# DESPICABLE
# SPECIES

# DESPICABLE SPECIES

## On Cowbirds, Kudzu, Hornworms, and Other Scourges

## *Janet Lembke*

*Illustrations by Joe Nutt*

The Lyons Press

Guilford, Connecticut

*An imprint of The Globe Pequot Press*

PRINTED IN THE UNITED STATES OF AMERICA

FIRST PAPERBACK EDITION

ORIGINALLY PUBLISHED IN HARDCOVER BY THE LYONS PRESS

Library of Congress Cataloging-in-Publication Data

Lembke, Janet.
    Despicable species : cowbirds, kudzu, and other scourges / Janet
Lembke : illustrations by Joe Nutt.
        p.   cm.
    Includes bibliographical references (p.209).
    ISBN 1-58574-199-X
    1. Symbiosis.   2. Parasitism.   I. Title.
QH548.L46   1999
577.8'5—dc21                                          99-15220
                                                       CIP

Permission from the following to quote copyrighted material is gratefully acknowledged:

Alfred A. Knopf: eighteen-line excerpt from "The Great Scarf of Birds" from *Collected Poems,
1953–1993* by John Updike. Copyright © 1995 by John Updike.

HarperCollins Publishers, Inc.: four-line excerpt from "Mushrooms" from *The Collected Poems*
by Sylvia Plath. Copyright © 1981 by Ted Hughes.

University Press of New England: eleven-line excerpt from "Kudzu" from *Helmets* by James
Dickey. Copyright © 1962, 1963, 1964 by James Dickey.

Viking Penguin: six-line excerpt from *The Odyssey* translated by Robert Fagles. Copyright
© Robert Fagles, 1996.

Several pieces have appeared in somewhat different form in these publications:

"The Natural History of Proteus," "Unfinished Business," "Heritage," "Prospect and
Refuge," "A Foot in the Door," *Brightleaf: A Southern Review of Books*, Vol. 1, Nos. 1–4;
Vol 2, No. 2.

"The Riddles of the Sphinx," *Oxford American*, Issue No. 29.

"Murmurations," *Bird Watcher's Digest*, Vol. 20, No. 4.

For three people
who are all at the farthest remove from despicability:
Harriet Divoky
Jean Wilson Kraus
JoAnne Powell

# CONTENTS

# ACKNOWLEDGMENTS

As always, many people have contributed information and recounted, with the proper glee and shudders, anecdotes of their own encounters with despicable species. Some of these good people are fully named in the stories. Others who deserve a share of credit are listed below:

- Bob Marlin; Nick Lyons; Southern writers Franklin Burroughs, James Kilgo, John Lane, Janisse Ray, "Prospect and Refuge"
- Franklin Burroughs, Ruth Roberson, "The Barkings of a Joyful Squirrel"
- Grace Evans, Meg Rawls, "The Natural History of Proteus"
- Jeffery Beam; Bob Marlin; Michael Waldvogel, North Carolina Cooperative Extension Service, "Legs"
- Nick Lyons, "Blood" and "Unfinished Business"
- Tom Glasgow, cooperative extension agent, New Bern, North Carolina; F. Timothy Edwards, landscape architect, South Carolina Department of Transportation, "Heritage"
- Sara Mack, "The Wisdom of Nature"
- Staff and patrons of the Augusta County Library, Fishersville, Virginia, questionnaire
- Field-trippers who came to Great Neck Point under the auspices of the North Carolina Maritime Museum, questionnaire
- Liz Lathrop and the Oriental Birders, questionnaire

Nor should my entertaining husband, the Chief, and my supportive friends, especially Nick and my editor Lilly Golden at the Lyons Press, be ignored. To all of you and to any whom I may have overlooked, my deepest thanks. From now on, may you remain unbitten, unblistered, and unstung.

*The real, which is perfectly simple and supremely beautiful, too often escapes us, giving way before the imaginary, which is less troublesome to acquire. Instead of going back to the facts and seeing for ourselves, we blindly follow tradition.*

—Jean Henri Fabre

# DESPICABLE
# SPECIES

# LIVING TOGETHER, LIKE IT OR NOT

American Dog Tick (*Dermacentor variabilis*)

## An Introduction

How do we deal with the bad stuff? With all those disgusting, sickening, despicable, repellently alien lives that impinge on ours?

The world offers much to cherish and admire. Roses, oak trees, butterflies, ladybugs, bald eagles, dolphins, and puppies—these are among the thousand thousand things, great and small, that delight both spirit and imagination. They lend their fragrances to our lives, shade our summer days, bring fluttering color to the flowers, eat the pests that eat our gardens, soar to give us models for our aspirations, leap and roll with glistening power in sight and in mind's eye, and lick us, love us unconditionally. The subjects of our fondness are—only the mean-hearted would dispute it—Good with a capital G.

Other lives connect with ours in less all-out friendly ways. Some are neutral, asking only—if they ask at all—for the kind of attention that is easily given: a deer browsing at wood's edge, a red-tailed hawk perched in a roadside tree, swamp roses sprouting in a drainage ditch, June bugs thudding on doors and window screens. And there are lives so minuscule that we absolutely fail to notice them, like the pillow mites that subsist on the bits of skin we slough off as we sleep. Still other lives affect ours in ways that require work, as when oregano assumes its takeover mode in the herb garden or the neighbors' cats dance, leaving footprints all over the car. Then we weed or we wash. But the plants and animals causing events like these fall into the acceptable range—except when the puppy eats one of a pair of new shoes or the browsing deer is suddenly impelled to bound into the path of an oncoming car.

And then there are the lives that people love to hate: aggressive weeds, slithering or pulsating creatures that trigger the shudder factor, and the bugs, oh, the bugs in their teeming millions that ruin summer nights, crunch unpleasantly underfoot, deliver painful stings, or draw blood with their stabbing, sucking mouths. Sometimes, as with scraps of life so small we cannot see them—bacteria, viruses, dinoflagellates—they sicken and they kill.

Yet we're stuck on this earth together. And one small word

describes the immense and unimaginable array of alliances, conspiracies, collisions, and entanglements that mark our lives: *symbiosis*. It simply means "living together" but encompasses the not-so-simple fact that every plant and animal, and every other squirming, blooming, often invisible form of life, conducts its affairs, want to or not, in some kind of relationship with neighbors quite unlike itself.

In grade school, I thought that symbiosis was invariably a good thing, an always friendly conjoining of vital forces to achieve a common good. Back then, the lowly lichens were my introduction to the concept. I've since come to appreciate how quirky they are: gray-green, yellow, or reddish brown, leafy or crusty, they grow on trees, grow on rocks, and thrive equally well in heat or cold. And how oddly useful, for they not only serve as the food of choice to reindeer and caribou but also provide surprisingly flamboyant colors, purple and glowing rose-red, to the dyer's trade. And how utterly improbable! A fungus and an alga, the latter sometimes of the blue-green sort but often the green *Protococcus*, mingle in an inseparable union. More than that, the union amalgamates two distinct biological kingdoms, that of the Fungi and that of the Monera. The Monera, as immemorially ancient as the advent of life on earth, comprises one-celled organisms, like bacteria, that lack a nucleus, and, indeed, cyanobacteria—blue bacteria—is another name for blue-green algae. Each of a lichen's components helps the other, alga by producing vitamins, fungus by absorbing necessary water from the air, both by exchanging the discrete carbohydrates that each manufactures. Neither can exist alone. The situation is win-win all around.

But turn the natural world's kaleidoscope and look elsewhere: it's clear that relationships come in a hodgepodge of forms, some grand, some grotesque, with every possible permutation in between. Sometimes the relationships—and resemblances—are close indeed. Take, for example, the seven species of woodpecker that I observe in both my homes, the Shenandoah Valley and coastal North Carolina. Sometimes two species work for food—*tock-ta-tock-tock*—on the same tree. The food-gathering relationships here include the social mix of bird and bird, the birds' bark-chiseling or sap-seeking behavior on the tree, and the birds' predation on the insect cafeteria caught in the sap or hidden beneath the bark. More often, however, physically disparate actors are intertwined in a boggling array of situations and positions. Fungal threads snuggle by harmonious necessity with the roots of trees; behold, a crop of morel mushrooms at the base of a sweet gum. A ghost-pale isopod makes itself at home first as parasite, then as tenant in the menhaden's mouth; the fish suffers little while the hanger-on thrives. The HIV virus must live and replicate in the human body; as it does so, it entombs itself. All of these relationships are symbiotic, but win-win is hardly the rule. Instead, the business of living together throws friends, enemies, and bystanders onto the same patch of earth. Some get along famously or find neutral ground. Others, to ensure the replication of their kind, succeed in fighting to the death.

Science identifies several distinct styles of symbiosis. It is Burwell Wingfield, teacher of biology at Virginia Military Institute and expert hunter of morels, who names and explains these associations to me, and diagrams them with a symbol—Plus, Minus, or Zero—to indicate benefit, harm, or a neutral outcome for each for the two parties involved.

The first style is mutualism, a top-of-the-line Plus-Plus. It describes the kind of partnership in which two unrelated species work together, each supporting the other, because the lives of both depend upon the association. This is the lichens' win-win style, and that of the yucca plants and their specially adapted moths: plant needs moth for pollination, moth needs plant for larval food. And the relationship is obligate—it must occur, that is, in order for either partner to survive.

The second is protocooperation, the other Plus-Plus arrangement. The word was coined by Warder Clyde Allee (1885–1955), an ecologist with a particular interest in the communities of marine life-forms, and it refers to a rudimentary, always involuntary sort of teamwork. The association is passive; a common good is achieved without awareness or intent on the part of any symbiont. Mixed schools of fish, mixed flocks of birds engage in this behavior; when one species finds food or encounters a threat, the other species in the aggregation are alerted. Another protocooperative effort occurs between a tree and the sow bugs, centipedes, and other small creatures that live in the larder of leaf mold provided by the tree; as they feed, they churn and aerate the soil, helping the tree capture the moisture and nutrients necessary to its growth. Protocooperation also takes on darker guises. One that sorely afflicts people is the opportunistic association of the *Borrelia* spirochete (a relative of syphilis) with *Fusobacterium*, one of its Moneran kin; when the two occur on the same human gum at the same time, their joined forces cause the galloping form of gingivitis known as trench mouth. (If the sufferer is included in this symbiotic equation, then we're looking at parasitism, but more of that shortly).

Commensalism—which comes from Latin words that mean "together at the same table"—rates Plus-Zero in the symbiosis stakes. One party, usually the smaller, gains, while its larger host is unaffected. This is the relationship that occurs between the menhaden and its isopod (its "nursemaid" in folk parlance); after the juvenile isopod, lodged in the menhaden's gills, matures, it abandons its parasitic ways to become a boarder in the menhaden's mouth, where it helps itself liberally to the plankton soup swept in by its host. Commensalism is also the mode de vivre between our kind and the starling when the bird probes for insect food beneath the earth of a manmade clearing. And it's the kind of association seen in barnyard or pasture where cowbirds and cattle egrets tag along behind cows and horses to batten on the bugs that the larger animals have stirred up in passing.

A fourth style of symbiosis, the Plus-Minus version, takes two forms: parasistism and predation. Parasitism is by far the most common form of symbiosis on this earth. Always, one party gains, while the other forfeits resources and sometimes its life. Sometimes parasitism is obligate, sometimes opportunistic, as when a species takes temporary advantage of another's nest or cache of food. A parasite's host may be restricted to one particular organism (potato for the fungal potato blight, human being for the bacterial trench mouth). A two-host arrangement may also occur, in which the parasite, halfway through its life cycle, moves from one bed-and-breakfast to another (such as pig, then human being for the pork tapeworm). Or the possible hosts may be many, a diverse group of equally succulent and nutritive possibilities: any passing warm-blooded animal, like dog, deer, or person, is capable of satisfying an American dog tick's craving for protein. The parasites may even pile up in a scenario called hyperparasitism; the dog tick, for example, that sups on mammalian blood may itself be host to two kinds of internal microscopic organisms, each of which depends for its very life on tick juice. This arrangement demonstrates the fact that there are ectoparasites and endoparasites—those like the tick that work from the outside and those that work only on inside jobs. In every case, the host is larger than the invader and also serves as its only source of food.

⋅ Another kind of parasitism does not involve one creature munching directly on another but is rather a mooching arrangement—social parasitism. It occurs in a stunning array of variations, notably among the ants (though it comes in a human version, too: the legend of the man who came for dinner but stayed on for years). In every variant, one species associates itself intimately with another and gains benefits like food and shelter from the association. Two of the many apparitions of this phenomenon are cleptobiosis ("life-stealing"), in which one species lives near or among another and either feeds upon the host colony's refuse or robs its workers, and dulosis ("slavery"), a symbiosis noted by Darwin in *On the Origin of Species*, in which one kind of ant takes the workers of another kind captive and sets them to such necessary tasks as food storage, nest building, and caring for the slave-maker's brood. Ants and wasps also engage in an extreme form of social parasitism called inquilinism, from the Latin *inquilinus*, meaning "someone who inhabits a place not his own." In the case of insects, the mooching species moves into its host's nest; there, coercing the host to deliver vital services, it degenerates physically, becoming unable to care for itself in any matters other than reproduction.

As for the predatory form of the Plus-Minus symbiosis, the seeker of prey behaves in much the way as a parasite, though it certainly puts more muscle and greater agility into its work, sometimes hunting down its food and sometimes, like the angler fish and various spiders, enticing it. And the predator has greater choices than the parasite when it comes to taking advantage of whatever nutritious opportunity comes galloping, swimming, or creeping into range. Nor is the predator always larger than its prey (think of a lion bringing down a zebra, or a man shooting a deer).

Curiously, both parasite and predator may enhance the well-being of the prey species. For the sake of future meals, the former does not kill a healthy host. The latter tends to kill and eat the weakest individuals. Plus-Minus symbioses may seem repellent to us—the stab-and-grab attentions of lions or mosquitoes, the stealthy single-mindedness of a cat waiting to pounce, the cowbird and red-head duck dumping their eggs in another species' nest, the pestilential viruses afflicting throats and sinuses with the common cold—but in the long run they actually work to strengthen prey species and help them adapt to life in an unforgiving world. As Burwell Wingfield says, "The parasites and predators are merely trying to survive. In order for them (or us) to live, something must die, be it plant or animal. It is the nature of life."

The business of living together includes one Zero-Minus combination, in which one party profits not at all but the other invariably loses. It's known as amensalism—"being shoved away from the table"—and it takes two forms, competition and antibiosis. In both forms, resources—food, water, and space—are at stake. The bigger, stronger species helps itself, while the smaller, weaker species goes without, declining or altogether dying out. The competitive form of amensalism puts on a vigorous green show in our North Carolina vegetable garden every year. For every pea and bean seed planted, a hydra-headed batch of weeds springs up—jimson, bindweed, nut grass, witchgrass, rabbit tobacco, cranesbill, and several dozen others, all jostling and pushing. The harvest would fail except that I also move into a competitive mode, wielding rake and hoe. Antibiosis is less visible and more insidious. It is nothing less than chemical warfare, with the unaffected species secreting toxins that inhibit or kill the other party or parties within its reach. Think of antibiotics like penicillin, a bread-mold secretion, and the nitrogenous compound allantoin, excreted by maggots; both kill bacteria (and in so doing, add considerably to human comfort). The black walnut tree also engages in such tactics, secreting juglone, a substance that poisons plants trying to grow within its root zone. There is no profit in this gambit, except that, at the very beginning, when the sprouted black walnut seed puts down its first roots, it discourages competition; once established, however, the tree needs no toxin to maintain its dominance. Not long after moving into our Virginia house, we cut down the black walnut, a squirrel-planted volunteer, that had seized dominion over the small back yard. The ground beneath was bare of grass, not to mention weeds. It took five years, and the deaths of two live Christmas trees, before the juglone had dissipated enough for new greenery to take hold in the poisoned soil. Behaving at times like black walnut trees, people also practice antibiosis by spreading herbicides in the hope of a perfectly weed-free lawn (though what perfection there might be in weedlessness, I do not see).

The final incarnation of symbiosis, the bleak Minus-Minus version, is synnecrosis—a Greek term that means "dying together." In a situation akin to that of the feuding Hatfields and McCoys, mutual inhibition occurs between two species. And both may die because of the fast-as-a-bullet toxins released into the shared habitat. Burwell Wingfield illustrates this lose-lose arrangement with a general example from the strange and populous kingdom of the fungi: "Suppose the genus *Penicillium* contains a species X, which produces a particularly strong variation of the antibiotic penicillin. This might be toxic to a species of the genus *Streptomyces*. Suppose also that species Y of the genus *Streptomyces* produces a variation of streptomycin that is toxic to *Penicillium X*. If these two fungi growing in the soil encounter each other, both will react to the antibiotics, and that part of both fungal filaments in the vicinity will be killed." But, where win-lose scenarios may occur on both an individual and a species basis, lose-lose synnecrosis spells doom only for individuals. It does not portend the end of any species. The species have evolved their own survival strategies, have devised blind, unstoppable means of perpetuating themselves. Wingfield notes that synnecrosis takes place mainly between organisms living in the soil, but he speculates with a wry shake of his head that, given the ongoing savagery in places like Bosnia, Kosovo, Kurdistan, Rwanda, Somalia, and Ireland, it has spilled over onto mankind.

So here we are, every last one of us, living together, like it or not. As for the lives large and small that we don't like living with, what is it about them (or us) that drives us to profanity, lifts our swatting hands, or excites our shudders? Peculiarly, out of the whole breathing, thumping, chewing, squirming mass, *Homo sapiens* is the only one that makes value judgments about the others. (*Et quis iudicat iudices?*) Even when there's no longer anything to gain by being wary or timorous, we still respond with almost instinctive fear and loathing to a host of things that aren't like us. Sometimes, we flinch not at the reality but at the pictures provided on the Discovery Channel: North America, for example, is not home to the hyena, but the very thought of a scavenger that seems to laugh maniacally as it zeroes in on a meal of carrion evokes a reflexive shiver. The creature, behaving in a way that repels us, is relegated (by all but the most intent zoologists) to the status of a lowlife not deserving our notice, much less our regard.

The hyena's credentials for despicability are easy to understand. So are those of the phylum Arthropoda, especially its insects and arachnids like spiders, chiggers, and ticks. As Annie Dillard says of the insects, they "are not only cold-blooded, and green- and yel-low-blooded, but are also cased in a clacking horn. They lack the grace to go about as we do, softside-out to the wind and thorns. They have rigid eyes and brains strung down their backs. . . . No form is too gruesome, no behavior too grotesque." Nor is it simply the otherness of their looks that damns them; they also bite and sting and transport disease.

But, peculiarly, there is a dichotomy between the things that we despise and those that we both loathe and fear. The former often wreak bodily harm, while the latter, though they rarely hurt us, frighten us literally out of our wits, tossing us like toy boats on wild seas of unreason. Why do we simply dislike mosquitoes and poison ivy, which surely cause excruciating itches and blisters, but shudder at snakes, spiders, and centipedes, which would rather flee the scene than attack our kind? We flinch involuntarily not only at the imme-diate slinking, spinning, lurking presence of such creatures but at the very idea that these monsters occupy room on our planet. Imagina-tion afflicts us as desperately as does reality. The speed, the lack of feet, or their obscene abundance, the fangs and possibilities of venom, the attenuated scaly bodies or the plump hairy ones, the flicking tongues, the multiple eyes—all these characteristics sepa-rate the reptiles and arthropods from everything that's sensible and comely. They contradict our notions of the way things ought to be. A phobic reaction sets in: palpitations, sweating palms, a smother of doom that makes the sufferer gasp for air.

Genetic programming may dictate our fear. Not only humankind but all the primates recoil from things like snakes and other sudden apparitions in the trees or grass. The ecologist and natural philoso-pher Paul Shepard (1925–1996) has written:

> No monkey makes light of shapes in the gloom that might be
> leopards, snakes, or owls. Storms and lightning terrify them.
> That primal event of the murder story, the scream in the dark, is
> the primate's ultimate signal of panic. The dark side of imagina-
> tion generates fearful shapes and bad dreams, infecting shadows

with menace, even when the leopards are in fact far away. Such
fears are the price of intelligence and the consequences of exposure—the price of sleeping lightly in trees or on cliffs.

*H. sapiens*, the fiercest primate of all, has fully inherited the ancestral
timorousness. It is, as Shepard explains, a protective device that
makes for longevity. Our eyes, and those of our simian kin, tell the
story. We are diurnal creatures, active in the light of day, and once
we were all arboreal, living aloft in the canopy of rain-forest trees.
To enable this way of life, evolution equipped the primates with
both binocular and color vision, the first to allow accurate estimates
of distances to be traversed, the second to tell the difference
between bitter green fruit and that which has attained a succulent
ripeness. Birds have color vision, but when it comes to furry, infant-
suckling mammals, primates are the only ones—with the odd exception of a few rodent species—that are blessed with color vision. But
we also see shades of gray. Outside its fovea, the central area that
perceives the world in color, the primate retina possesses rod cells
that see only in monochrome. Thus, we are able to see not only that
there are shadows and nighttime darkness but also what moves—or
may at any instant materialize—within the dimness and the dark.
What is and what might be—the real leopard and the leopard of
imagination—live side by side in the vision of primates. And
because we have been designed to expect trouble in the shadows,
our actual safety is greatly enhanced.

With snakes, human aversions have been given extra strength
by myth and theology: the head of Medusa crowned with writhing,
hissing serpents and, worse, all hopes of Eden forever lost. As for
the spiders, I have seen it suggested by Richard Conniff that they,
too, may be the objects of an aversion "embedded in our genetic
memory." Certainly, they, too, are the subjects of myth and folklore:
the princess Arachne transformed into a spider because the perfection of her weaving insulted the divine spinster Athena and, on the
positive side, the solitary prisoner, like Scotland's fourteenth-century
Robert the Bruce, finding a friend in the spider who shares his cell.

The latter view may speak to the difficulty of human rapprochement with eight-leggers; only when there is no other company are spiders to be considered welcome or, at least, ungrudgingly granted tenancy of nearby space. I doubt that the "hundred-leggers," the centipedes, are ever welcome as they skitter across walls and come up suddenly through drains. They share to a hideous excess the spider's fault of too many legs. And they are small enough so that imagination insists on the possibility that they will travel across soft, defenseless skin and invade not just a sleeper's dreams but also the private parts. This perception of the centipede is, in its horrid way, more venomous than the creature's actual bite.

With snakes, spiders, and the whole gaudy, clacking, creeping, many-shaped, and quite uncountable army of the planet's bugs, it's fairly easy to understand the shudder factor. But what of the other living things that people despise: the cowbirds, the poison ivy, and the myriad other bugbears like rats, slugs, crabgrass, jellyfish, and viruses? A good part of despicability comes, of course, from the all-too-real bites, stings, and blisters and from the invisible but equally real frissons and terrors that rise from the collision of other lives with ours. Another clue lies in the very word *bugbear,* which popped up in English in 1581. It referred in its earliest days to things that were spectral and scary—a goblin or a ghost. It still denotes things, real or imaginary, that give rise to a gamut of stressful emotions, ranging from a twinge of annoyance to disabling fear. And *bug* itself, the fourteenth-century word at the heart of *bugbear,* originally designated a phenomenon, usually supernatural, that inspired disturbing physical symptoms, like clammy hands and thumping hearts— symptoms, striking sudden as a lightning bolt, that in themselves exacerbated fright. *Bogeyman* and *bugaboo:* a Black Sabbath of ghoulies and ghosties and long-legged beasties shelter also within these two terms. And to this day, anything we do not understand becomes a bugbear, deserving in its otherness our suspicion and dislike. With a sigh for the idiosyncrasies of humankind, however, it should be noted that things that are bugbears to some are beauties to others. The Japanese relish eating kudzu and keep caged crickets as pets, the Chinese consider spiders lucky, and English householders set out nest boxes for starlings. (I've never heard of anyone, however, who admires the tick.)

A species' despicability may also be determined, at least in American minds, by its failure to contribute one jot or tittle to the well-being of our kind. "What is not useful is vicious," wrote Cotton Mather (1663–1728) with the moral certainty of a Puritan who knew beyond all doubt that God had granted man dominion over the fish of the sea, the fowl of the air, the cattle, the creeping things, and every herb bearing seed upon the face of the earth. In this view, a species need not go so far as to harm people; its malice resides in the bare fact that it exists but leads the lily's idle life, neither toiling nor spinning on behalf of those who dominate. The corollary is that the living things that manage to tend to their own instinctual business without getting in our way nonetheless deserve our condemnation and our swats because they do nothing to help us.

This is also the view—*because they are there*—that makes some people shoot at tree swallows flying over the pond just around the corner from my North Carolina home and that leads others to poach carnivorous plants, the sundews and Venus flytraps, from a nearby national forest. It may be said, of course, that there's sport in the first and money in the second. But both reflect relationships turned topsy-turvy. In a world in which the odds were more fairly distributed, the relationship of swallow and gunner, plant and digger would be Zero-Zero, both present in the same place but neither affecting the other in any way. Both gunner and digger would be nothing more than innocent observers—should they even deign to notice bird or plant. But when rifle and spade are wielded, observer turns into predator, and the symbiosis becomes Plus-Minus, with the predator always winning and the prey uprooted or killed outright.

In the stories that follow, the often peculiar relationships of our kind and the species we despise are looked at in their natural, mythic, and literary incarnations. It will be noted that almost every species that we scorn or loathe is superlatively abundant—starlings in their uncountable legions, kudzu in sprawling green tangles that have seized millions of acres, hornworms in every tomato patch and tobacco field, squirrels up every tree. Some of these species are exotic imports that have thrived with grand excess in the United States because the natural controls—the predators and parasites—of their native lands are absent here. Others are indigenous and have always been at home in the New World. But no matter their origin, all are far more ancient in evolutionary terms than we, and all are with us until the end (be it bang or whimper). They may indeed continue to be here, a crawling, flying, leaping horde of survivors, well after we're gone. Howard Ensign Evans, an entomologist who nostalgically remembers stomping tobacco hornworms with his bare feet on his father's farm, offers these words of caution (the italics are mine): "Swatting a mosquito is allowable, considering that mosquitoes are so well able to flood the earth with their kind. *But swat it respectfully:* the mosquito is a product of a million years of evolution." Sue Hubbell, a notable science writer and champion of hexapods, goes so far as to wonder "if we are justified in acting as though we have a *right* to live in an insect-free world." But swat and spray and slaughter as we will, the things we find despicable are well-nigh extermination-proof—with the possible exception of the one species that is peculiarly fitted with the power to self-destruct. As Pogo the 'possum says, "I have seen the enemy, and he is us." But more of this matter in the book's final story.

Along the way, the questions rise like weeds in gardens or mosquitoes from a swamp. Some questions are specific, concerning the lives of particular despicables: How does a centipede manage to run without tripping itself? And how does a cowbird know what it is when the first birds it sees are those of another species? To questions like these, answers may often be found. It turns out that many of the things we despise are, in their own sly ways, quite remarkable.

Other questions are far more obdurate; nor do I propose to attempt to resolve them. The hope is rather to stimulate reflection.

What, for example, might be the rights of that running centipede or the new-hatched cowbird, of a rat or a snake, of jimsonweed or crabgrass? Do rights exist, except as a concept of human devising? To what extent are our lives interlocked with those of the things we despise? To what extent are all the lives on earth interlocked one with another, even at second, third, and fourth removes? Are there lives that the world can do without? Species, genera, whole families whose removal would not cause some cataclysmic wobble in the planet's spin? Why treasure endangered species but scorn the commonplace? Is rarity valuable in and of itself? Might intentional extermination of a species ever be justified? How much is enough? And most vital of all, what means shall we give our children so that they may learn to recognize and delight in the world beyond their finite skins? These questions and others each need a reasoned response, not one that smacks of eco-think, which has a flatulent tendency to come from the gut, not the brain.

Answers are important because of the dilemmas inherent in our attempts to manage our natural surroundings. I think of the good intentions that did in not only the salt-marsh mosquitoes on Florida's Merritt Island but also *Ammodramus maritimus nigrescens*, the dusky subspecies of the seaside sparrow. In the name of human comfort, the marshes were drained to eliminate the mosquitoes' breeding grounds. The hoped-for consequence was achieved: a drastic drop in the mosquito population. But in the wake of that achievement, like an unexpected aftershock, a small bird with a chunky bill, a dark-colored back, and a pale breast heavily streaked with black became extinct because its food supply had disappeared.

We're stuck here, like it or not. So, it behooves us to think hard and well about how we fit into the scheme of things and how we should behave ourselves so that the scheme continues to cohere. As Paul Shepard has said, "What problems and characteristics are in fact uniquely human cannot be clear until we know more about what is not human, and, in discovering what we are not, gain a fresh appreciation of the rich and diverse otherness of the natural world in its own right."

Here, then, are tales of despicable species—mirrors in which we might take a good, long, quizzical look at our most peculiar selves.

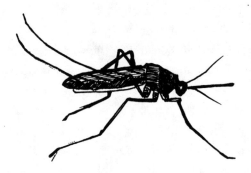

Mosquito (*Culex pipiens*)

# PROSPECT AND REFUGE

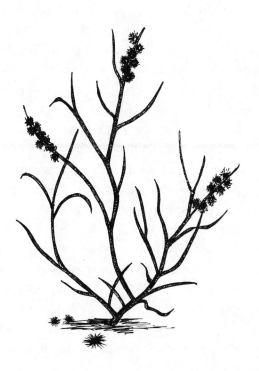

Sandbur (*Cenchrus tribuloides*)

## Sandburs

G RASS: imagination conjures an endless panorama of green—sweet new-mown lawns, cattle-cropped pastures, wild prairies where leaves and stems rise tall, then lapse in the wind like waves on the sea. And other colors gleam there, too, the red kernels of Indian corn and the well-sung amber waves of grain. Grass: the word itself rustles and whispers and summons Walt Whitman. The reality of any grassy expanse, be it wild or tame, may be just as appealing as the visions that rise in mind's eye—Spartina, sea oats, tall oats grass, wild rice, maize, rye, bluegrass, red fescue, purple love grass, and a hundred others native to the New World. Grass, nonetheless—as all of us who touch earth know—has its grosser manifestations, like crabgrass, an introduced species, that intrudes in the lawn and insistently nudges aside more modest greenery, and like the native quack grass and nut grass that rage through vegetable patch and flower garden with the swarming, single-minded thrust of a Mongol horde. And at the shore and in disturbed places like roadsides and abandoned fields there grows a kind of grass that is perfectly in balance with its own world but must, by any human reckoning, be considered truly vicious.

Cenchrus tribuloides, "thornlike millet"—I meet it first on Ossabaw Island, Georgia. The group of plants commonly known as millet produces nutritious seeds, which are readily eaten by people, animals, and birds. Cenchrus is Greek for "millet," but this thornlike imposter has simply stolen the millet group's good name and is related only insofar as both belong to the Poaceae, the huge and multifarious grass family, which includes not only lush coverings like bluegrass and fescue, hay crops like timothy, and the economically important cereals that comprise the amber waves, but also gathers in some prickly mistakes (if nature can be said to make mistakes). The only similarities between C. tribuloides and true millet lie in the size and roundness of their seeds. And there all visible resemblance stops.

Ossabaw Island—what a peculiar and time-warped place! It lies off the Georgia coast only twenty short miles from Savannah—

twenty-five thousand acres of sand, tidal marshes, and freshwater ponds, of live oaks swathed in Spanish moss, and huge, fan-leafed palms that rustle and rattle in the sea wind. And it's altogether a more exotic place than any I've seen before, a place in which a peg-legged pirate or even a dinosaur might not seem amiss. I'm here, though, with a gaggle (or perhaps a pride) of writers, editors, and photographers whose special interest is the natural world, particularly as it manifests itself throughout the South. We're here to talk specifically about how being Southerners, by birth or by choice, colors our experience of the land and all that lives thereon. For the weekend of our conversation, we are encamped in somewhat seedy elegance in the palatial Mediterrranean-style house—at least eighteen bedrooms, most with attached baths—that was built as a winter hideaway by Henry Torrey, the Midwestern industrialist who bought the entire island in the 1920s. His daughter, Sandy West, now in her eighties, still lives amid call bells for maids and pantries stocked with glassware and china for dozens of guests. The maids no longer exist, but the guests still arrive, bringing their own food and their own agendas, usually something to do with the arts. In the past several decades, their number has included composers, sculptors, novelists, environmental activists, and the woman who designed Barney the dinosaur. We are simply one small and recent group in a long line of individuals and organizations using the big house as a conference center—and gawking at the strangeness of the sea-island landscape. As is eminently fitting, Ossabaw was sold to Georgia in 1978 and has become the state's first Natural Heritage Preserve.

In 1995, however, the National Trust for Historic Preservation put the island on its list of the country's eleven most endangered places. Long before the 1920s and the incursion of a personal fortune, Ossabaw had lost its claim to being a pristine sea island. From the time that people—the Indians first, then the European explorers and colonists, and later the planters and their slaves—set foot on the sandy beaches, the land and its denizens have been unnaturally altered. Marshlands have been drained and filled, and tide-interrupting causeways built; some woods have been cleared for growing cotton and raising cattle; a web of dirt roads, seventy-five miles altogether, crisscrosses the dunes and swales. An auto graveyard even decorates the island's North End. Much of the original sea-island flora and fauna remains, of course, but has been joined over three hundred years by foreign things. Amid the woods live the gray squirrels and the birds, woodpeckers and warblers, that have always been here, but introduced animals also lay claim today to a not inconsiderable amount of space. White-tailed deer, armadillos, and a great slew of feral pigs, some three thousand of them in many colors from black to rusty red, move with skittish caution through the open underbrush. And Sicilian donkeys, descended from five brought in as pets for Henry Torrey's children, have made themselves wildly at home in the fields and woods. Behind the palatial house, peafowl, geese, black-and-white Muscovy ducks with red wattles, and three horses wander at will, honking, squawking, neighing. One huge coal-black pig dubbed Lucky has become a permanent hanger-on, dining well on kitchen scraps. Half-grown wild piglets hover on the fringes of the action hoping that some food will be spilled for them. I see Sandy West out back in the early morning refilling feed and water buckets. Despite her years, she works with youthful verve and vigor.

But some things on Ossabaw are immemorial. The name is said to mean "yaupon holly bushes' place," and the yaupons still grow here in thriving numbers. Nor do they at all resemble the flourishing but spindly shrubs I know on the Carolina coast; they have attained the status of true trees and boast solid, mottled gray-green trunks that may measure as much as a full foot in diameter. The live oaks, some growing naturally and others planted more than two hundred years ago, are as gnarled as the trees in any enchanted woods conjured by the Brothers Grimm. I see one oak, huge and decaying but still almost fully decked out with glossy dark green leaves, that first poked its head up through the soil some six hundred years ago, around the time that Columbus set sail. Snakes—rat snakes, king snakes, and black racers, rattlesnakes and copperheads—lurk and slither everywhere as they always have, while alligators haunt the sloughs, bald eagles soar overhead, and shorebirds—willets, yellow-legged sandpipers, and the tiny, hard-to-tell-apart sandpiper species collectively known as "peeps"—run and peck through the sands at water's edge.

And here on the beaches and dunes—like a thorn on a rose, like a serpent's tooth—*C. tribuloides* also makes itself at home. The sands of Ossabaw, and those of its sister sea islands, like Saint Catherines, Blackbeard, and Sapelo, were not designed with human comfort as a prime consideration. Beachcombing is best accomplished in a pair of shoes or boots, and even then there's no escaping the long-spined burs, which will insinuate themselves into the waffle treads of the soles. That's how I first discover them. After the morning efforts to define Southern nature writers (if they are definable), we make an early afternoon excursion to South Beach. And here I am, sitting on a handy tree stump to watch the various acrobatics of a kingfisher and an immature bald eagle, but when I cross my legs, hoisting left ankle over right knee, the hand used for the hoisting is right smartly punctured by the burs stuck in my soles. Burgrass! Sandbur! Sandspur! The common names for this common plant hiss and spit in their own right and make for excellent invective. The half-hour ride back to the palatial house is spent gingerly extracting needle-sharp sandburs from the soles of my shoes. I save some for further study by sticking them on my shirt, where they cling unbudgeably till they are put into a pint-sized plastic freezer bag. And even that is not enough to keep them from pricking and stabbing; though the bag has been folded around them, the rigid spines, packed as thickly as those on a sea urchin and each at least a centimeter long, have readily pierced three layers of tough plastic.

After the outing, I go to the back yard and take pictures of the half-grown wild piglets that are hanging around hoping for crumbs.

When they cotton to my presence, they glance at me with quick suspicion and scoot. Then people gather again beside giant yaupons to continue exploring this entity dubbed the Southern nature writer. The discussions ramble and rage. Some of the talk is profoundly silly, especially in the attempts to define ourselves in relation to Westerners: Westerners see things, Southerners see characters; the West is geological, the South biological. But two of our number come close to definitions that have the ring of truth. One says, "We are an endangered species. We still have a sense of place." And the other, "There are things we see in a landscape—prospect and refuge. We as writers are nurtured by a sense of the past." He will tell me later that the notion of prospect and refuge is not original with him but rather comes out of a theory about how our feelings of beauty and fitness in landscape reflect our deep Darwinian past, when we needed both a point of vantage and a place of concealment if we were to succeed, on the one hand, in eating and, on the other, in not being eaten.

Prospect and refuge, vantage and safe shelter—I latch on to these, along with a picture of Southern writers being rocked like babies in the cradle of the past. But it's not only the past that nurtures us. We are nurtured also by trumpet creepers, longleaf pines, alligators, and a folk memory of ivory-billed woodpeckers. Sustenance comes, too, in beaches and salt marshes on the flat-as-a-flounder coast, it comes in Appalachian peaks and hollows—and all of it well steamed and stewed in a cauldron of hot, humid summers. There's much to consider, from eagles and azaleas to feral pigs and the peculiar thornlike millet.

As I learn later, the genus *Cenchrus* consists of about twenty-five species, most of them native to the New World's tropics. It is indeed a common plant—a common scourge—fond of sandy soil and quick to injure the unwary. Nor are all of its models confined to warmer regions; three are widely found in North America, and one—*C. longispinus*, "long-spined millet"—spreads across the continent. This long-spined sort may be found in disturbed areas, like roadsides and vacant lots, from New Hampshire to Oregon, south through the Carolinas and Georgia, and west through Texas into New Mexico. Its usual mode of growth is to sprawl, hugging the ground unobtrusively in what might seem an effort at going unnoticed, at hiding its great piercing power. *C. tribuloides*, its thornlike cousin, is the species usually damned as a "sandspur," and, from Staten Island to Florida, it is to be found with truly pestiferous abundance on the dunes and beaches of the Atlantic coast, including those at my home beach, Great Neck Point. "Those miserable little monkeys"—that's how a neighbor describes them. How I have missed a painful encounter in the years of my residence beside the wide and salty river Neuse, I do not know—fool luck, perhaps. Of the trio of pseudomillets thriving in North America, the third is *C. incertus*, "uncertain millet," which may be named for its wont to start out as an annual plant growing in a tall upright clump but then to dig in, when fortune brings the right weather and nutrients, and transform itself into a perennial.

Burgrass! Sandbur! Sandspur! These stubborn plants inspire not just ordinary folk but also botanists to lose their objectivity. Usually, the latter deploy sobersided technical terms like these used in *Gray's New Manual of Botany* to describe the prickery, stickery seed pods of *C. tribuloides*: "Spikelets 1-flowered, acuminate, 2–6 together, subtended by a short-pediceled ovoid or globular involucre of rigid connate spines." But listen to these judgments in two field guides. With laudable restraint, the one identifying shore plants calls thornlike millet "Unpleasant to walk on barefooted!" The other, a volume devoted to grasses, goes all out to say this of the long-spined variety: "Horribly spiny flower clusters, extremely painful to step on." Granted, these guides were written to satisfy popular curiosity, but only in describing *Cenchrus* species do they forfeit objectivity and utter a loud, sharp yelp. It's fair to suspect that both attest to excruciating personal encounters with unpleasant, horrible things that look like tiny morning stars bristling with a hundred spikes.

I spy another difference between Southerners and Westerners. While the former would surely rank sandburs high on any list of despicable weeds, the latter might nominate puncture vine, which has reached the status of a pest in places like Arizona, New Mexico, and West Texas. With a habit of prostrate growth and a nutlet with two stout, sharp-pointed spines that jut out like the horns of a steer, it is capable of doing serious harm not only to bare feet but to bicycle tires and the mouths of grazing livestock. The formal name of this well-defended plant is *Tribulus terrestris*, "earth thorn," and it belongs to the Zygophyllaceae, the caltrop family, which includes other thorn-studded species like the star thistles and pests like the knapweeds, along with such gardeners' favorites as bachelor's button and dusty miller. But puncture vine is not a grass, nor is it native to the New World but came as a stowaway from the Mediterranean.

The sandburs have been here from the beginning. The burs are really nothing more than coverings for seeds. Like chestnuts protected within a densely prickery pod, sweet gum seeds within a thorny ball, and cockleburs within a horrid palisade of barbed spines, the seeds of *Cenchrus* hide behind armor. Bur and seed together constitute the fruit. But why build such a fortress? Two reasons present themselves, and the first is protection: to keep the fruits from being devoured and digested, never to have a chance at sprouting. It has been noted that cattle and sheep avoid eating these plants, even when they are young and easy to chew. The second reason is transport: to stick tightly in fur, hair, flesh, and shoes so that the seeds and, with them, the next generations of *Cenchrus* are dispersed as widely as possible. Prospect and refuge—these are not only human goals but those of *Cenchrus* also: ensuring the future and shielding its tender imminence.

As we leave Ossabaw, I see a tiny pig—red as Georgia clay from snout to tail—streaking across the road that leads to the dock and the boat that will ferry us back to the mainland. Along with the sandburs, it encapsulates for me the secret of Ossabaw. Not just the people who write on the intricate peculiarities of nature but every last one of us needs some old-time prickliness, along with a good dash of upstart, feral notions, in the underbrush of our imaginations.

Razorback Hog

# THE BARKINGS OF A
# JOYFUL SQUIRREL

Gray Squirrel (*Sciurus carolinensis*)

Gray Squirrel

WHEN I press Dick Coleman, my doctor, for the reason he finds squirrels the most despicable species on earth, he says, "Because they're stupid, that's why."

I've always thought of Eastern gray squirrels as fat, sassy tail-twitchers, as admirably fearless acrobats (I have no head for heights), and as infinitely clever thieves able to get the better of any bird feeder. With great wit and no little accuracy, a friend calls them the monkeys of North America. Squirrels are also good to eat—if they're cooked so that they retain some of their juices rather than being fried until the meat is nothing but dry, leathery strips wrapped around tiny bones. Brunswick stew—hurray! And I think of other squirrels, not our familiar gray species nor the tree-dwelling kind in general but rather the playful creatures that romp through human imagination—among them, Rocky Squirrel, the bucktoothed side-kick of Bullwinkle Moose in the 1960s cartoons, and Beatrix Potter's classic Squirrel Nutkin, an English red squirrel, brother of Twinkle-berry and neighbor to Peter Rabbit and Mrs. Tiggy-Winkle the hedgehog. And at Mary Baldwin, a small liberal arts college for women located in Virginia, the sports teams, volleyball to soccer, call themselves the Squirrels. Altogether, be they four-legged or human, real or imaginary, squirrels are entertaining and inspirational, not to mention delicious. In just what ways can they possibly be stupid?

"They pile up nuts," Dick says with a *hmpf*, "and they bury them. Then they forget where they put them. Every year, those durn animals tear up my yard looking for nuts."

So squirrels can wreak costly havoc in suburban landscaping. I've also been told that, sneaking in through vents beneath the eaves, they can tear up an attic as efficiently as they do a lawn. Neither of these reasons, however, sits at the top of the list for despising them. The complaint heard most often comes from back-yard bird-watchers, who sputter and fume about another kind of expensive wreckage: the ongoing, undeterrable theft of bird seed and the consequent destruction of feeders. As it turns out, Dick has complaints in that department, too: "Squirrels! They mess with the sunflower seed put on the ground by the woodpeckers. Squirrels, chipmunks, doves, it's a zoo under the feeders."

What kind of creature is this gray squirrel that causes such condemnation and outrage in humankind?

A rodent, that's what. First cousin to the flying squirrels and ground squirrels; kin at only a slight remove to mice, rats, and voles, beavers and porcupines. All are members of the order Rodentia, a Latin word that means "gnawers." And all have notable incisors, enamel in front, dentine in back, which are not rooted in the jaw but continue to grow throughout the animal's life. Lest they become so long that they keep the animal from feeding properly, such teeth must be frequently filed down, a task accomplished, of course, by chomping into whatever's handy—a tree for dam building, the husk of a succulent nut, the rafters and stored goods in an attic, the bird feeder in the back yard. (Mice, which always make themselves at home in our North Carolina trailer during the winter, have planted their incisors not only into candles and soup-can labels but also into the stove's pink fiberglass insulation, which they've used to line the nests located in sundry drawers. And once upon a time, a gray squirrel sank its teeth into my tomatoes, still green on the vine, but more about that later.) Gnawing sharpens the teeth by quickly wearing away the soft dentine and leaving a well-honed cutting edge on the hard anterior enamel.

In the next zoological subdivision, that of family, the squirrels belong to the Sciuridae, which means simply "squirrelkind." The Latin word for squirrel is *sciurus*. It occurs in the work of Martial, the Roman poet, and that of his contemporary, the naturalist Pliny, in the first century A.D.; the squirrels to which they refer, however, are European species, not the gray squirrel, a native of North America that was introduced into the Old World only in the last hundred years.

Squirrelkind is further divided into three groups: tree, flying, and ground squirrels. The last group denotes the burrowers—the chipmunks and woodchucks of the Eastern United States, the prairie dogs of the central plains, the marmots of the Rockies, and at least twenty species of striped ground squirrels. (Burrower though it be, I have, however, seen a woodchuck huff and chuff its way up a tree to make a desperate—and successful—high-rise escape from my dog.) The flying squirrels are huge-eyed creatures of the night that leave their cavity nests after dark to glide in search of nuts, grains, blossoms, insects, and occasionally birds' eggs. My favorite early observer of the natural scene, John Lawson, surveyor general to the lords proprietors of the Carolinas, published this description of airborne squirrelkind in 1709:

> He has not Wings, as Birds or Bats have, there being a fine thin Skin cover'd with Hair, as the rest of the parts are. This is from the Fore-Feet to the Hinder-Feet, which is extended and holds so much Air, as buoys him up, from one Tree to another, that are greater distances asunder than other Squirrels can reach by jumping or springing. He is made very tame, is an Enemy to a Cornfield (as all Squirrels are) and eats only the germinating Eye of the Grain, which is very sweet.

Today, while the Southern flying squirrel thrives, the Northern species, not uncommon in the northern United States and Canada, is so hard to find below the Mason-Dixon line that it has been listed in several states as threatened or endangered.

The tree squirrels of the East are three in number: Eastern red, Eastern gray, and fox. The fox, the giant among them, is formally named *Sciurus niger*, "black squirrel," for the black face it sometimes displays on its otherwise gray self; many that are resident in the North, however, wear rusty pelts. The pipsqueak is the red squirrel, *Tamiasciurus hudsonicus*, which means "Hudson's housekeeper squirrel"; its Western cousins, the spruce squirrel, *T. fremonti*, and the chickaree, *T. douglasii*, also bear scientific names that honor intrepid nineteenth-century explorers—respectively, Colonel John Charles Frémont, who shared an expedition to California with Kit Carson, and David Douglas, the Scottish plant collector, who is honored also in the common name of the Douglas fir. The Eastern red squirrel, a highly territorial beast, chatters at any creature, human or otherwise, that invades it chosen realm. I do not see it on the low-lying Carolina coast, but in the wooded town park where once I walked the dog during our Virginia winters, its dry, querulous bark was a constant as we made our rounds. The middle-sized gray squirrel—the one deemed stupid by my doctor—is formally known as *Sciurus carolinensis*, the Carolina squirrel, but it ranges widely in the Eastern United States, from New England southward. And it seems, in my observations at any rate, that as one goes farther south, the species gets noticeably smaller. It is large and bold in Connecticut and New York; I've seen typical specimens, many of them melanistic, all nearly as big and furry as Persian cats, romping through the Brooklyn Botanic Gardens. But in the Shenandoah Valley, though still hearty, the species is reduced in heft by perhaps a quarter, and on the Carolina coast, it's a puny little thing. Some that I see often in our riverside yard belong to a clan with a genetic predisposition for bushy blond tails, but all the gray squirrels that sport in our sweet gums and loblolly pines are such little things, more bones than meat, that I'd hesitate to put them in a pot of Brunswick stew.

In the seventeenth century, when Europeans first began to colonize these wild shores where prodigious hardwood forests and piney woods stretched ever westward, gray squirrels were even more abundant, more ubiquitous than they are today. But between that time and the earliest years of the twentieth century, they were almost extirpated. Their numbers had been so reduced that some doubted the species would survive. Several reasons account for the disastrous diminishment of gray squirrel populations. One was loss of habitat as the unbroken aboriginal forests fell to the ax. A second reason, just as important and equally fatal, was that noted by John Lawson: like its flying cousins, the gray squirrel was, is, and ever shall be an "Enemy to a Cornfield." The animal's sweet tooth was nearly its undoing, for bounties were widely handed out by sore-beset farmers for the carcasses that shooters brought in. In 1748, Peter Kalm, the Swedish plant collector, noted in his journal that in Maryland everyone was required to kill four squirrels yearly and to present their heads "to a local officer to prevent deceit."

Kalm also reported an imaginative rumor, circulating among Pennsylvanians:

> Squirrels are the chief food of the rattlesnake and other snakes, and it is a common fancy with the people hereabouts that when the rattlesnake lies on the ground and fixes its eyes upon a squirrel, the latter will be as if charmed, and that though it be on the uppermost branches of a tree it will come down by degrees until it leaps into the snake's mouth. The snake then licks the little animal several times and makes it wet all over with his spittle so that it may go down the throat easier. It then swallows the whole squirrel at once.

But snakes, no matter how hearty their appetites, could never have accounted for the demise of many squirrels.

Along with dwindling forests and slaughter for money, I see a third reason for plummeting populations: the eternal human fondness for sport, in which people go hunting not to obtain necessary victuals but rather to show off their marksmanship. *Bang!*—the Carolina parakeets and the sky-darkening flights of passenger pigeons fall to earth forever. *Bang!*—the Eastern elk are extirpated, the shaggy, thunderous herds of bison also disappear from Appalachian valleys and from the Great Plains as well. *Bang!*—the gray squirrel came within a whisker of joining them in whatever heaven is reserved for animals.

It may be, however, that squirrel shooting—for whatever reason, money or sport—helped Americans beat the British in the Revolutionary War. In the early days, when this country was still a collection of colonies, its people developed long, lean rifles for hunting small game. Slimmed to 40- to 50-calibers, using lighter balls and less powder than traditional muskets, these flintlocks could outshoot the typical British weapon three yards to one and not only that but do so with winning accuracy.

The squirrel as denizen of the woods, the squirrel as fancier of corn, and the squirrel as target pop up frequently in the writings of John James Audubon, who wielded a gun as skillfully as he did a paintbrush. In one of his eclectic "Episodes," which are short, wonderfully pictorial—though frequently overinventive or downright mendacious—essays on people and events, he notes that, like the cunning raccoon, the squirrel knows "when the corn is juicy and pleasant to eat." In another, he contrasts the bleakness of camping on the granite coast of Labrador with the pleasures of doing the same thing in Mississippi, where, despite heat and mosquitoes, "the barkings of a joyful Squirrel, or the notes of the Barred Owl, that grave buffoon of our western woods, never fail to gladden the camper." Then, in an Episode entitled "Kentucky Sports," Audubon's talent for invention enters jauntily when he speaks of another kind of barking:

> *Barking off Squirrels* is a delightful sport, and in my opinion requires a greater degree of accuracy than any other. I first witnessed this manner of procuring squirrels whilst near the town of Frankfort. The performer was the celebrated Daniel Boone. We . . . followed the margins of the Kentucky River, until we reached a piece of flat land thickly covered with black walnuts, oaks, and hickories. As the general mast was a good one that year, Squirrels were seen gambolling on every tree around us. My companion, a stout, hale, and athletic man, dressed in a homespun hunting shirt, bare-legged and moccasined, carried a long and heavy rifle, . . . which he hoped would not fail on this occasion, as he felt proud to show me his skill. The gun was wiped, the powder measured, the ball patched with six-hundred-thread linen, and the charge sent home with a hickory rod. We moved not a step from the place, for the Squirrels were so numerous that it was not necessary to go after them.

The fiction here is Audubon's claim to acquaintance with Boone, fifty years his senior, who would have been doddering, or close to it, at the time of this purported encounter. But the squirrel-hunting method that he describes is nothing less than gospel truth. The hunter did not set each squirrel directly in his sights but rather aimed at the bark just below the point at which the animal clung to a tree. When the ball hit, the bark was "shivered into splinters." It was the mighty concussion that killed the squirrel, which then fell undamaged to the ground. The barking technique was designed to avoid shivering the squirrels themselves into splinters. And so a day of hunting would proceed, squirrel after squirrel after squirrel, as many as the gunners wished. (I wonder how many ended up in pots of Brunswick stew.) Dick, my doctor, would surely have lauded their prodigiously successful efforts.

After hitting a low point in the early 1900s, *S. carolinensis* has recovered nicely from onslaughts of sportsmen like Audubon and Boone. The new lease on life, however, is not so much accounted for by a loss of human interest in hunting for the sake of hunting as by the imposition of hunting seasons and limits and, more, by the protection and restoration of Eastern woodlands, the gray squirrel's necessary high-rise habitat. State parks and game commissions, national forests, private organizations promoting conservation—all have served to pull back the joyful barker from the brink of doom.

The wanton squirrels that blithely eat the farmers' corn, frolic through the trees in plain sight of shooters, or dig endless divots out of Dick Coleman's yard—are they stupid? Hardly, nor does stupidity figure at all in the equation. The animals are simply predisposed to be there, collecting, eating, storing food, be it given by nature or planted by man. The cleverness of squirrels, though, is well worth noting.

And clever they are, demonstrably so. A two-part documentary film called *Daylight Robbery*, made several years ago in England, shows gray squirrels outwitting the most contortive human attempts to thwart their raids on bird feeders. In that land, the gray squirrel is an invader, an exotic import that, since its arrival a century ago, has taken over much of the native red squirrel's territory. Squirrel Nutkin, his brother Twinkleberry, and their small, talkative red-brown kind have suffered conquest and a great reduction in number. So it's not just raids by gray squirrels—"tree rats," as some would have it—that rile the human population but their displacement of a species hallowed by Beatrix Potter and as lovable, if not so mythic, as Eeyore, Piglet, and Winnie-the-Pooh (not to mention Rat and Mole). The feeders built to foil attack by marauding foreigners were equipped with a multitude of ingenious devices: baffles that tilted and windmills that spun under the squirrel's weight, lines strung with rotating disks, tubes and tunnels, and a rocket-shaped sled that had to be mounted and ridden before the goal could be won. En route to the feeder, the squirrel sometimes had to make gargantuan leaps between one station and the next. The feeder itself was placed so that no access was possible other than that through this obstacle course more daunting than any in boot camp. Did victory elude the squirrel? Of course not. Winning through would take several attempts, but the bird-seed prize was inevitably seized, gnawed, and digested. Gray squirrels are not only clever; they possess a fierce determination and a considerable aptitude for problem solving.

Two decades ago, I watched a gray squirrel work its way through the mechanics of a special problem—that of taking one of my green tomatoes up a tree. I lived then in the basement apartment of an 1840s house built on the side of a steep hill in the small Virginia town of my growing up. The apartment's location in the basement does not, however, mean that it is subterranean; instead, through windows six feet high, the living-dining room gives a fine view of the steeply downsloping back yard. And in that yard (where once I found kudzu stealthily sprouting up and stretching out), I'd made a garden to ease my annual craving for homegrown tomatoes, red-ripe and sweetened by the sun. The garden attracted thieves. One, not surprising, was my landlady, but I hadn't expected the squirrel. I became aware of its activities one noon as I stood in the kitchen fixing a double-thick BLT for lunch. The kitchen overlooks the side yard, but its window is also a six-footer that frames a tall tree. I no longer recall what kind of tree it is—it's still very much there, and I could go look—but its identity is not so important as the memory of the blue jays that built a rickety stick-nest in the lower branches one spring and let me watch them hatch and fledge their young. And certainly not so important as the memory of the astounding acrobatics executed by that squirrel as it attempted to carry a fresh-picked green tomato up the trunk. Time after time, it made a flying leap upward, grabbed the bark with its cling-fast claws, and tried for the top at full speed. Time after time, the green tomato, a good bit larger than the squirrel's head, acted as a weight that loosened the claws, swung the animal around one hundred and eighty degrees, and sent it lickety-split back to the ground. To the rescue: cleverness, determination, and a keen ability to solve food-related conundrums. And how was the job of getting tomato from ground to the leafy heights accomplished? The squirrel ascended the tree backward—tail first, tomato last. Without futile preliminaries, similar ascents were made the next day and the day after. It may have been the one of the smallest circuses on earth, but what grand entertainment, and all for the price of a few green tomatoes! The depredations of the landlady (my tomatoes but *her* back yard) were far more serious.

By now it must be clear that squirrels, at least to my way of thinking, are anything but despicable. And I'm sure that even those thousands of people who truly, deeply detest them will acknowledge that they play a not insignificant role in the scheme of things. In fact, they're part of a win-win symbiosis: just as trees serve squirrels, providing them with food, shelter, nesting materials, exercise, and safety from earthbound creatures (or those, at least, that do not carry guns), so squirrels serve trees by practicing reforestation. If not for those nuts that they store but forget about or fail to find despite the most intensive digging, the Eastern United States would be less thickly covered with black walnuts, the many kinds of oaks, and a multitude of different hickories, including the pecan.

Villains? Vaudevillians? Something else? Whatever gray squirrels really are, we tend to see them mainly as their lives impinge on ours. In that respect, I must grant that on occasion they're nuisances, teasing dogs with malice aforethought, bombarding passersby with sticks and acorns (the very missiles that Squirrel Nutkin, grown irritable with age, hurled down from his tree). But if the critters invade and tear up attics, if they work mischief on yards and feeders, it's surely because we ourselves have presented them with opportunities that their inborn squirrelness cannot resist. Problems like those of Dick Coleman—the dug-up lawn, the zoo under the feeders—aren't problems at all to the teeming, tail-flicking hordes of S. carolinensis that wreak the damage. They're simply taking full advantage of the circumstances in which they find themselves.

Dick's wife, Jan, has taken to trapping and transporting squirrels. Alas, though that strategy bears short-term satisfaction, it's tantamount to commanding the tides not to roll in. Remove one squirrel, another—maybe two—will take its place. But, for what it's worth, I have a cure for squirrel-wrought devastation. Bird feeders should, of course, be banished altogether. Then, lawn and bordering beds of shrubs and flowers should be converted into pavement or, more aesthetically pleasing, a Japanese garden unsuitable for burying nuts: rocks, carefully raked sand, greenery that does not furnish squirrel food, and a pond with lily pads and bullfrogs. The damage worked by the frogs would be minimal, and if their population should explode, well, frogs' legs are almost as good to eat as Brunswick stew.

Ah, Brunswick stew! By way of a coda to this inquiry into the nature of the gray squirrel, here is the old-fashioned but eminently usable recipe handed on by my grandmother, Jannette, for whom I was named (and, yes, my name sports all those extra letters, but for reasons of laziness I do not use them). Her father, a medal-winning marksman, spent much time in the field, and to commemorate his exploits, his wife, my great-grandmother, designed and worked a large piece of filet crochet that shows him in deerstalker and hunting breeches, rifle on his shoulder and hound dog at his side. The piece, mounted on brown velvet and framed, is now displayed in my upstairs hall.

End of divagation. Here's the recipe. For the conservative palate, chicken may be substituted.

## JANNETTE EAST'S BRUNSWICK STEW

*3 squirrels, skinned and cut into pieces*
*1 gallon water*

- Boil till meat comes off bones. Chop meat fine after removing bones. Put back into the same water. Add:
  *¼ pound lean country bacon, chopped fine*
  *1 quart Irish potatoes, sliced thin or cut in squares*
  *½ quart lima beans, fresh or dried soaked in water and split*
  *1 quart tomatoes, peeled and cut*
  *1 large onion, chopped*
  *½ teaspoon Worcestershire sauce*
  *salt and pepper to taste*

- Boil gently for 5 hours. As water boils down renew with hot water and stir often to prevent burning. Add:
  *6 ears corn*
  *½ pound butter*

- Cook until corn is done and butter is melted.

NOTES FOR THE CONTEMPORARY COOK:

1. Cooking time may be reduced from 5 to 2½ hours. The secret is to keep the ingredients at a constant and *gentle* simmer.
2. Cut the kernels off the corn and add them, not the cobs, to the stew. Frozen corn may be substituted.
3. The butter may be eliminated, for the country bacon will provide enough fat for smoothness and flavor.

4. As accompaniments, serve my grandmother's favorites: biscuits and butter, a tossed green salad, and, for dessert, devil's food cake or lemon meringue pie.

5. As Audubon might have said, *Bon appétit.*

# MURMURATIONS

Starling (*Sturnus vulgaris*)

European Starling

APRIL: a sudden eviction has occurred in the new living quarters located in a tall pine snag near the wide and salty river Neuse. For the last two years, the snag has served as a high-rise condo for woodpeckers. Excavating a fresh new hole midway down the weathered, barkless trunk each year, a pair of pileated woodpeckers has nested deep within the dead wood. Early this April, yellow-shafted flickers took up residence in the pileateds' old quarters, and red-bellied woodpeckers drilled out a completely new apartment in the uppermost level of the snag, at a juncture that is marked by the silvery stubs of three broken-off branches.

It was, in fact, the red-bellied male that alerted me to the new apartment in the first place. For several days I'd heard him drumming and trilling, trilling and drumming. I knew what bird he was by his cry, which is not the whistle or skriek typical of other woodpeckers, and certainly not the lunatic cackle of the pileateds, but rather a rapid, throaty, sweet-ringing music. Of course, I went to see if I could find him. And there he was, clinging to the snag, propped on his stiff-feathered tail, mere inches from a hole with freshly chipped edges. As I watched, he slipped inside. During the next week, I listened and watched, found the year's other woodpeckers, and kept an eye on the whole community. The pair of pileateds spelled each other; sometimes his head, sometimes, hers, bold and red-crested, would be visible in the entrance to their nest. The flickers were cagey, seldom allowing themselves to be seen; their comings and goings were given away mainly by small, downy feathers caught on the rough wood of their doorway. All the while, the male red-belly put himself on display near the tip of the snag.

It may have been the red-belly's musical presence that attracted the rude attentions of the small, swart, pushy stranger. Or perhaps the stranger's dark eye was simply drawn to a handy-looking hole. Whatever the reason, eviction ensued. It was swift and merciless, nor could it be appealed. The stranger was a starling, of course, a bird notorious for throwing out legitimate tenants and moving right in. Within an hour, he began furnishing the cavity with pine straw.

His diligence soon lured a mate, who helped him put the finishing touches on interior decoration and then settled in to lay her eggs. As his kind does (and some other kinds as well), he guarded her jealously and at the same time copulated with other females every chance he could get. He may even have found a second mate and established another household.

Late May: the secretive flickers leave no trace of their nesting failure or success, but the pileateds fledge at least three young, two of them red-mustached males. After the red-bellies' dispossession, I often saw them drumming near the snag and heard their trills; the appearance of young birds testifies to a triumph over the forces of darkness. As for those forces, the soot-brown bodies of fledglings mingle with black adult bodies to crowd the three silvery branch stubs at the top of the snag. And there they perch, announcing themselves with wheezes and sighs.

*Sturnus vulgaris*, the European starling—this is the bird that North Americans love to despise. Indeed, we seem to despise it above all other avian species, even the house sparrow and the rock dove. (The latter is the euphemistic name applied by ornithologists to ordinary pigeons.) The starling regularly dispossesses birds that people consider more attractive—not just woodpeckers but other cavity nesters, like bluebirds, titmice, the golden prothonotary warblers, and the great crested flycatchers, which are feathered in cinnamon and lemon. It even competes for space with cavity- and box-nesting wood ducks. Nor are its raids on the territory of other species confined to the natural world. It also drives city folk to desperation by soiling buildings as thoroughly as pigeons do; it enrages farmers, impoverishes them, by wreaking havoc on fruit and grain crops. It behaves like a ghostly intruder, invisible and frightening; it has figured at least once in rumors of diabolic practices. It provoked my peaceable, nature-loving father into picking up his gun.

My father's long-ago market-garden farm in Ohio: I remember that he toted a shotgun as he led me to a crab-apple tree halfway up the dirt lane, past the storage barn where horsedrawn ploughs, harrows, wagons, and hay balers were kept, along with a mess of wooden barrels for making vinegar. The season must have been late fall or early winter, for the tree was leafless. The branches, however, were far from bare. Black birds no bigger than my child-sized fist had perched themselves wing to wing on almost every available limb and twig. They were thick as cockle-burs; they hunched and glowered; softly, they jeered. I knew they were bad birds before my father introduced them. "Starlings," he said. "Stay back, honey. I'll show them." And he fired. Darkness fell from the tree in small feathered clots. But most of the flock simply wheeled upward with a slap of wings and flew away. After counting the fallen, my father explained that starlings showed no mercy, working intolerable damage on his cherries and plums, pears and peaches—the dang birds invited being shot. The air did not clear. I knew even then that my sense of something bad resided not so much in the birds themselves and their unlovely irruption into a child's awareness as in my father's lethal response to their presence. It wasn't until I was older than he'd ever lived to be that my brother, who'd read his farming daybooks, made me privy to more information: with that old shotgun, an Iver Johnson .410, he had not only regularly blasted down starlings but kept meticulous records on the number of victims. Worse, he'd com-pounded the crime by mounting the same kind of determined campaign against house sparrows.

Starlings, however, are not protected anywhere in the United States. As one species in an unholy trinity that includes the house sparrow and the pigeon, the bird is considered open game, up for slaughter in any season. State laws may apply to all three, nonethe-less, with a hunting license required for would-be bird plinkers, or a special permit for scientists who aim to trap, collect, and study them. But these pariahs—who springs to their defense? Will animal-rights activists ever mount campaigns to protect them? The success of any such movements would surely be doomed at the outset: the unlovely three are too many to count, while their partisans are few. What good are they anyway?

In particular, what good are starlings? Apart, that is, from being good to themselves in the highly successful perpetuation of their own squat, sooty kind? But, though the bird does not, and never will, muster much human support, it has also proved useful to us, albeit in tiny and peculiar ways. It is ornamental: in Europe, people intentionally attract starlings to their yards by erecting nest boxes. It helps anglers: in Britain, the feathers, taken from birds shot in their winter roosts, are using in tying fishing flies. It is edible: the bodies of the plucked birds are sent to the Continent and turned into paté. And *S. vulgaris*, though vulgar indeed, has been downright inspirational: not only did it provoke two—not one but *two*—compositions from Mozart, who kept a caged starling as a pet, but it has also provided an image for W. H. Auden and served as a muse to John Updike, from whom it called forth a fine poem. But more later about the music and the poems, and about a more recent starling project that has been awarded a grant form the Minnesota State Arts Board.

What good are starlings to the greater world, the world that encompasses both them and us? Are they required in the scheme of things? A necessary presence without which the links connecting all life on earth would be irremediably severed? Would their absence create a slow leak in earth's substance? Would it make a hole, a void, into which everything else would be sucked, vanishing forever?

All the starlings on earth—and they are legion—got off to an innocuous start. The family, with twenty-four genera and more than one hundred species, is the Sturnidae, which means simply "starlingkind." The family name comes from the Latin *sturnus*, the word that Romans used to designate the starling most familiar to the Western world, the bird that science has denominated *vulgaris*, or "common." The family includes the mynahs along with a varied batch of drab, compact, slim-billed birds that bear a close family resemblance to the European starlings we know (which are arguably the drabbest of the lot) and also many others that sport colorful feathers—rose, violet, glossy purple, vibrant orange, yellow, green, or white—that people use for their own adornment. Some of the common names speak of this splendor: golden-breasted starling, violet-backed starling, red-winged, red-browed, and rose-colored starlings. Some of these birds wear crests, others grow wattles, and the several Asian species, in particular, feature patches of bare skin around the eyes. Though most species nest in holes, some are choosy, preferring holes located in cliffs or behind waterfalls, while others like to dig out their own cavities in muddy banks or drill them, woodpeckerlike, in dead trees. A few, however, actually build nests that are cup-shaped or domed. Many starlings lay spotted eggs, though eggs laid by members of the genus *Sturnus* are generally spotless and pale. And, except for the Antarctic, starlingkind has made itself at home all over the world. Our bird, the European starling, may be found not only throughout Western Europe and the United States, including Alaska's Arctic north, but also in places as far-flung as Iran and Siberia, Egypt and the Madeira Islands, not to mention all the regions in between. It has been introduced into South Africa and Australia. It was the very first species I saw as my plane came in for a landing in New Zealand (the second and third species were also birds that had been transported for reasons of sentiment from their native lands: the house sparrow and the mallard).

European starling—that is the occasionally ornamental, sometimes useful, peculiarly inspiring, ubiquitous, plaguey bird that will inhabit this story from here on. Its everyday name might seem a chiding acknowledgment of the source of North America's infestation. But, no, the name is not reserved solely for exasperated use by people living on the west side of the Atlantic; it's heard throughout Europe, too. Another geographic term would have been more accurate: the starling we know best, along with its numerous kin, probably originated in southeast Asia. And there, in its earliest incarnations, it was likely an arboreal bird that sipped nectar, ate fruit and pollen, and snacked on the insects that were also attracted to the feast in the trees. Some of the Asian species still lead elevated lives, but members of the genus *Sturnus* have notably descended to the ground. Though they have hardly forsaken their passion for fruit, their dinners more often consist of grubs and other insect delicacies harbored in the soil. Clever birds, they long ago spotted and learned to exploit an underused subterranean larder.

Part of the European starling's success in colonizing our country coast to coast has to do with its ability to find buried treasure. Its preferred locations for foraging are open places with short vegetation, like pastures grazed by livestock and closely manicured lawns. The technique is this: insert closed bill into ground, open mandibles, probe for and find grub, shut bill on prey, and withdraw it forthwith. Meanwhile, between captures, a feeding bird will strut on the grass, engage in leapfrog with comrades, lift off and up for a couple of feet, wheel, and come down for a quick landing before reinserting bill in ground. The neighbors two doors up the street in my small Virginia town provide another sort of meal for starlings; at precisely 8:20 every morning, they cast slices of white bread upon their neatly mowed back lawn. And, oh, the commotions! At least two dozen birds descend before the bread hits the grass. Left, right, left, right, they march briskly to the bread, grab it, and tug it apart, one bird to each corner of a slice. Then, like jugglers, they toss the tidbits in the air, catch them, toss again, prancing and leaping all the while, and when they finally tire of the game, either swallow or fly off with this fluffy white food they didn't work for. Later in the day, they may return for a more ordinary bout of probing for invertebrates in our neighbors' back yard.

But the starling could not have colonized—no, conquered— the country had it arrived on these shores with the earliest settlers, nor could it have gained so much as a clawhold for the next two or three centuries. The dense forests in the East, despite natural clearings, and the tallgrass prairies of the Great Plains gave little quarter to ground-feeding birds. It was ax and plough that created habitat suitable for the species. Or, to be less euphemistic, habitat was created by the people who set fires to clear away vegetation, who wielded tree-felling axes and grass-cutting scythes. Denuding the land, we made it starling-friendly. Nor is it likely that starlings would have settled in without some help. Though the European starling has ever been a migratory bird, breeding in Norway and Britain, for example, and wintering on Spain's Costa Brava, its seasonal movements cover limited territory. If it hadn't been for a man-made magic carpet, the Atlantic Ocean would have presented a daunting natural barrier to the species' westward expansion.

At this point, William Shakespeare figures in the tale. He is not, however, among those who have found the starling inspirational. Instead, he bids one of his characters to employ it as an instrument of mockery. The locus is *King Henry IV, Part I*. Henry considers Edmund Mortimer, the Earl of March, to be a traitor who gave up in battle and so "wilfully betray'd the lives of those he did lead to fight." He will not pay ransom for such a treasonous man. But the younger Henry Percy, known as Hotspur, protests, saying that Mortimer fought nobly, indeed was wounded, and suffered an honorable defeat. The king is not persuaded. Angered, frightened by the political threat that Mortimer poses, he enjoins Hotspur not to speak of Mortimer again. But immediately on Henry's departure from the scene, Hotspur tells his uncle, Thomas Percy, Earl of Worcester, that he shall speak to the king of Mortimer. And not only speak but find ingenious help in doing so. The indignant Hotspur tells his uncle that King Henry

> . . . said he would not ransom Mortimer;
> Forbade my tongue to speak of Mortimer;
> But I will find him when he lies asleep,
> And in his ear I'll holla 'Mortimer!'
> Nay, I'll have a starling shall be taught to speak
> Nothing but 'Mortimer,' and give it him,
> To keep his anger still in motion.

Here Shakespeare notes the starling's stellar gift for mimicry. And though the Bard does not say so outright, the bird that Hotspur intends to enlist as a torment to Henry will be a caged bird, one caught young, just fledged perhaps, and steadily tutored thereafter by regular repetition of the enraging word it is supposed to imitate.

"I knew I was the only person at the barn—my car was the only one in the parking lot," says Debra, who is married to my younger son. "When I went in, the horses nickered—this is normal. Then I thought I heard voices, multiple voices, and distinct snatches of speech just beyond my hearing comprehension. I actually walked through the barn and called out a couple of times. Even at midday, one doesn't want strangers hanging around a stable."

Because the study of art history does not yet provide Debra with much remuneration, she earns some money at a stable, exercising horses, grooming them, teaching riding, caring for tack, and just plain mucking out. The labor is not purely physical, however: she has also constructed the Heirloom Index, a genealogical data base for Egyptian Arabian horses of the stock owned by nineteenth-century pashas and khedives or collected in the desert, then bred in both Egypt and England, and exported all over the world by Lady Anne Blunt. But Debra's greatest benefit from the job is not the wages nor the satisfaction from having created a widely consulted data base, not to mention the glad chance to be outdoors in all seasons. Just as the children of a teacher in a private school often attend that school without fees, her own Egyptian Arabian, a stallion named Ibntep of Rose, is boarded there free of charge. Ibntep is a wonder horse, survivor of a broken leg, but that is another tale. He takes part in this story as the prime reason that Debra entered the barn that day: to ride him out and put him through all his elegant paces so that he could stretch his long silver-gray legs. But the voice of the impossible intruder gave her pause.

"The voices were still speaking when I returned to the door," she says. "I looked up then and saw a starling perched on a rafter, a single bird that continued to speak in human tongues as I watched and listened. Now and then it would punctuate its monologue with its own call and the calls of other birds. Starlings always remind me of those birdcall tapes and records—the sound is garbled and scratchy and skips from one song to another. But I was definitely hearing human phonemes."

As she stood there looking and listening, relieved that no one was rustling the animals or stealing tack, she wondered why the bird didn't imitate horses instead of people. Then she realized where this bird had learned its mumbles and its murmurations; she knew how it had picked up the sounds, if not the content, of human conversation. "Looking up at the bird reminded me of all the times I've driven down country roads and seen starlings lined up on the telephone wires. They tend to sit next to the poles when there are only a few birds, probably because there's less sway in the line. And that's where the leakage is greatest. I know a lineman who used to joke about all the things he heard when he rejoined phone lines."

So, among their many other failings, starlings are eaves-droppers.

At least once in history, starlings have been the models, not the imitators. My cousin Bess relates her own story, told to her by grown-ups after the event—and after they had realized that noth-ing ailed her. As a toddler, she spent most of her days and nights with a taciturn nursemaid on the third floor of a palatial stone house. Her bedroom lay under the eaves and overlooked a court-yard (I know that room; I've slept there, too). But despite her isola-tion from the rest of the household, she was a talkative child. The problem was, no one could understand her, for she spoke in whis-tles, screeches, and warbles. Worry spread upstairs and down that the child had a speech impediment or, worse, a deficient brain, and it persisted until the day that someone noticed the birds scrabbling, fussing, and probably nesting under the eaves. "A little kid repeats what it hears, and what I heard was bird," Bess says. "Of course I spoke starling."

In that respect then, starlings are like little kids, adept at repeating what they hear. And their stellar talent for mimicry was a subject for comment long before Shakespeare's time. The Roman naturalist Pliny the Elder, writing in the first century A.D., mentions a starling and several nightingales that belonged to Nero as a boy and to his stepbrother Britannicus, son of the emperor Claudius. The birds, given into the care of a trainer, were taught to utter words, then sentences of increasing length, in both Latin and Greek. Taking his feathered pupils into a quiet, private room, where no other sounds could interrupt and possibly influence the lessons, the trainer repeated words and rewarded success with food. Pliny remarks, as well, on other species that can be taught to talk—parrots, magpies—and declares that a magpie kept from repeating, repeating, repeating its favorite words will dwindle and die.

Pliny's record is one of the earliest to note the European starling's abilities as a vocal acrobat. (For ordinary, untutored starling-talk, the Romans used an onomatopoeic verb: *pisito*, I wheeze, I sigh, I hiss.) Pliny is far from the first, however, to make mention of the bird. In the Western tradition, Homer's *Iliad* may well have that honor, twice bringing an image of starlings—*psares* in Greek—into the fray of battle and both times associating them with crows. In Book 16, the Greek warrior Patroclus, maddened by grief at the death of his best friend, strikes at the enemy's front lines like a hawk stooping at full speed to scatter frightened crows and starlings. In Book 17, the Trojans in turn relentlessly pursue the Greeks; Aeneas (who will sail westward after Troy's fall to found Rome) and the crown prince Hector charge the Greek army's rear, and the soldiers flee before them like a cloud of cawing, shrieking crows and starlings.

Starlings were rarer in Homer's day, and also in Shakespeare's, than they are at present. (The Bard is only waiting in the wings. He'll reenter after the story makes one or two more turns.) But whether starlings were more highly regarded in ages past than they are today, as a boon to some few, a bane to everybody else, is a question to which all the answers have been lost in the fogs of time. It is known, however, that the English word that designates the species is venerable indeed. The oldest English texts, dating back to the early 700s A.D., use the form *staer*. And the Lindesfarne Gospels, about A.D. 950, offer two *staras* instead of the canonical pair of sparrows in Matthew 10:29. Later, in the fifteenth century, *The Book of St. Albans*—treating of matters important to the nobility, like hawking, hunting, and heraldry—compiles and codifies venereal terms, particularly the group names that might (or might not) be important to hunters; here, familiar terms, like a school of fish and a host of angels, mingle with those that are decidedly inventive or outright strange, like a skulk of foxes and a husk of hares. The terms pertaining to birds show much imagination: a pitying of doves, a siege of herons, a murder of crows, a "murmuracion of stares." (It was definitely murmurations that Debra heard, not the sharper *pisito*.) In the early days, the suffix *-ling* was sometimes tacked on, sometimes left off (it may be a diminutive, though no one knows for certain, but it has since glued itself to the rest of the word, from which it is now inseparable). A Northumbrian poem of the fourteenth century sings of a sparrowhawk that flew by the "sterling," and a tale of King Arthur, written in the mid-1400s, brings in a battle image that might have come from Homer: "They smote in among them as a falcon among starlings." In 1667, the bird's gift for mimicry astonished Samuel Pepys, who wrote in his diary of a "starling which . . . do whistle and talk the most and best that ever I heard anything in my life." The bird's penchant for imitation continues to astound; in the 1980s, one Margarete Corbo wrote *Arnie, the Darling Starling* about her adventures with a garrulous bird, whom she found as a naked nestling and cared for fondly until its death four years later. She learned of its talent for mimicry when, one day, it suddenly shrieked the name she'd given it— "Arnie! Arnie!" But of her views, more later.

Mimicry: from all that's been said above, it may seem that the starling is confined to imitating talk, whistles, and telephone conversations, to Greek and Latin sentences or the endless hollering of "Mortimer." Not so: the bird's agile tongue finds stimulus in music, too. And so Mozart discovered to his amazement and delight. The place, Vienna; the year, 1784: soon after the composer had completed his Concerto for Piano and Orchestra in G Major (K. 453), he heard the happy little tune that informs the last movement played back to him out of the blue. Or rather, not out of the blue but out of the bill of a caged starling, which he bought immediately from a nearby shop for a sum that would amount to about twelve dollars in U.S. currency today. In this case, composer inspired starling; the bird had listened well while the man had fingered the piano keys and put the piece together. A surviving description notes, however, that the bird's mimicry was not perfect, for it sang a G sharp instead of the G flat that had been written into the music. Mozart, enchanted, forgiving the bird, commented, "That was lovely." Not much later, in 1787, the roles were reversed: starling inspired composer. The man also knew how to listen well and, using what he'd heard, wrote a sextet for strings and two horns that one critic has described this way: "Full of wrong notes, structural defects, uncalled for cadenzas, the piece represents a playful, practical-joking vein in Mozart of which his biography gives ample evidence, but which his music seldom exhibits." The critic does not mention Mozart's pet, but here there is no doubt that the music imitates the bird. Throughout, the notes are embellished with murmurs, wheezes, whistles, and sighs, and the very last note is a loud, ungainly, completely appropriate squawk. The piece (K. 522) has been nicknamed *ein musikalischer Spass*, a musical joke. And what a merry jest it is!

In 1880, less than a century later, Cassell's *Natural History* delivered a pronouncement that all too soon came to seem like a joke: "The Starlings are found only in the Old World, where they form a very large and natural Group."

Reenter Shakespeare, the man behind the man who provided European starlings with their magic carpet. In 1890 and 1891, only a flash of time after Cassell had made his definitive assignment of starlings to the Old World, flocks totalling eighty to a hundred birds altogether were brought across the Atlantic and released in New York City's Central Park. The importer was Eugene Scheiffelin, the leading light of an acclimatization society. Such societies, with the laudable aim of aiding and abetting the exchange of plants and animals from one part of the world to another, were founded by the dozens in the nineteenth century. In the case of starlings, this aim was realized with measureless success. Scheiffelin's starlings were not the first, however, to have been brought westward. Earlier attempts had been made at settling the species in the New World, notably in California, where a small flock languished for several years in the 1880s before it disappeared. What the motive for that particular introduction might have been, I do not know. But the reason for Scheiffelin's importation has lasted as stubbornly as the birds that he released in Central Park: Shakespeare. In an admiring, entirely well-meant obeisance to the playwright's memory, he planned to bring to the United States every avian species that Shakespeare mentioned. And birds certainly inhabit the plays and poems—birds like the thrush, jay, and lark of *The Winter's Tale*; the sweet-singing nightingales and wrens of several venues; and, in *Macbeth*, the darkly ominous "maggot-pies and choughs and rooks." One hallowed Shakespearean species—the tiny wren that we know as the winter wren—was native to North America as well as to Europe; another—chanticleer, the rooster, along with his hens—had been most successfully imported and established by the Spaniards who explored all the Americas decades before Shakespeare set pen to paper and centuries before Scheiffelin thought to honor the Bard. But where today are the New World's nightingales, and where our song thrushes, skylarks, and crowlike choughs? The only introduced Shakespearean species that have taken hold are the starling and the house sparrow (Scheiffelin brought in the house sparrow, too, but he was only one of many sponsors of that most successful immigrant). And the starling has not just taken hold but taken as its rightful fiefdom more territory than anyone could have foretold.

The timeline for success reads this way: 1890, release in Central Park. 1920, a sure claim staked out to the mid-Atlantic states along the eastern side of the Appalachians, from Chesapeake Bay down through the Carolinas. 1930, except for Maine, occupation of the United States west to—and sometimes across—the mighty Mississippi. 1940, Maine taken over, along with much of New Brunswick; the birds' western limit now marked by the Rockies, with incursions into Mexico and Manitoba. 1950, another push west, this time to California's coast range, and another push north, into the southernmost regions of British Columbia. 1960, a large new claim established, from the southern tip of Canada's Georgian Bay west to the Pacific, thence south to Baja California. And still no end to the great grab: land seized in the Northwest Territories and Alaska, some of it at latitudes above the Arctic Circle. By 1993, a modest count numbered the offspring of the hundred birds released by Scheiffelin at a quite immodest two hundred million.

A simple formula describes the birds' explosive expansion in a single century: starlings increase in number with a growth in the human population. As we develop places for ourselves to live, the starlings see an open invitation and move right in. Symbiosis—that's the proper word to denominate the unwitting interaction between *Homo sapiens* and *S. vulgaris*, and its form is commensalism, in which one species gains but the other receives nothing whatsoever, no reward and no penalty. And in the involuntary arrangement between starlingkind and humanity, another element comes into play: try though we do, we can't shake off the association. Wherever we go, the starling follows. The only places in the continental United States in which the bird is uncommon are exactly those places that discourage us—the stony, treeless peaks and ridges of the Rockies, the sere landscapes of northern New Mexico, and Alaska's high tundra.

But commensalism accounts for only some of the starling's success. The bird has inner resources to call on: an almost omnivorous appetite, an eye for the advantage, a flair for adapting instanter to changing circumstances, a profound wariness, and a hair-trigger instinct for self-preservation. I'm also confident that the bird possesses an intellect greater than that found within most feathered skulls. But it's the appetite that most amazes me. Or rather, the anatomical acrobatics that allow the starling to alter its dining habits according to seasonal rhythms in the abundance or scarcity of certain foods. The facts about this biological feat first came to my attention through the 1993 booklet *European Starling*, Number 48 in the series of monographs issued on North American birds by the American Ornithologists' Union. In the warmer months, *S. vulgaris* dines mainly on a grand cafeteria of insects, along with other invertebrates like snails, earthworms, and spiders. When cold weather sets in and insect life huddles quiescently in winter quarters, plants provide the bird's primary diet; the list of preferred items includes both wild and domesticated fruits—plums, cherries, holly and sumac berries, hackberries, the applelike fruit of hawthorns, the small blue-black drupes of the black gum. But vegetables are harder than meat for a bird to digest. Since aboriginal times, however, the starling's gut has taken that inconvenience into account. During the frigid months, not only does the gizzard grow larger, but the gut also lengthens, and so do the villi—the tiny, hairlike intestinal protuberances that help absorb nutriments. And vegetables speed through the bird's gut at a clip far faster than that of meat.

But my greatest reward for latching on to Number 48 in the AOU's series has lain not so much in acquiring such arcane snippets of starling lore as in talking with Paul Cabe, the scientist who wrote it. And getting together with him to talk about starlings was yet another example of what I've come to think of as the small-world phenomenon, which is the oddly recurrent concatenation of the past with the present, the wonderfully unexpected awakening of dormant connections. When I ordered the monograph from a specialist in ornithological books, I was unaware of my earlier acquaintance with its author. But my memory was jostled by his familiar surname and, more, by his mention of collecting starlings in the Virginia county where I spend each winter. I'd first met Paul a good fifteen years before reading his work, had met him indeed when he was just a boy, not a slender, bearded man with a splendidly shaggy, home-barbered mop of chestnut-brown hair. He's now halfway through his fourth decade and teaches courses in biology and environmental studies at Saint Olaf College in Northfield, Minnesota. His parents—running a local business, performing frequently in summer theater and winter play-readings, restoring a quirkily Victorian house in the small town of my growing up—have forever been part of my days. One of his sisters is also a leading light of the county's bird club. Back home during his Christmas break, Paul sits at my kitchen table munching on a sandwich and answering a flock of questions between bites. With him is his slim, sloe-eyed and dark-haired, very pretty wife, Leigh Ann Beavers, who is an ardent starling partisan in her own right. She calls them "one of the prettiest birds we have."

"Yes, but," says Paul, "starlings are like certain kinds of music, loud and raucous, very hard to listen to."

"Yes, but," Leigh Ann responds, "there are melodies underneath."

Paul is by trade a population geneticist, not an ornithologist. And it is he around whom a suspicion of witchcraft crystalized. The factual basis for this canard may be found in his choice, years before, of a subject for his dissertation, which investigated starling populations in different areas of the United States, comparing them and seeking to uncover genetic variations from group to group. Why starlings? Because of their abundance and ubiquity, they'd be easy to collect for the samples needed, and, yes, he had obtained the permits necessary for legally taking birds, even those of the pestiferous sort. He hoped to "find signatures on a geographic scale that showed how they expanded across North America." And given a mere hundred ancestors for the millions on tens of millions of birds now peppering the New World, it seemed possible that the species might show evidence of population bottlenecks. A bottleneck occurs when the gene pool available to a species as a whole becomes limited because a founding group is small or a residual population, like that of the black-footed ferret or the Florida panther, is tiny. These isolated groups do not have access to genetic variety. In Paul's estimation, however, "One hundred individuals is not a small bottleneck." Sufficient variety was imported with the birds. Yet they themselves had the potential for creating new bottlenecks as small flocks with little variation in their genes migrated hither and yon, not so gradually spreading from the Atlantic to the Pacific. Starlings like to move; they migrate readily, with flocks that swell, diminish, then swell again as birds come and go. Checking on possible differences among populations, looking for signatures, Paul collected birds in Virginia, Vermont, Colorado, and California. Mathematical predictions of what should have happened turned out to be predictions of what had actually happened: the New World's birds showed only an insignificant decrease in genetic variation. Had geographic expansion left a signature in current populations? Apparently not; his work suggested that any differences so created would be ephemeral, for each population of European starlings in the United States is much the same as any other. While birds like fox sparrows that stay put and live in geographic islands, one in this valley, another in the next, show high degrees of difference—high, that is, compared to other kinds of birds— among their separate communities, only a low degree pertains among birds that are quick to travel. Paul says, "Starlings move around so much that any differences there might have been are erased."

It was en route to these conclusions that Paul found himself suspected of involvement in diabolic rites. For his studies of genetic differences among starlings, he took tissue samples from dead birds and stored them in liquid nitrogen till testing time. The testing procedure that he used was molecular—"a fairly old, late-sixties technique"—that looks at particular enzymes and examines the differences in the DNA-directed structures of their proteins. One balmy afternoon Paul decided to pursue his tissue-taking labors on the back porch of the Victorian house that his parents had barely begun to renovate. As his father, Tom Cabe, describes it, the dwelling was nothing more than a "tumbledown, haunted wreck, surrounded by ailanthus trees and overgrown boxwood," and the family was living in the nearby carriage house, actually a garage with a second-floor apartment that had been built on the foundations of a nineteenth-century stable. The porch was a better locale, airier and brighter, for the task of dissecting starlings. The scene that met the eyes of a child exploring the neighborhood and peeking through the boxwood was that of a grown man working with the fiercest, most minutely focused attention on—oh, no!—dead birds. And not just dead but black! Horror-struck, the child-spy ran back to the aunt she was visiting and babbled that she'd seen a bearded man doing rituals behind the Cabes' house. But what else could she have thought at the sight of someone picking apart those little, dark-as-midnight carcasses?

Alive, the birds are beautiful. Leigh Ann and I agree on that. Their old feathers shine glossy black, with a faint iridescence like that of oil shimmering on dark water. (According to Debra, my art-historian daughter-in-law, birds see more of the color spectrum than people do, and so the iridescence must appear to them as a "brilliant, dazzling array of moving color.") And their new feathers, donned after the yearly summertime molt, are decorated with delicate, lacy white spots. For the spring breeding season, the dark gray-brown bills of all the adults, male and female, turn to a rich golden yellow. And their legs—always eager to strut and march, prance and cavort—are cinnamon-brown.

"Each bird is an individual," Leigh Ann says. "If you see them perched on a wire, each one is doing something different—preening, stretching its wings, lifting its bill, whistling, taking a nap. No two are alike."

This view of starlings is one for which the Minnesota State Arts Board has awarded her a grant. She is a printmaker, working mainly with intaglio monotypes. To achieve them, she uses various methods to incise Plexiglas and copper plates with lines into which ink is forced; the tops of the plates are then wiped clean and run with paper through a press. Sometimes she enhances the images formed by the lines with monoprinting, a technique that adds ink to the surface of the plate in order to create complexities of shading and tone. Her recently funded project celebrates the starlings introduced to Central Park by Eugene Scheiffelin a little more than a century ago. Eighty birds? One hundred? Because no one knows exactly how many were released, Leigh Ann has settled on a happy medium of ninety, with each portrayed in a way unlike that of its eighty-nine companions. The time-consuming preliminary task of photographing starlings doing almost everything they can do has been painstakingly accomplished. The fine limited-edition intaglios have been made. And Leigh Ann has had a sponsored show of ninety starlings in ninety different modes of display.

Little did Scheiffelin know what he had wrought for the world of art when he imported stocky black birds with raucous voices and antic habits. Not just the visual but also the verbal arts have benefited from his act of homage. Not that starlings haven't been noticed immemorially by poets, and in our time, they have furnished W. H. Auden with a tender and hopeful simile. In the prologue to "On This Island," he offers a prayer to Love and asks that longing bring a man's thought "Alive like patterns a murmuration of starlings,/Rising in joy over wolds, unwittingly weave." "Joy" seems the meet word to describe the instinctive speed and synchronicity with which a large flock lifts and wheels, but Auden's starlings rise over the rolling, open uplands of his native England. It is John Updike who has noticed descendants of Scheiffelin's birds and proceeded not only to sing them but to let them name his poem "The Great Scarf of Birds."

The place is a golf course, in Updike's Ipswich I would think, for it lies on the edge of a salt marsh. The time is autumn, leaves turning orange and red, the geese flying south. The "trumpeting" of geese induces the poet and his golfing companions to look upward. And they see this:

> As if out of the Bible
> or science fiction,
> a cloud appeared, a cloud of dots
> like iron filings which a magnet
> underneath the paper undulates.
> It dartingly darkened in spots,
> paled, pulsed, compressed, distended, yet
> held an identity firm: a flock
> of starlings, as much one thing as a rock.
> One will moved above the trees
> the liquid and hesitant drift.

The starlings landed and thickly covered the grass of the fairway. But as the poet watched, one bird rose. Instantly,

> . . . lifting in a casual billow,
> the flock ascended as a lady's scarf,
> transparent, of gray, might be twitched
> by one corner, drawn upward, and then,
> decided against, negligently tossed toward a chair:
> dissolving all anxiety,
> the southward cloud withdrew into the air.

It does not diminish this poem as art to say that it is also an act of science: its description of starling behavior shows an acuity that might be envied by any naturalist.

"What good are starlings?" I ask Paul Cabe. "Or, to put it otherwise, how bad are they?"

He admits that they can be pests. At least, they're seen as such, notably in urban settings. And, showing a special fondness for cherries, they can strip orchards bare. As for their effect on other birds, he says, "Starlings probably depress the population densities of medium-sized woodpeckers and other cavity-nesting species. The great crested flycatchers are hard hit. But starlings won't drive any bird to extinction."

Nor will starlings come to cause more widespread ruination than they do now, with every orchard failing, every city thickly, hopelessly mired in droppings. The general despicability of starlings seems to me largely a matter of perception. How to improve their public image? How to turn each one into a "darling starling," like the garrulous Arnie? That bird's keeper, Margarete Corbo, obligingly provides us with a theory: "Since we never consider as a nuisance those creatures with whom we choose to share our lives, my solution would be known as Adopt a Starling." She adds that we should cease trying to rid ourselves of Arnie's brethren in favor of giving all bird lovers starlings that "shall be taught to speak nothing but 'Mortimer.'" Alas, despite its brave support for this most unpopular of birds, the theory lacks all practicality.

It occurs to me then to ask Paul a subversive question, a question that pushes the bounds of ecological correctness: "Suppose, just suppose that the cosmic plug were somehow pulled and all European starlings went the way of the dodo. Are starlings necessary?"

Paul replies, "I think about this sort of question a lot. It comes up frequently in environmental studies and conservation biology. My opinion is that for most species, extinction will not have cata-

strophic community or ecosystem results. Other species will probably shift their abundance, but nothing will disappear or fall apart."

And what might a world without starlings be like? A world
without starlings probing the soil of pastures, golf courses, and the
median strips of four-lane highways? Without starlings listening to
conversations through the high wires along every street and country
lane? Without starlings not just barging into woodpecker holes and
bird boxes but also invading attics, eaves, and clothes-dryer vents to
build their nests? Without starlings marching, prancing, and fighting
over bread or lifting into air as lightly as a lady's scarf?

"I think we would have fewer urban birds," Paul says. "I have
never been able to think of any native species which would do as
well in cities."

February: once again, winter recedes. The starlings shine glossy
black, and their yellow bills advertise the advent of the breeding
season. My annual sojourn in the valley of Virginia is nearly done;
I'll soon head back to North Carolina's wide and salty river Neuse.
There in the woodpeckers' condo-tree, I'll see another round of
winner-takes-all. No sense making book on the winner's identity. But
the red-bellied woodpeckers, the great crested flycatchers, the
pushed-aside others have grand instincts for beating the odds.
They'll be calling, nesting, and replicating themselves this year, next
year, and in years to come.

And another question, quite unlooked for, surfaces: What if the
cosmic plug were pulled on us? Is *Homo sapiens* necessary?

I'm willing to bet that if we weren't around, clearing land and
raising cities, there'd be fewer starlings.

But it's my daughter-in-law Debra who has the last word. In
addition to horses she takes care of small animals, particularly those
that have suffered injury or abandonment. Her skills were long ago
gained by rearing starlings and pigeons that had fallen from their
nests. So starlings are old acquaintances. "I've felt a smile creep over
my face," she writes, "listening to a starling go through its scratchy

repertoire of songs, and I honed my bird-listening ear figuring out which species it was imitating."

It's her ornithomorphism—her ability to see humankind in the image of a bird—that most catches my attention. She calls starlings the "clowns" of the bird world and its "gaudy gypsies, with voices pretending to be many of their neighbors." And she adds, "They are raucous, rude, wildly adaptable, murderous, thieving, irreverent mockers. Are they birds or humans? All of us have a little starling inside us, some more than others."

Amen.

# THE NATURAL HISTORY OF PROTEUS

Small Lobose Amoeba

Scaled Cyst Stage

Zoospore

Filose Amoeba

Stage with Extended Flagellum

Large Lobose Amoeba

*Pfiesteria piscicida*

NORTH Carolina's wide and salty river Neuse flows seventy-five feet from my front door. For fourteen years I've watched it, for fourteen summers set my crab pots and pulled a gill net out from the riverbank. At times, many of the menhaden in the net would show oozing red sores in the anal area, and some of the crab shells would be deeply pitted, as if they'd been burned with a red-hot poker. I kept the fish as bait for the pots but tossed back damaged crabs. The trouble, however, was occasional, with a low incidence of apparent disease. Then, in September 1995, I saw a strange glint on the river—sunlight reflected off a thousand thousand floating, bobbing silver carcasses. The water close inshore was covered with dead fish.

Fish kills have occurred for many reasons since fish first came to swim, hundreds of millions of years ago, in the waters of the world. But only in the last decade has it been recognized that the sores, the deformities, the dying may signal the presence of a shape-changer. As infinitely agile as Proteus, the Old Man of the Sea, it lives, feasts, and replicates itself beneath the surface of the waves. And at one and the same time, like Proteus, it is powerful, dangerous, prophetic.

Homer has given us the mythic history of that ancient Proteus, and the old Greek tale has once again found new life in English. Feisty and eloquent, in lines that surge like a swelling, lapsing sea, it is recounted in Robert Fagles's recent version of *The Odyssey*. The venue for the story is the salty deeps of the Mediterranean and Pharos, the island located a hard day's sail from the Egyptian coast. There, the Old Man of the Sea regularly comes ashore with his retinue of seals and curls up in a cave to sleep away the sea-washed afternoons. And there, Menelaus has been marooned for twenty days. After leaving the smoke-blackened ruins of Troy and pausing in Egypt on his way back to Sparta, he has come this far only to find that heaven is holding back the winds. Sails limp, rations nearly gone, stamina running low, he and his crew have almost lost hope for a day of safe return. Then one of the god's daughters offers help.

Ambush Proteus, she says, catch the "immortal Old Man of the Sea who never lies"; he will instruct you how to cross the "swarming sea" and reach the shores of home. And she gives Menelaus a warning to

> muster your heart and hold him fast,
> wildly as he writhes and fights you to escape.
> He'll try all kinds of escape—twist and turn
> into every beast that moves across the earth,
> transforming himself into water, superhuman fire,
> but you hold on for dear life, hug him all the harder!

Proteus does just that: he shifts—*zip*—with pelting, dazing speed from god into lion, serpent, panther, wild boar, torrential cataract, towering tree, and back to god again. Parting at last with the information Menelaus longs for, how to call forth the necessary, sail-filling winds and set course again, the Old Man of the Sea also prophesies the past, present, and future of the other Greeks who fought at Troy. And more, he tells Menelaus precisely what act of hybris so angered the gods that they stranded him on Pharos: in his haste to go home, the lord of Sparta had failed to make the proper sacrifices before he embarked from Egypt. He will, however, be given a chance to retrace his route and make amends.

Split-second transformations, resounding prophecy, a warning against arrogance: these are the main elements in the mythic history of Proteus. They are present also in the natural history of the shape-changer that ranges the creeks, estuaries, and sounds of our coast from the Chesapeake Bay to the southern tip of Florida. With striped bass, bluefish, and blue crabs, with speckled trout and sea horses, these waters swarm as actively as the Homeric sea. (They host bottle-nosed dolphins, too, the very species that escorted the Greek ships across the Aegean sea to Troy.) And the Proteus within the deeps and shallows here, though older than the idea of time, rose only an instant ago into human ken. First reported in 1985, it was identified in the early 1990s by JoAnn Burkholder, an aquatic botanist at North Carolina State University, as a dinoflagellate, a microscopic organism and one unlike any that had ever before swum into our awareness. A marine creature, it seems to prefer brackish and saline habitats, though it has been known to range off-shore in the full-strength salt of the Atlantic ocean. Some scientists believe that, in any of its habitats, it may assume as many as twenty-four fluent shapes during its life cycle; others opt for a maximum of eight. The precise number of stages is not so important, however, as the fact that some of them are toxic, killing fish as surely as the dynamite still used by a few scofflaw fishermen.

The discovery of this protean dinoflagellate and its predatory habits have been widely noted by the media: TV news reports and documentaries; National Public Radio commentaries; articles in magazines as various as *Outdoor Life, U.S. News & World Report,* and *Sports Illustrated;* and, not least, a 364-page book ominously entitled *When the Waters Turned to Blood.* Written by Rodney Barker, published in 1997, it sensationalizes politics and personalities but scants the creature's natural history. The author also offers considerable misin-formation. The title, for example, implies that the presence of the dinoflagellate in an active state is signaled when our sounds and rivers turn red. No; the phenomenon known as a "red tide" is due entirely to other microorganisms. And he writes that the word *dinoflagellate* is pronounced with a short *i.* Not so; the *i* is long, as in *dire.* The most egregious error lurks in his statement that the organ-ism belongs not to the plant kingdom but to that of the animals. In reality, it inhabits neither. Biologists, though they bicker about the precise number, have added three or four more kingdoms to the list. Joining the classic duumvirate of Plantae and Animalia are the Prokaryotes (sometimes called the Monera), single-celled organisms that lack a nucleus; the Protista, single-celled and other microscopic organisms with a nucleus; and the Fungi, from shiitakes and chanterelles to the causative agents of athlete's foot. With all other dinoflagellates, the Proteus in our sounds, estuaries, and brackish creeks dwells amid the Protista, the "very first ones," a kingdom in which diatoms, simple algae, and the slime molds also hold sway. It's a kingdom far more primal than those of plants and animals.

This creature nearly as old as the sea has been formally named *Pfiesteria piscicida* (pronounced Feast-EAR-ee-ah pis-kih-SEED-ah). The genus, which was assigned by Dr. Burkholder, honors the late Lois Pfiester (1936–1992), an algal biologist who specialized in freshwater dinoflagellates. Dr. Burkholder's original choice for the species name, *piscimortis* or "fish death," was invalidated by the rules of scientific nomenclature, and Karen Steinhalter of the University of Florida came up with the far more vigorous descriptive term *piscicida*, which translates into "fish killer." At least two species of *Pfiesteria*, including *piscicida*, have been identified in the coastal rivers of North Carolina, and two more in the Chesapeake Bay and its tributaries.

*Dinoflagellate*, an ecumenical word that combines the Greek *dinos* with the Latin *flagella*, means "whirling whips" and describes the threadlike filaments, usually two, that the organisms use to propel themselves up, down, and sideways through the water. Many species are protected by an outer membrane of cellulose called a theca; others, only thinly covered, are termed "naked" or "unarmored" dinoflagellates. Fish killer, an armored species, takes the standard form in several of its incarnations. Then—*zip*—it changes. The typically ovoid, daintily flagellated single cell becomes larger and assumes the shape of an amoeba with arms like the points of stars. *Zip*—it loses definition and spreads into an even larger amoeba, flattened and lobed. *Zip*—this stage turns itself into a minuscule spore that drifts in the water or into a ball-like cyst that sinks to the bottom and waits. Nor is shifting into these shapes and a slew of others the only way in which fish killer exhibits a stunning virtuosity. Sometimes it eats flesh; sometimes, like a plant, it feeds itself by photosynthesis. The latter is accomplished through a kind of swift, silent mugging: fish killer is a specialized thief—a kleptochloroplast, in formal terms—that steals chlorophyll from its rightful owners. In some incarnations it may be toxic; in others, benign. It is equally artful, equally devious in matters of reproduction, able to perpetuate its own kind either by sexual means or vegetatively through cell division. There is no set order for any of these changes. Fish killer spins through myriad transformations in accord with the varying signals delivered by its environment. It is, for example, the natural secreta—the flaking tissues, the excrement—of passing schools of fish that trigger a carnivorous frenzy. *Zip*—with a sudden twist, this microscopic ambush-predator changes from sleeping cyst to star or flattened sphere and releases a toxin that makes fish lethargic and injures their skin so that they are no longer able to maintain their internal salt balance. Open, sometimes hemorrhaging lesions may develop. The tiny predator feeds then on the blood and sloughed tissues. And when the fish die, fish killer reverts to a benign phase and feeds on the carcasses.

Fish killer's quick changes, its multiple disguises are not unique among dinoflagellates; some freshwater species have been observed to have as many as three dozen life stages. Though its style of predation seems extraordinary, other *Pfiesteria*-like dinoflagellates almost certainly employ similar techniques. And the creature's general noxiousness is shared by a sizable group of Protista. These species are known collectively as harmful algal blooms, a term that's been condensed into an easy-to-pronounce acronym, HAB. Several varieties of HAB cause the notorious red tides that occur in coastal waters. Not all red tides bloom red; some may be brown or yellow-green. But no matter the color, every one is the work of a species of dinoflagellate, like *Gonyaulax polyedra*, *G. catenella*, or the unarmored *Gymnodium breve*. Many of these dinoflagellates contain potent neurotoxins. Clams, oysters, and scallops that feed on them are able to store the toxins in their flesh without ill effect, but should a fish swim through contaminated waters, it will suffer paralysis and die. Should a human being eat shellfish harvested from tainted waters, or even breathe red-tidal surf or spray flung into the air, the central nervous system will be attacked: sickness, paralysis, sometimes death. It's *G. breve*—the "naked, short" species—that flouted the common belief that red tides occur only in sultry summer: riding the Gulf Stream, it swept up the North Carolina coast from Florida in

October 1987, and persisted till the end of December. Coastal waters were closed to shellfishing; trawlers and dredges stayed in harbor; seafood markets shut their doors. Since then, the red tides of the Southeastern coast have kept themselves elsewhere—off Texas and Florida, for example—but nothing guarantees that these lethally colorful HAB species will never return to the Carolinas and the river Neuse. Another sort of HAB, probably a dinoflagellate (though it may be a diatom, with a siliceous skeleton) is believed to be the cause of ciguatera, or tropical fish poisoning. Fish like dog snappers and some species of amberjack eat smaller fish that have fed on this toxin-producing protist, and people who innocently dine upon affected food-fish suffer serious consequences to the central nervous system.

It is speculated that *P. piscicida* may also inflict harm on human beings. Nausea, loss of memory, lesions on the skin—these and other possible effects have been reported. Cause and effect, however, have not been firmly linked, nor have fish killer's particular toxins been identified. But tutored by my own experience, I can firmly state this much: cuts from sharp fish scales and puncture wounds from the spines of fish and crabs are as septic as a dog's bite or that of a man. It's likely that many—not all, but many—of the infections blamed on fish killer are triggered instead by failure to cleanse the wounds and apply antiseptics. I suspect, too, that the human propensity to place blame elsewhere thrusts fish killer into the role of a scapegoat. Nonetheless, if someone should fall into water that abounds with dead fish, these are the recommendations: bathe as soon as possible, washing off with soap and uncontaminated water or a solution of one part household bleach to ten parts water.

What is fish killer's role in the world? Much has been said about the interdependence of all the great and small lives on earth. On the Carolina coast, I see the concept illustrated by myriad symbiotic relationships: the yucca fertilized by a specially adapted moth; the isopod that lives in the mouth of a menhaden and uses its host as a food-catching device; the aquatic grasses that serve as nurseries for new-hatched mullet and red drum, as food for mallards and canvasbacks; the black tupelos that entice migrating birds with their high-energy fruit and are then rewarded by avian dispersal of the seeds. In these relationships, at least one party needs the other to survive. The common metaphor is the "web of life"; the picture is that of intricate, often messy interconnections—a sticky fabric spun all higgledy-piggledy by a drunken spider—in which anything affecting one strand reverberates through all the others. Damage leads to the endangerment of living species, breaks mean their extinction, and all of us—moths, isopods, fish, plants, birds, people—are the poorer for destruction of any part of the web. But how does this conventional view of the natural scheme accommodate the nefarious HAB species? What is their place in the tanglement of things? What can their purpose possibly be?

*P. piscicida* and its kin are opportunists. They have a place on the low end of the food chain, yes, and in the lab, fish killer has been seen to serve as supper for slightly larger aquatic organisms, like rotifers and copepods (not to mention the journalists of the late 1990s who have feasted gluttonously on its newly recognized presence). No other life-form, however, is dependent on fish killer for its own survival. The ecological role of any HAB may best be understood if it is viewed as an organism subtly, ingeniously adapted to take advantage of its habitat. Like the European starling that long ago seized the chance to exploit a subterranean larder of grubs and worms not accessible to other birds, like the kudzu vine that grows in soils too poor in nutrients to support other vegetation, the HAB species have made themselves at home in niches that are otherwise unoccupied or underused. Indeed, with one huge exception, all earth's living entities, both great and small, are niche dwellers. The exception, of course, is we the people, who scrambled out of our original hunting-gathering niche only a brief ten to twelve thousand years ago when we managed to domesticate our plant and animal food. Since then, we've been roaming ever farther afield, exploring the planet, flying to the moon, and en route discovering a few creatures that are something like ourselves and teeming multitudes that are not. The first category is filled primarily with mammals and marsupials; the second with aliens, some of which, like goldfish and ladybugs, are humanly acceptable, while others, like rattlesnakes, ticks, tapeworms, and P. piscicida, are only to be cussed, avoided, and killed when possible. The aliens are numerous and wily, and they know their niches far better than we do.

But why should the HAB species in general and P. piscicida in particular—these aboriginal protists that have swarmed almost forever in this planet's waters—seem to have become newly malevolent? Or not malevolent, for they are not capable of forming wishes, but rather explosively predatory where once they'd seemed quiet and well behaved?

Considering these questions, I am reminded of the sea changes that the passage of millennia has worked in the Nereids. According to classical myth, these saltwater nymphs, daughters of the sea god Nereus, sported with Poseidon and Proteus in the waters of the fish-friendly sea. Riding bow waves along with the dolphins, they escorted Achilles' ships to Troy. And I like to think that as they gamboled in the surf and swelling combers, the sea glowed not just with phosphorescence but with the playful holiness of their presence. Between those times and ours, however, the position of the Nereids in Greek folk belief has been altered. They are no longer seen as bright, good-natured spirits but have assumed a dark, demonic character. A friend tells me that on the island of Rhodes he met an old man who said he'd encountered the Nereids one night. How did he respond? "I drew a cross on the ground and a circle around it," he told my friend, "and I stayed in the circle till dawn." Precisely why the Nereids' luster has been extinguished, my friend did not say.

Like the daughters of Nereus, fish killer has lived in the world's waters well-nigh onto forever. And like those sea nymphs it has become an object of fear. Though it has always fed itself by stealing chloroplasts to enable photosynthesis or by engaging in chemical warfare to acquire meat, its populations now burgeon, becoming more ravenous and lethal every day. What accounts for this great burst of insatiability?

Enter prophecy. The Greeks knew that it might be used not just to predict the future but also to tell the truth about the present. The Proteus in our streams and sounds, captured at last and identified, delivers a twofold truth. Its first component is the nutrient overload—the pollution—in coastal waters. Corn and soy farms, hog and poultry operations, pulp mills and phosphate mines, military-aircraft-repair depots, towns with water-treatment plants, rural waterfront communities with septic systems, the weedless and chemically lush riverfront lawn of my neighbor three doors down— all these and more variously release fertilizers, heavy metals, animal and human wastes, and other unnatural ingredients into the water. The bottoms where fish killer is wont to hide till prey comes swimming by are also paved with new detritus, from dead algae to mud washed down the river from raw residential developments under construction two hundred miles upstream. Like a tonic, like the kiss of a prince, this broth—particularly rich in phosphorus and nitrogen compounds—has waked the organism from its slumber. *P. piscicida* has always caused fish kills, but now, at century's end, its lethal work is seen with increasing frequency.

The second part of fish killer's truth is that of the increased stratification of its aqueous habitat. In the summer, when the dinoflagellate is most predaceous, the water in estuaries becomes layered: freshwater from rain and runoff flowing downstream meets wedges of denser saltwater working their way upstream, and the two pile up, one over the other, rather than mingling. This phenomenon is natural. The problem is that it's being unnaturally intensified, and in two ways. First, factors like farming, urban paving, and growing development of the watershed lead to a great increase in runoff (and to the likelihood of floods, especially in times of nor'easter or hurricane). Then, with private and commercial development and the consequent removal of vegetation and with the steady loss of wetlands, there has been a concomitant decrease in the ability of the soil to soak up water and release it slowly. When rain and runoff cannot be stored, they enter the water in all-or-nothing spurts. And when the summertime flow downstream is robbed of freshwater because none has been held back for steady release, then saltwater intrudes with greater force. But come a storm and a burst of heavy runoff, the flow of freshwater will also be enhanced. And when large amounts of freshwater become layered with large amounts of salt, fish killer especially rejoices.

Yet a third element in fish killer's transformation from peaceable microbe to terrorist was posited in 1997 by a physician working with river-related illnesses in coastal Maryland. He made the logical speculation that copper is implicated. Copper, added to animal feeds to prevent spoilage, forms salts when it is leached from manure piles, particularly on factory farms producing hogs and poultry. With other agricultural runoff, the salts then enter waterways, where they kill the algae and cryptomonads on which the dinoflagellate feeds in its benign state. Then, cupboard bare of its customary food, it must look for other prey.

Clearly, we've made the life of *P. piscicida* far easier by offering it an increased supply of food and better furnishings for the niche it's filled since time out of mind. And with these new comforts, the organism's noxiousness has been exacerbated. Its populations explode. Dead fish—menhaden, flounder, croaker, spot, a host of others—rise and float on the surface in great rafts. Battening on their flesh, fish killer ceases to reproduce vegetatively and becomes a sexual creature. But when it is sated or disturbed, it encysts itself, sinks to the soft bottom, and once again waits.

The Proteus of estuaries and sounds has given warning: we have been as neglectful as Menelaus of our responsibilities. We have not given due respect to nature's invisible forces. And our particular hybris has consisted of making heedless assumptions about water's ability to flush away chemical insults.

But the glare of nationwide publicity has at last, after more than a decade of denial and foot dragging, embarrassed the government of North Carolina, where fish killer's toxic work has been most evident, into setting up clean-water standards and authorizing funds to monitor the rivers and sounds and reduce their burden of pollution. Maryland and Virginia, where either fish killer or a closely related dinoflagellate has only recently been discovered, have been far less sluggardly; they sprang instantly into action, closing the streams in which it was found to both commercial fishing and recreation. Those waters have since been reopened but will be closed again should fish killer reappear in lethal form.

The moral is that where there's evidence pointing to fish killer's presence, the most useful approach is that of exercising caution and common sense. Here on the wide and salty Neuse, when the water flows free of floating carcasses and the shells of the blue crabs are clean, my neighbors and I are safe in setting crab pots and pulling out our nets. To our good fortune, such benignity is the river's customary mode.

As species, *P. piscicida* and the others in its genus are well-nigh immortal. Try as we will, our best efforts can in no way cause their endangerment, much less push them into extinction. And should some eruption of chance, some unimaginable accident, exterminate the creatures but leave all other life on earth intact, other equally hungry opportunists would move into the vacant niche. We do, however, have the power to damp fish killer's appetites, to reduce its ravages. Like Menelaus, we still have a chance to clean up our own act and make amends.

Planozygote

# BLOOD

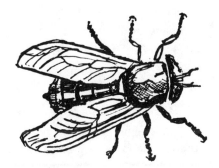

Black Horsefly (*Tabanus atratus*)

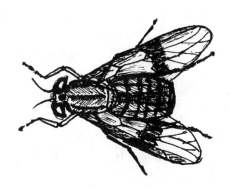

Deerfly (*Chrysops vittatus*)

## The Tabanids—Deerflies and Horseflies

I N mid-May, beware. Stay clear of coastal Carolina's deep green
shadows. Forgo the walks along the hedgerows and tangled
edges of Great Neck Point. Shun the woods and the rush-
guarded margins of the creek. Otherwise, as surely as blue jays will
steal eggs and hornworms infest the tomatoes, deerflies will also
make an instinctive response to spring's stern requirements.

The ravenous hordes descend without a sound. They circle
their quarry, then land, piercing flesh the instant they alight. Nor do
they cease to strike until their hunger has been satisfied. And, like
the legendary Amazons, every last one of this raging multitude is
female. The reclusive males loll idly elsewhere, feasting on pollen
and plant juices, committing careless acts of pollination while they
wait to fertilize the eggs of their own kind—eggs that can develop
only after the females are sated with blood. But after four weeks, no
later than the middle of June, the future of the species will have
been assured. Except for the occasional marauder, the deerflies disappear—until next year.

Coastal Carolinians aren't the only ones who are plagued by
these winged tormentors. A friend, an ardent angler who fishes out
West when he can, writes, "The only despicable creature I know is
the deerfly, which punctures my right hand, just as I'm trying to cast
a fly; I dearly love to squash them slowly." In his opinion, though
other forms of life, like snakes, qualify as thoroughly unlikable,
"Deerflies are really bad-ass." No region of North America offers
escape from deerflies; everywhere, from the East Coast to the West,
from the Rio Grande north into the Canadian Rockies, their various
species—all of them in the genus *Chrysops*, the "golden-eyes"—home
in on unprotected flesh and drink their blood meals. Their common
name refers to their preferred source of food: white-tailed and mule
deer. But lacking deer or other ungulates, they are perfectly satisfied
to feast on human victims. The head and neck are their favorite
zones of attack, and some, landing on the hair, will even try to burrow down to the scalp.

With all other true flies, from common houseflies and robber flies, bot- and blowflies to the daintier dance flies and flower-loving flies, the golden-eyes are members of the order Diptera, insects with "two wings." Most other insects boast two pairs of wings (though some, like silverfish and book lice, have none at all). In the Diptera, the second pair was anciently modified into a set of small organs, called halteres and shaped like narrow clubs, that are used as flight stabilizers. But where many kinds of flies vibrate their wings, zooming from here to there with a drone or a buzz, the golden-eyes opt for stealth and glide in on their prey with the silence of a moth. Their suborder is Brachycera—the "shorthorns"—a term that refers to their relatively short antennae. And the family is that of the Tabanidae, which they share with the physically larger, equally bloodthirsty horseflies; the family's name springs from *tabanus*, the Romans' generic word for any horsefly. That fly, a worldwide scourge, is represented by more than one hundred species in North America alone. *Tabanus americanus, T. punctifer, T. atratus, T. trimaculatus*—American horsefly, sting-bearing horsefly, black horsefly, three-spotted horsefly—the scientific name for each species usually refers to the insect's appearance or point of origin. One particularly vicious sort is commonly called a greeneye or greenie. Unlike its smaller deerfly cousins, which prefer a victim's upper reaches, it homes in on ankles and legs.

Oistros—that's the general word that the Greeks used for flies of this kind. And, in a time before time, it was just such a fly that inflicted divine retribution on the maiden Io, who had, without intent, attracted the lordly fancy of Zeus. But his wife, the lady Hera, took exception to putting up with yet another bout of philandering, and summoning the easy magic at her command, she turned Io into a cow. More than that, she sent an *oistros* winging after the now transformed object of her husband's lust. And *oistros*, often translated as gadfly, was nothing other than an aboriginal horsefly, perhaps the first one ever in the world. According to Aeschylus in his play *Suppliants*, this is what Hera and her *oistros* accomplished: when the fly found the cow (grazing placidly, no doubt, in some green meadow), it bit her, drawing blood, and bit her again, again, again until her hide was running red. Io, flystung and stampeded, ran for her life. She hurled herself the length of Asia, through cities and towns, through pastures and deserts. And in her mad flight, her body plundered by the winged drover's sharp bites, she transformed the landscape: valleys sank in her tracks, and mountains rose, rivers welled forth. At last she came to an Egyptian oasis sacred to Zeus. The people there were horror-struck at her appearance: a flyblown, half-human piece of patchflesh, here woman, there cow. Zeus saw her, of course, and took pity, restoring her to fully human form after he'd duly performed an act of insemination.

But the gods have left the world now and retreated into myth. It is Hera's *oistros* that has stayed behind, and its progeny, the tabanids—the legions of hairy horseflies, the myriad deerflies with dark or golden eyes—are with us today, still inflicting their bites, still feasting on our blood so that their kind shall endure.

How do they find us? The female tabanid, be she horsefly or deerfly, is not alerted by body heat, as ticks and chiggers are, but rather responds, like *oistros*, to the movement, the slink or stride or gallop, of the quarry. The harder Io ran, the faster and more furious was *oistros*'s pursuit. In the female tabanid's zeal for prey, she has been discovered to respond as well to other moving things, like cars and trains. And she's capable of speed; someone with a passion to know such things has clocked her flight at regularly more than twenty-five miles an hour. When prey appears, she homes in soundlessly. Nor is it only warm-blooded mammals that are chosen for a meal; some species of horseflies feed on reptiles, including alligators.

When contact is made, blood is instantly tapped through an incision made by the fly's mouthparts, which resemble a tiny wedge with scalpel-sharp edges. Blood welling freely, the fly soaks it up with the labellum, a spongelike segment of the lower lip. After such a meal, the deeply pierced victim often continues to bleed; livestock that are frequently attacked may forfeit their lives to the loss of blood. Meanwhile, the fly has hastened forth on silent wings to find a mate among the nectar-sipping males. (The enormous compound eyes of male tabanids touch along the midline, giving them the one-eyed aspect of a Cyclops.) After her eggs are fertilized, the female fly lays them on aquatic or streamside plants with leaves that overhang the water. And when the larvae hatch, they dine voraciously. Deerfly maggots eat only vegetables, like decaying wood, leaf mold, and beached pondweed, while their horsefly kin feed solely on meat, devouring such creatures as snails, worms, insects, and even frogs. Then, appetites satisfied for the nonce, they burrow in the mud and pupate; all of them overwinter, and some species go down for a long sleep that lasts two years. In spring, the adults emerge. The horseflies are large, robust, and furry, with eyes that often show an iridescent blue or green; the deerflies are smaller, their soft bodies striped tan and charcoal gray, their wings bearing dark calligraphic patterns, with eyes plain brown in some species but in others shiny gold or green zigzagged with red. On emergence, the females' stealthy and relentless quest for blood begins again.

Other biting flies exist: the midges, especially the no-see-ums, and the blackflies, scourge of Maine, that are bold and ravenous enough to sneak into a victim's sleeping bag. I feel almost fortunate that in Carolina I must contend every year only with several kinds of deerfly. One is *Chrysops vittatus*, the "banded golden-eye"; these are the fearless raiders with dark eyes and dark-patterned wings that haunt the woods and hedgerows. The other is a piratical species, no bigger than a housefly but with gleaming golden eyes, that works over the water near shore and strikes me as I fish the gill net. But I must swear and swat only for those blessedly brief four weeks in spring. The problem could be worse: in some circumstances, deerflies are vectors for eye diseases or tularemia, a plaguelike bacterial disease that afflicts wild animals, especially rabbits and rodents, but is transmitted from these hosts to human beings by fly bites.

And the problem could be even more dire. Some flies are

unapologetically more bad-ass than deer- and horseflies were ever designed to be. Compared to these prodigies, the tabanids are mere amateurs at the painful exploitation of prey. Consider the blowflies, the Calliphoridae, or "beautiful robbers," a name that encompasses both their parasitic behavior and also the gleaming metallic colors on the bodies of the family's typical members, the bluebottle and greenbottle flies. The larvae of most calliphorid species dine on dead tissue, but some, like the screwworm—*Cochliomyia hominivorax*, the "spirally twisted man-eating fly"—devour living flesh. Screwworm flies, once ubiquitous pests of cattle in the United States, were thought to have been eliminated through the release of laboratory-sterilized males and the subsequent reproductive failures in badly infested areas, but recently, in 1997, they were back at work in Texas, stowaways that came silently, deviously over the Mexican border under the hide of a dog.

(It's hard to admit that flies can be of service to us. So, I'll do it in a parenthetical whisper. The maggots of some calliphorid species have been used since time immemorial to cleanse wounds, a task they accomplish by eating dead, germ-laden tissue. Unknown to themselves, they also encourage healing by excreting a substance called allantoin, which is used as an active ingredient both in expensive over-the-counter cosmetics and in the kind of medicinal cream that patients undergoing radiation treatments for cancer may apply to affected skin to allay the effects of literally being cooked by high-intensity X-rays. The secret: allantoin not only fights bacteria but works to stimulate cell growth.)

In the bad-ass department, consider also the large tribe of botflies, one of which specializes in humankind. That one is *Dermatobia hominis*, "skin-creature of man," which is not equipped to deposit its eggs beneath the human hide and so enlists the help of other insects, like mosquitoes and biting flies. It lays its eggs on these carriers; when they bite, the warmth of the prey causes the eggs to hatch, and the newborn maggots then insinuate themselves into the bleeding wound. This tribe includes a gaggle of families, including the Gasterophilidae, the "stomach-loving" horse botflies; the Cuterebridae, the "skin-boring" robust botflies; and the Oestridae, the family of bot- and warble flies that is named (with scientific inexactness but a proper focus on the torment caused) for Hera's *oistros*. "Warble" has nothing to do with sound but rather refers to the subcutaneous swellings made in the flesh of cattle as the larvae of the ox warble fly grow and feast on the products—the pus and decayed flesh—of the irritation that they cause. The larvae of another oestrid develop in the nostrils of sheep. As for the gasterophilids, they lay their eggs on equine lips or nostrils, and when the horse licks them into its mouth, they travel to the stomach or intestines, where they hatch into maggots with a hunger great enough to debilitate the animal. In all these families, the larvae leave their nurseries within the host and drop to the ground to pupate.

Information about *Cuterebra*, the genus for which the cuterebrids were named, comes from my daughter Elisabeth, a veterinarian, who has seen the ravages its larvae cause in cats and dogs. (She also supplied the facts about the screwworm's dog-assisted reentry.) Nominally a parasite of rabbits and rodents, the skin-borer fly may sometimes jump those bounds in seeking to perpetuate its kind, sometimes affecting people as well as wild animals and pets. And the flies are large, often growing to twenty millimeters in length, while the maggots are downright huge, measuring in at twenty-five millimeters—more than an inch! A monster like that leaves behind a deep pit in its host's flesh when it drops out to pupate. My daughter names the species as the most despicable she knows.

"Supposing," I say to her, "there were no such thing as *Cuterebra*?"

"I'd see fewer cats and rabbits with holes in them, that's all," she replies.

I tell her that I can't imagine how they fit into the earth's puzzle pieces or what their purpose is, beyond ensuring a future from now to eternity for the skin borers. And I say, "*Cuterebra*—looks like the world could do without them."

"The world *could* do without them," she says. "But they're parasites. They cannot do without the world."

The same holds true for all the predatory and parasitic flies— the screws and bots, the deerflies and horseflies, the thousand detestable others. This May, on the first day that the bad-ass hordes come swarming in on silent wings, I'll mark my journal with a marginal sketch of a fly with dark patterns on its wings. I'll start counting down on the Four-Week War, and for twenty-eight days I shall swat and squash them with a pleasure equal, at least, to that of my fishing friend. And I shall remember that I don't need them, but in their lust for what I and my kind can supply, they do need us.

# A FOOT IN THE DOOR

Destroying Angel (*Amanita virosa*)

The Fungi

WITHOUT any warning, without the slightest sound or motion, my childhood repossesses me. The ambush begins in the entrance hall of Rowan Oak, William Faulkner's 1840s plantation house in Oxford, Mississippi; it overwhelms me, whisks me away as I stand in the doorway of the writer's study scanning the outline, penciled on the walls in red and black, of his last book, *A Fable*. But I'm not really there. Instead, in a most physical way, I've been transported to the Shenandoah Valley in the era of World War II; I've been returned to my grandmother's high-ceilinged brick house, where I spent the war years while my father served overseas with the Army. There, summer and winter, the air smelled, as it does at Rowan Oak, of furniture polish, brass polish, and mold—most of all, mold.

That pervasive, primeval mustiness emanated from my grandmother's books—books in the breakfront, books on the mantel and in the walnut secretary, books in the Wernicke case that had long ago lost the glass doors protecting its shelves. *The Princess and the Goblin; The Water-Babies;* Bulfinch's *The Age of Fable* and collections of myths edited by Gayley and Guerber; a leather-bound Latin lexicon used by my great-uncle at Washington and Lee in 1888; and first editions, once my mother's, of *The Wonderful Wizard of Oz* and *The Road to Oz*—these are among the titles I remember or, in the case of the myths and lexicon, have appropriated and shelved among my own books to this day. Those volumes still bear the immemorial staleness on their foxed pages.

That smell haunts Faulkner's grand house. Barred by velvet ropes from entering the rooms, I cannot read the titles of his books, but I can see their full buckram or leather bindings, and the gilt lettering on their spines. Though it stands to reason that few, if any, are the same as those in my grandmother's smaller collection, the Mississippi books and hers are definitely kin. They are related not only by their production in Victorian times or the early 1900s but also by the fact that both libraries have been similarly affected by enduring summer after summer in the land of mockingbirds, gentility, and heat combined with high humidity. The cloth and paper of these volumes are filled with fusty memories of dripping, sweltering summers, and, without the drying effects of air-conditioning, they receive a new infusion of moisture every year. Of course, more moisture promotes more mold.

There's a technical term for the action of mold on books, the foxing and splotching, the greenish patches that sprout on the covers, and the eternally dank smell: tropical deterioration. One does not have to be in the tropics, however, to note these phenomena; all that's needed is a cozy steam bath of warmth and humidity— behold! Last summer, on the sticky, hot Carolina coast, the wide, nylon-webbing collar of a dog four years dead came to light as I was cleaning out a dresser drawer (why, against all the urgings of good sense, do we keep these reminders of heartbreak?); the collar, originally brown, was coated thickly with turquoise mold. And not only nylon and paper are attacked but also fabric of all kinds, leather, lumber and living woods, bark, leaves, seeds, and insects both dead and alive. So are shower curtains, rubber duckies and spandex shorts, electrical insulation, kerosene-based aviation fuel, and just about everything else in the whole wide world. Nothing of organic origin is immune.

But molds are hardly confined to the warmer places on the earth. Some are drought resistant and enter a dormant state until the rains finally come. Others live only in the water as parasites on aquatic life. Like goblins or the tiger that hides under the bed, most work at highest efficiency in the dark. And a varied group known as psychrophiles—cold-lovers—is even known to thrive in the winter-long darkness of both the Arctic and the Antarctic. As for their local success in surviving near-freezing conditions, anyone with a refrigerator knows what happens to the leftovers that have migrated back into the nether regions. Mashed potatoes turn pink; cream cheese grows gray-green fur; brown specks blossom abundantly in the applesauce. Wherever molds appear, they take over, nor can they be easily dislodged.

The reason for these silent and ubiquitous encroachments is, of course, perpetuation of the species—as many as 250,000 of them according to several estimates. The molds are inescapable; everywhere across the planet they help themselves to an over-spilling cornucopia of opportunities that satisfy their appetites and enable them to reproduce. And together they constitute a biological kingdom, the Fungi, until recently thought to be an odd and populous offshoot of the plant kingdom but now recognized as a distinct realm of its own. Fungus and mold—the words applied to the realm's inhabitants are well-nigh interchangeable, though the former comes from a Latin transformation of a Greek word for sponge and the latter first fell bluntly from an Anglo-Saxon tongue.

Mold or fungus (use the term you prefer), the kingdom's populace falls into five sometimes overlapping classes. It was thought until recently that a sixth group, the slime molds, was part of the realm, but it has been reassigned to the kingdom of the Protista, which it shares with organisms like the diatoms and dinoflagellates. The remaining five all have polysyllabic Greek names: Phycomycetes, or weed fungi, which includes the two divisions Mastigomycotina and Zygomycotina; Ascomycetes; Basidiomycetes, and Fungi Imperfecti, also known as Deuteromycotina. The suffixes -mycetes and -mycotina simply mean "fungi," and the rest of these tongue twisters denominate the five respectively as whip and yoke (the two divisions of the weeds), skin, pedestal, and imperfect or secondary fungi. Each tribe performs a multiplicity of jobs. One species of whip, for example, attacks fish in aquaria—the dreaded "ich"; another causes potato blight. The whip that infests potatoes is *Phytophthora infestans*—"plant corrupter," and it alone was responsible for the wholesale putrefaction in the fields that led to Ireland's five gaunt years of famine, beginning in 1845. The yokes may be blamed for pin molds on food and leather. The skins, so called for the sacs in which they encase their spores, include baker's and brewer's yeast, the splendidly edible morels, and the cup fungi that grow on wood. The pedestals tend to have a base supporting a cap; these are the typical toadstools and mushrooms, though the group also encompasses bracket fungi and the no-neck puffballs that balloon straight out of the ground. Some pedestals form fairy rings, some glow with cold luminescence, some are hallucinogenic, still others produce toxins that kill. The imperfects lack the capability for sexual reproduction (note the taxonomical value judgment here); several species of imperfects are used as agents for ripening cheeses, like the blue-veined Roquefort and the green Gorgonzola, while others play humanly counterproductive roles, causing ringworm and athlete's foot. And within each tribe dwell species that are strictly parasitic, bound to feed on the tissues of a live host; strictly sapro-phytic, obliged to eat dead matter; or opportunistic, switching from one mode to the other in order to take advantage of the nutrients available, be they alive or dead.

All of the fungi, no matter their tribe, seem to plague rather than please us. Some animals, to be sure, draw benefit from their existence; several species of ants and beetles assure themselves of a steady food supply by farming fungi, and the endangered red-cock-aded woodpeckers could hardly drill their cavity-nests in old-growth pines if not for the molds that soften old heartwood with rot. As for humankind, we do honor the outright edibility—raw, sautéed, or stuffed—of a few fungi, and appreciate their invaluable assistance in the preparation of foods like bread, cheeses, and the famous sauces Worcestershire and soy. The molds are friends not only of the gourmet but of the tippler, too, for it is they that stimu-late the fermentation that results in hard cider, stout, and lager, wines both sweet and dry, Bombay Sapphire, Jack Daniels, and all other forms of potable alcohol. (As for other forms of service to mankind, a parenthetical acknowledgment should probably be made of *Empusa musca*—"fly-digester"—a mold that's parasitic on the com-mon housefly.) And when it comes to sickness or health, where would we be without drugs like penicillin, ergotine, and strepto-mycin? By their name, you can tell that all of the -mycin antibiotics are derived from fungus, for *mykos* is the Greek word for mushroom, with the hard *k* softened to *c* in the suffix.

But there's nothing soft in human perceptions of the Fungi and their work. We know their devastations well: Dutch elm disease; American chestnut blight; mycotoxins killing cattle and poultry; myriad fungal infections decimating crops; ergotism, or St. Anthony's fire, leading to loss of human limbs; histoplasmosis affecting human lungs; and a hideous host of other damaging or fatal ills, including mycotismus, which is the fancy clinical term for mushroom poisoning. The everyday words that have been used in English for a thousand years demonstrate an ineradicable human suspicion of the Fungi, and an aboriginal dislike, not to mention terror at the threats they pose to crops and, so, to human life. Listen to the no-nonsense curtness, the thud and hiss of these terms: blight, blotch, bunt, mold, mildew, rot, rust, scurf, smut, and wilt. With the exception of the very last, all are ancient Teutonic words that hit the ear like imprecations. Wilt, the odd word out, first rose on American tongues in the mid-1800s. It certainly provides a graphic description of what happens to tomato vines set in mold-infested soil. Mold enters through their roots and rises, clogging their water-conducting vascular systems; vines shooting sturdily upward one day are drooping and wizened the next, and dead the week after. Northern Europeans and Americans, however, are hardly alone in their appraisal of the harm that Fungi do. The Romans nominated a god of rust, Robigus, and, lest grain be blighted before it could be reaped, attempted to appease him with an annual festival, the Robigalia, held on April 25. And, should a Roman suffer poisoning from eating fungus, the natural historian Pliny (A.D. 23–79) offered a long list of antidotes, from radishes and twice-boiled cabbage to lily roots and the lees of wine. But no remedy, natural or otherwise, kept the emperor Claudius from succumbing in A.D. 54 to mycotismus. Unwittingly, he ate toxic fungi especially selected for him by his wife Agrippina. Of that occasion, Pliny remarked that, as things turned out, a worse poison had been left behind: Claudius's son and successor Nero.

Mushrooms—in his 1597 *Herball,* the English botanist John Gerard called them "bastard plants" and "earthie excrescences" and pronounced that "some are very venemous and full of poison, others not so noisome; and neither of them very wholesome meate." Gerard's contemporary Shakespeare went even farther in *King Lear* by assigning a prime cause to the Fungi, and especially to the rusts, bunts, and stinking smuts that swell and rot cereal crops and can also poison people who eat infected grains. Edgar, the rightful but dispossessed heir of the Earl of Gloucester, acts the madman, staggering from a hut and howling his credentials as a lunatic: "This is the foul fiend Flibbertigibbet: he begins at curfew, and walks till the first cock; he gives the web and the pin, squints the eye, and makes the harelip, mildews the white wheat, and hurts the poor creature of earth." So, it's Flibbertigibbet's fault that mold, doing little good but much harm, has silently helped itself to worldwide dominion and engages in its stealthy enterprises not just between curfew and cockcrow but from sunup to sundown as well.

There's no stopping the foul fiend, no escaping his excrescences, the Fungi. For, they are surely the planet's most successful reproductive strategists. Asexually, a single cell may attenuate itself, branching and budding to produce a mycelium—the threadlike food-obtaining structure—that reaches out for feet, yards, acres; each portion is genetically identical to every other. We've all seen the tatters, the frayed gauze of various mycelia clinging to damp ground, rotting wood, and those back-of-beyond refrigerator experiments. This kind of cell may also divide and send out spores, each of which contains the potential for a faithful clone of its parent. Then, the organism may leave behind the celibate life and become a fully sexual entity. The rusts, of which there may be four to five thousand species (not many as molds go), are characterized by a modus vivendi called pleomorphism—the adoption of multiple forms. They are also characterized by dependence on alternate hosts—gooseberry or currant and pines, for example, or barberry and wheat. In its sexual phase, on the currant or barberry where it overwinters, a rust contains gametes. These are sex cells properly called pycniospores, or "close-packed spores," each with a single haploid nucleus—that are analogous to the egg and sperm of animals. (Why only analogous, not the same, is an amazing matter that I'll come to shortly.) After sexual fusion takes place, aeciospores—"assaulting spores"—are formed, with each containing two nuclei, one from each pycniospore; these aeciospores establish a mycelium. From that network, crop after crop of urediospores—"blighting spores"—arises and becomes a windborne plague, leaving the first host to infect the second. After the damage is done, teliospores—"final spores" (that aren't truly final)—are produced; some may germinate forthwith, some sleep through the winter, but in all of them the two nuclei fuse. The germination of teliospores leads to the last phase of the cycle, the sporidia—"small spores"—which are also called basidiospores, a word that means "pedestal spores." Wind bears the sporidia back to the alternate host—from wheat to barberry, pine to currant—where they grow into mycelia, from which arise the picnia that produce the sexually active pycniospores.

Sex among the Fungi, what a peculiar phenomenon! Or so it may seem to us who are accustomed to think of mating in terms of male and female, each with its own gender-specific set of genes. But in the voiceless, covert kingdom of the molds, a multiplicity of genders, or mating types, holds sway. It is known that each has not one but two sets of genes, though how they work is still a mystery. And each mating type may contain a humanly bewildering array of variants. Sexual fusion is impossible only between types possessing an identical array. The possibilities for fusion, though, are boggling. If one type has twenty variants, and the other two hundred, then twenty times two hundred combinations can occur, for an orgiastic total of four thousand. Why such complexity? One notable mycologist has pointed out that for species divided into male and female,

only half the population is available to each gender for the establishment of a fertile union, but with multiple variants the molds have
maximized their chances for meeting a suitable partner in the subterranean dark. Nor do fungal peculiarities stop with this huge
enhancement of the odds in the mating game. When male and
female animals pair, sperm unites with egg, and the single sets of
chromosomes in each—the haploid sets—fuse into a diploid set,
from which the embryo develops. But when molds mate, fusion is
delayed, and the nuclei remain separated until the moment that a
fully mature specimen puts forth its spores. The reason for delay
may have to do with retaining the possibility for multiple partners,
for fusion with other wandering mycelial threads, and thus for assuring the various perfect species of genetic diversity on a truly grand
scale.

Success is the Fungi's reason for being. The envoys of Flibbertigibbet care nothing about ripening cheeses or blighting elms, feeding ants or causing the tender skin between our toes to itch and
peel. Instead, as Sylvia Plath puts it in her poem "Mushrooms," "*Our
kind multiplies:/We shall by morning/Inherit the earth./Our foot's in the door.*"
Multiply they have, and not just the pedestal mushrooms about
which she writes, but every last member of the realm. More than
that, their mycelia, if not their feet, have reached far beyond the
door; they've gained the equator, they're found at the poles, they've
entered almost every cranny in between. And it is not unimaginable
that if an asteroid sends our kind the way of the dinosaurs, or if
we're suicidal enough to envelop ourselves in the mushroom clouds
of nuclear warfare, the Fungi will survive whatever cataclysm wipes
us out.

Meanwhile, they scavenge as they've always done. They, more
than the selective vultures, more than the picky hyenas and hagfish,
are angels of decay. They not only consume the things we treasure,
like books, tomato vines, and old dog collars, but are able to ingest
every bit of the world's organic trash. In the vastness of their hunger
lies usefulness and also something to admire.

And something else arose from the moldy books at Rowan Oak, and the musty fabrics: they took me to another time, another place, another recognition—part grudging, part not—that there are lives above, below, and quite beyond those of my kind.

Fly Amanita (*Amanita muscaria*)

# THE CREATURE WITH NINETEEN LIVES

Opossum (*Didelphis marsupialis*)

## Common Opossum

ONCE upon a time, in another life three decades ago, I lived with husband and school-aged children in a big house on a large lot with woods and New England stone walls. And my husband, who was not in any way an animal lover, promised the children that they might keep as pets whatever wildlife they happened to catch on our grounds. Out, then, with the collecting jars, nets, and a Havahart trap. The results: uncountable fireflies, monarch caterpillars, a bullfrog or two, a ring-necked snake, a half-grown cat, and an opossum. The cat and frogs were let go, the snake (a fragile species) died, the caterpillars were fed until they pupated and later burst forth as butterflies, and the 'possum made the father of my children rescind his promise. They put it in a fifty-gallon aquarium, though, and kept it for a week, feeding it table scraps and listening to its hisses of complaint.

I remember the catching and caging of the 'possum. It had a long bare tail, coarse gray body fur, hairless black ears, and a white face that culminated in an eraser-pink nose. I remember its teeth, too—long, pointy things covered with a green scum that may have been algae. I also remember thinking and saying that it wasn't really the kind of animal that made a decent house pet. Hey, children, let that critter go. They did after that long week, and I thought about 'possums rarely after that. Only recently, when someone named the opossum as her top candidate for the most despicable species on earth and clinched the nomination with a wide-eyed shiver, did the image of this animal resurface. Now it won't go away. The only way to exorcise its presence in my mind is to investigate it and uncover its secrets.

What could be despicable about the common opossum? Those green teeth, all fifty of them. The coarse, tatty-looking grayish fur. The long, tough, quite naked terminal appendage, which is monstrously rodentlike. Then, given the chance (a loose-fitting lid), it plunders garbage cans as eagerly as any 'coon. And my neighbor Mo Wixon, aged seventy-seven and wise in the ways of critters and plants, once went to war against 'possumkind. He was raising red worms, mighty popular with fishermen, in a bunch of old refrigerators turned on their backs. He says, still irritated by the memory,

"Those 'possums would get in there and just gobble up my worms. I had to knock a couple of 'em in the head."

Looks and habits apart, the animal is often seen either as a creature of Walt Kelly—Pogo the 'possum, eternally adorable and wise—or as a primitive life-form notable for its stupidity and thus the butt of jokes like this one: Why did the chicken cross the road? To show the 'possum it could be done. 'Possum lives in light verse, too. My grandmother's scrapbook contains this ditty, "In the Georgia Woods," which was published by the *Atlanta Constitution* a decade or so after the beginning of the twentieth century:

> The snarlin' o' the 'possum,
> > The boundin' o' the buck,
> Barkin' o' the squirrel,
> > An' the rabbit's foot for luck,
>
> Like a fiddle's music
> > When it knows the tune to play,
> But a rifle's ringin' echo
> > Beats a fiddle any day.
>
> An' the whirrin' o' the partridge—
> > Then there's music in the air,
> An' 'possum served for supper
> > Beats a hotel bill o' fare.

Then, along with raccoon, jay, and other rural creatures, 'possum also inhabits at least one folk song, "Bile Them Cabbage Down." In one rollicking verse, raccoon, on the ground below a persimmon tree, directs 'possum, who is already up there, to "shake them 'simmons down." With that, it's fair to say that the opossum does not suffer from a picky appetite but relishes fruit along with other delectables like fishing worms, insects, birds' eggs, birds themselves, mice, and the kitchen scraps found in garbage cans. And it sometimes behaves like a four-footed vulture, eating carrion as it can, but more about this predilection shortly.

'Possum also inhabits the historical record. In 1612, Captain

John Smith, describing the wildlife of the Virginia colony, had this to say: "An Opassum hath a head like a swine and a taile like a Rat and is of the bignes of a Cat. Under her belly she hath a bagge, wherein she lodgeth, carrieth, and sucketh her young." Nothing remotely similar had ever been seen in Europe; so he borrowed the animal's Algonquian Indian name, wrote it down as best he could, and gave it to the world.

And so began the wondering European references to "'possum." Nearly a century after Captain Smith's day, John Lawson, gentleman surveyor, traveled the Carolinas and wrote of all that he saw, from eagles and polecats to oaks and 'simmon trees and on to the Indians, their manners and customs. The resulting book, *A New Voyage to Carolina*, was published in 1709 as an inducement to potential colonists to cross the ocean and settle in ever fair, ever fertile "Summer-Country." And Lawson was always keen at noting the peculiarities of the things that he came across: a hollow tulip poplar that once gave a man and his furniture ample living space; the Indian habit of feeding a child roasted "Rearmouse"—a bat, that is—as a cure for pica; the "excrementious Matter" that forms the rattles of a rattlesnake's tail; and the edibility, yea or nay, of every bird he came across. He commented also on the edibility of 'possum, which he sampled happily in South Carolina when his party ran short of victuals: "The Weather was very cold, the Winds holding *Northerly*. We made ourselves as merry as we could, having a good Supper with the Scraps of Venison we had given us by the *Indians*, having kill'd 3 Teal and a Possum; which Medly all together made a curious Ragoo." His detailed description of 'possum shows clearly that he found the living animal to be a whole sight more curious than this gamey stew:

> The *Possum* is found no where but in *America*. He is the Wonder of all the Land-Animals, being the size of a Badger and near that Colour. The Male's Pizzle is placed retrograde; and in time of Coition, they differ from all other Animals, turning Tail to Tail, as Dog and Bitch when ty'd. The Female, doubtless, breeds her Young at her Teats; for I have seen them stick fast thereto, when

they have been no bigger than a small Raspberry, and seemingly inanimate. She has a Paunch, or false Belly, wherein she carries her Young, after they are from those Teats, till they can shift for themselves. Their food is Roots, Poultry, or wild Fruits. They have no hair on their Tails, but a sort of a Scale, or hard Crust, as the Bevers have. If a Cat has nine lives, this Creature surely has nineteen; for if you break every Bone in their Skin, and mash their Skull, leaving them for Dead, you may come an hour after, and they will be gone quite away, or perhaps you meet them creeping away. They are a very stupid Creature, utterly neglecting their Safety. They are the most like Rats of any thing. I have, for Necessity in the Wilderness, eaten of them. Their Flesh is very white, and well-tasted; but their ugly Tails put me out of Conceit with that Fare. They climb Trees as the Raccoons do. Their Fur is not esteem'd nor used, save that the *Indians* spin it into Girdles and Garters.

My friend Mo tells me that he's also out of conceit with that fare and would prefer hotel food any time. "I've seen people singeing off the hair in a bonfire before cooking the beast," he says. "But when I was a child on the farm, lightning would strike a horse or cow, and it'd just lie there. And every time you'd walk by the old dead cow, there'd go a 'possum out the debris chute, or the fantail, so to speak. It was hard to eat them after that."

As it happens, John Lawson has fed us a Ragoo of fact and sup-position, along with mention of some typically eighteenth-century bone-breaking and skull-mashing (no animal-rights groups in those days). Lawson was off the mark on the location of teats and "paunch" but right about the retrograde pizzle and the ugly tail. He was dead wrong, as shall be seen, about the animal's stupidity. But his statement that 'possum is the "Wonder of all the Land-Animals" is hardly an exaggeration.

For the common opossum, sometimes called the Virginia opos-sum, is indeed a wonder. Formally, the animal is known as *Didelphis marsupialis*, the "marsupial double womb." That double womb is an anatomical phenomenon in the same odd league as the retrograde pizzle. But first, a look at the place of 'possum in the scheme of things. 'Possum belongs to a most peculiar order of mammals, the Marsupialia, which might be translated as "pouch beasts" for the pocketlike flap of skin—*marsupium* in Latin—that covers the nipples of females in many, though not all, species. Marsupials first appeared in the Cretaceous period; the oldest fossils come from North America and date back some eighty million years. Fossils of an animal almost identical to the present-day *D. marsupialis* go back a not inconsiderable sixty-five million years. As marsupials evolved, some of them were vulpine and some hyenalike, with bone-crushing teeth, while others grew as large as a modern-day rhinoceros. The carnivorous, jaguar-sized *Thylascosmilus* that once prowled what is now South America, sported long, powerful canine teeth, which it used for killing prey. In fact, it filled the niche occupied elsewhere by the sabertoothed cat and became extinct when the land bridge between North and South America was reestablished about ten million years ago and the sabertooth, a more canny predator, moved south into the stomping grounds of *Thylacosmilus*. Marsupials have since made themselves at home in two parts of the planet. Some 170 present-day species are native to Celebes, Timor, New Guinea, and, most notably, Australia, famed for its koalas, wallabies, kangaroos, bandicoots, Tasmanian devils, and a varied bunch of forest-dwelling creatures called possums. Those possums, noted by eighteenth-century European explorers like Captain James Cook, came to share the name of the earlier discovered North American 'possum because of their general kinship as marsupials; they are otherwise not closely related. Australian possums belong to the Phalangeridae, the phalanger family, so named for its phalanges—its toes—which have evolved so that the first and second digits on the hind feet are opposable and can be used to grasp branches as the animals travel through their arboreal habitat (the North American opossum can do this, too). Some phalanger species have developed flaps of skin between their fore and hind limbs, which can be spread to help them sail through the air in the manner of a flying squirrel.

The New World houses only two marsupial families, totalling seventy species, more or less. They include such creatures as the rat opossum, or selva, which has no pouch; the four-eyed opossum, which wears two large white spots over its eyes; and the yapok, or water opossum, which leads a semiaquatic life and so has been blessed with webbed toes, oily fur, and a pouch that can be tightly closed to keep its offspring dry. All these live in Central or South America. The only one of the seventy that has ventured north is *D. marsupialis*, the common opossum. Though it ranges as far below the equator as Argentina, it has made itself thoroughly at home in not only the United States but southern Canada as well.

But no matter the land it travels through, *D. marsupialis* (called 'possum from here on) is indeed a most surpassing wonder. It raises a sense of wonder, too, in anyone who contemplates the inventive ways in which the marsupials—also called metatherians, or "early beasts"—diverge from the placental mammals, which are sometimes dubbed eutherians, or "good beasts." (There seems to be some taxonomical snobbishness at work here.) The main features that bring marsupials into the mammal camp are, of course, warm-bloodedness, at least some body hair, and mammary glands that produce milk for suckling the live-born young. (There's an exception, however, to prove the rule: the kind of mammals known as prototherians, or "first beasts." These are the duck-billed platypuses and the echidnas, or spiny anteaters, sole members of the order Monotremata—the "single holes," for their combined intestinal-urinary-genital tract, or cloaca. They lay eggs but suckle their young after hatching. Because the females have no nipples, the young absorb milk through the skin.) Another characteristic that 'possum shares with many eutherian mammals is that it gets about on four legs. Most important, like every other living creature, mammal or not—no matter whether they fly or swim, wriggle or leap, skitter or walk—'possum lives to perpetuate its kind, to keep a basic 'possumhood alive on earth. But from there it goes its own peculiar way.

Take that naked, crusty-looking tail, which stretches out for half the creature's total length. It's prehensile. Despite common wisdom, it's used hardly at all as a hook for hanging upside down but serves rather as an anchor to steady the animal as it moves about on the branches of trees. The tail is also handy for toting materials for building the nests in which this mainly nocturnal animal sleeps during the daytime; 'possum scoops up grass and leaves with mouth and front feet, then pushes them rearward to the tail, which curls around them like a carrying strap. Add to the prehensile tail the clawless, opposable digits on each of the hind feet: 'possum ranges through its arboreal habitat with the greatest of safety and ease. It nonetheless seems to prefer ambling about on the ground and is more often found on terra firma than aloft. Should 'possum encounter a copperhead or rattler in its perambulations, it falters not, nor does it flinch, for it is a creature endowed with an extraordinary resistance to snake venom. Then, contemplate 'possum's mouth, which can open more than ninety degrees, almost as great a gape as a rattlesnake's, and look, too, at the fifty teeth—far more of them than are granted other kinds of mammal. 'Possum's skull is specially constructed to handle the stresses created when so many teeth crunch down (John Lawson and friends must have discovered that it was far from easy to mash such a hard head). Its brain, however, is on the small side when compared to the brains of what are sometimes called the "higher mammals." Its body temperature ranges far lower than that of most mammals—94.5 degrees Fahrenheit, as compared to the human 98.6, the mouse's 100.2, and the 101.5 of cows and pigs. 'Possum has thus been relegated to the lower ranks of mammalkind. Such seeming underendowment may help account for pronouncements, like Lawson's, that 'possum is a "very stupid Creature." And its pizzle is indeed placed retrograde, just as Lawson said, with the scrotal sac located in front of the penis. What's more, the penis is forked.

Ah, but with good reason. The animal's most dramatic divergence from other kinds of mammals lies in its reproductive arrangements. Note, to begin with, the double womb that led to the naming of the whole New World family, the Didelphidae, to which the common opossum belongs. Each female has two uteruses, each served by its own vagina. The male's two-pronged penis is supported by a two-pronged cartilage, the didelphid equivalent of the penis bone found in every male mammal except for the primates. This arrangement makes very good sense because, with a fork in each vagina, the chances for fertilization are maximized.

One female common opossum may carry as many as two dozen embryos, not all of which make it to adulthood; an average litter numbers seven to nine. Gestation takes about thirteen days, with the walls of the yolk sacs of the developing young attached to the mother in a rudimentary fashion that is nonetheless placental, for it allows for the transfer of nourishment from mother to fetuses. But when the young are born, they arrive in the world not through the double vaginas but through a birth canal that develops for the occasion between the vaginas. The babies, for all that they are blind, naked, and tiny as grubs, come equipped with clawed forelimbs that enable them to scrabble to the mother's pouch and latch for dear life on to her teats, which number thirteen. Not all succeed in reaching this destination; others may be out of luck because all available slots are already occupied. But those that do attain pouch-dom are kept in place by nipples that expand in the mouths of the newborns to form a firm bond. My friend Mo comes up with a bit of folk wisdom: "When they were raiding my worms and I knocked them over the head, there was one had four-five-six babies in her pouch. You know, they pull on the tit so hard it elongates. It was a good three inches." He's right about the length but wrong about the reason—it is maternal biology, not juvenile suction, that accounts for the three inches. (The expansion of the nipples also allows the newborn young of pouchless marsupials, like the selva, to hang on tight.) The babies will then stay in the mother's pouch for four or five weeks before they first crawl out. They are not completely independent then, for weaning does not take place till one hundred days after birth. But once out in the world, they sometimes cling to the fur on their mother's back and ride along as she meanders through her night's work. These little hitchhikers have also been known to return to the pouch for protection and milk.

And still no end to reproductive wonders. The female 'possum may remate within a few days of the birth of her young. But if these young have reached her pouch and are suckling, then the new fetuses will enter a state called embryonic diapause, a sort of suspended animation, in which fetal development stops until the pouch is once again empty and ready to receive the next generation. Usually, two litters are born each year.

'Possum may not seem bright. Beady-eyed and long-tailed, it lumbers along with apparent lethargy. There's also the notorious business of playing 'possum to which John Lawson refers: when confronted with danger, the animal goes belly up in an almost fatalistic fashion but creeps away as soon as danger disappears. It can maintain this deathlike quiescence for up to six hours. And there are those teeth, that bristling mouthful of green and pointy teeth. But Mo says, "An opossum is not near as vicious as people would think—hissing like a snake. You can just pick it up."

There is, however, one aspect of 'possum that may be considered truly despicable, especially by people who own horses or keep stables. 'Possum is the definitive host, the host sine qua non, of a protozoan, *Sarcocystis falcatula*, or "hook-shaped fleshy cyst." It is sometimes known as *S. neurona*, "found in the neurons," for the site it inhabited when it was first identified in horses. 'Possum ingests the protozoan when it eats intermediate hosts, like grackles and cowbirds; the microorganism takes hold and matures; 'possum then sends it into the world by way of feces. 'Possum is not affected in any way, but should a horse eat contaminated grain or graze in a pasture where 'possum has trod, it may become infested by the protozoan and suffer a disabling illness called equine protozoal myeloencephalitis, EPM for short, which causes brain lesions, convulsions, and sometimes death. 'Possum is not aware, of course, of its role as vector. All that hippophiles can do is try to eliminate the marsupial from the premises.

Given the intricacies of 'possumhood, it's unwise at best to call 'possum stupid. Consider its haunts. Think of its nineteen lives. It lives in story, song, and jingle-jangle verse, in joke and comic strip. It dwells in the notes of explorers and the digs of paleontologists. More significantly, it spends the rest of its lives in the tangible world, ever reconnoitering woods, fields, and yards, ever rambling in its own good time across the continent. It sequesters itself within the carcasses of livestock and wildlife. And were Mo still raising red worms, it would surely visit his refrigerators, there to fill its belly and stay fit for the overwhelmingly important business of bringing more 'possums into the world.

That's precisely what it has done with noteworthy success for, lo, sixty-five million years. John Lawson and others to the contrary, no animal that has kept its kind going for so long can possibly be considered stupid. To be sure, it has no knowledge of matters like time and longevity, nor does it care. But our own come-lately species, which knows and cares, might find another wonder here: any animal blessed with such an indomitable ability to survive, come glaciers or global warming, is a very clever animal indeed.

# THE RIDDLES OF THE SPHINX

Tomato Hornworm (*Manduca quinquemaculata*)

## Hornworms

A mild June morning at Great Neck Point on North Carolina's wide and salty river Neuse—perfect for picking green beans. Five-gallon bucket in hand, I go to the vegetable garden and set to work. Three gallons along, when I stretch and look up, scanning the rows of beans and adjacent tomatoes, there it is, only an arm's length away—a hornworm. I can see its jaws move. It's chomping vigorously, implacably on the leaf of a Better Boy vine. In its wake lie the shorn stumps of other leaves.

Anyone who fancies homegrown, sun-ripened tomatoes (the only kind worth eating) is well acquainted with hornworms. Where tomatoes grow, there also flourish great, fleshy, spike-tailed eating machines. Nor do the vines attract only the kind known as the tomato hornworm but cater as well to the appetites of several closely related caterpillar species that also favor chowing down on the Solanaceae, the nightshade family, which includes potatoes and eggplant among its domesticated members. The caterpillar now devouring my plants—a pudgy thing as big around as my thumb—is a tobacco hornworm, *Manduca sexta*, the "six-spotted glutton," so named for its fierce appetite and the six orange spots on each side of the abdomen of its moth. The giveaway to its identification is the "horn," the stiff, backward-jutting spine on its nether end. That of the tomato hornworm, *Manduca quinquemaculata*, the "five-spotted glutton," is green with a black edge, but this one glows red as a burning ember. I know that the rest of its body is grossly corpulent and bright green, the better to hide amid the leafy fodder as it munches its way toward metamorphosis. And its sides are marked with seven white slashes, while those of the tomato hornworm are patterned with eight white patches, each shaped like a streamlined L. Today, however, I can see only the caterpillar's tail, not the markings on its body, for as it was eating away at my vines, something else was eating it: tiny spindles wound with white thread protrude from its skin and nearly cover it—the cocoons of braconid wasps. The green beans are patient; they will wait. I break the leaf from the vine and carry it inside, hornworm and all, to take a photograph.

Braconid: the name means, roughly, "short stuff" and refers to the minuscule size of the wasps in this family. The larvae of some braconids are parasitic on the larval stages of other kinds of insects. Braconids are found worldwide. Many species of the tiny wasps, most of them stingless and all belonging to the genus *Apanteles*, the "imperfect ones," are native to the New World, and they are fierce natural predators on the pests of New World plants like tobacco and tomatoes; other braconids have been introduced from Europe to help control the ravages of insects on cash crops like cabbages. The species at work on my hornworm turns out to be *A. congregatus*, the "assembled-congregated-swarming imperfect ones," so named for the numerous cocoons often seen on a single host. The adult wasps lay their eggs on the victim's skin. When the eggs hatch, the larvae burrow into their host, feasting until it's time to chew back up to the surface, on which they then spin their delicate cocoons. Devoured from the inside out, infested caterpillars invariably die, though they are ravenous, their great jaws working furiously, until the end. The wasps may not emerge from their pupae until well after the host's death. Nature is not only red in tooth and claw but also relentless in jaw and ovipositor.

When I return to the vegetable garden after taking *M. sexta's* picture, memory diverts me from the tedium of picking beans, thousands of beans. Once upon a time, at my children's behest, we kept tomato hornworms, the species with a green spine, under glass for observation. Oh, the see-through wonders of glass: guppies and angelfish in an aquarium, woodland plants—with occasional insects and frogs—in a terrarium, and tomato hornworms in the largest possible size of applesauce jar. The point of the exercise was, of course, to watch the caterpillar transform itself into a moth. (We kept smaller caterpillars, too—monarchs, black swallowtails, Polyphemus moths—but they were housed in appropriately smaller jars as we awaited chrysalis or cocoon and, later, wings.) For every hornworm, the big jar would be furnished with four or five inches of soil from the garden. In, then, with the caterpillar, and on with the lid, punched full of air holes. The lid would be removed, usually more than once daily, for the insertion of a fresh batch of tomato leaves to satisfy the hornworm's gargantuan appetite. And when the body showed rolls of flesh and the green skin seemed ready to burst, the chewing would stop, the great jaws become still at last. But the rest of the caterpillar would go into feverish action, loping around, around, around in its cage of glass. For days on end, it would explore perimeter and surface. Then, according to some internal cue, it would tunnel into the soil and there slough its skin. I learned later that there is a timetable for these events: four preliminary molts; then, when the creature's weight reaches seven grams, it starts its restless peregrinations and on the fourth day burrows and undergoes a fifth molt from worm to pupa. We could not always see the dark-brown pupa and its long tongue case, curving like the handle of a tall pitcher. We never saw the emergence of its moth (though we succeeded every time with monarch, swallowtail, and Polyphemus). Does hornworm require a long period of pupation? Did our hopes outrun our patience? Or did the kitchen counter simply need to be cleared of an in-the-way applesauce jar? At this remove, I do not know.

The experience of my husband, the Chief, with hornworms was of entirely another order. He never thought of keeping them as live zoological specimens, no indeed. He murdered them, and with malice aforethought. Nor was he unlike other sons of North Carolina farmers, who sent their children into the brightleaf tobacco fields as soon as they were five or six years old. Some summers he was sent off to his uncle's farm, which offered precisely the same kind of outdoor entertainment. There and at home, he and the other young'uns cropped the sandlugs—picked the dirt-covered bottom leaves, that is, the leaves too close to the ground for a grown man to gather easily. The young'uns cropped the higher leaves, too, and topped the plants, cutting off the blossom stems that jut high above the leaves. They'd come out of the fields covered with gummy tobacco sap that turned black as it dried. The Chief's aunt would then dole out what she called "weak, dirty tobacco hugs"—light, brief, arm's-length pats made only after a search for the least gummy place on the recipient. After this greeting, the kids would be made to take off their shirts and scrub down before they were allowed back in the house. But cropping and topping were only two aspects of the job. Another was pest removal. Like other tobacco fieldhands of tender years, the Chief was given a Mason jar partly filled with kerosene and dispatched to walk the rows inspecting them for hornworms; each one he found would be plucked off and plopped into the oily fluid. When the jar was crammed with fat green bodies, he'd dump them at field's edge and set off again with a refill of kerosene. Though kerosene alone would have done the trick, that wasn't the end. Because the youngest children weren't allowed to handle fire, an older hand, aged ten or so, would strike a match and send the kerosene-soaked hornworms off in a great, leaping, popping, pyromaniacal blaze of glory. But that was more than half a century ago. Nowadays, pesticides do the worms in.

As the adage puts it, beauty lies in the eye of the beholder. So does the opposite, and try as I may, my eye discovers no beauty in hornworms. Goodness knows, they are many and multifarious, devoted not just to tomatoes and tobacco but to a huge array of garden crops, shrubs, deciduous trees, and conifers. From grapes to hydrangeas and honeysuckle, catalpas to cayenne peppers, nothing in the plant kingdom is safe. The caterpillars of most species come equipped with dorsal horns at the end of the abdomen; the others sport only a token bump or show no protrusion at all. Most pupate in the ground; a few, however, spin their cocoons amid leaf litter. The zoological family to which they all belong, horned or not, comprises no fewer than 124 species, which are hardly confined to the Americas. One species is endemic to the fabled Galápagos. Another is mentioned in the Hippocratic medical literature of ancient Greece, in which instructions are given for concocting a remedy for a suppurating womb from the dried and pulverized *kampai* of the *tithumallis*—the caterpillars of the spurge hawk moth.

Tobacco Hornworm (*Manduca sexta*)

These spurge-devouring caterpillars are described as wearing a *kentron*—a spike on their rumps, and that's the clue to their identity.

But ugly though I find hornworms, I can hardly deny my fascination with the colors and patterns on their bodies. And if I were still smoking, I might even envy the tobacco hornworm's gift for remaining unaffected by nicotine, which turns many human beings into raving addicts and kills most bugs when it's used as an insecticide. All the nicotine ingested by the tobacco glutton is simply, quickly excreted. But, oh, the bodies and habits of all hornworms are repulsive—the squishy obesity, the incessantly moving jaws. Their only challengers in the race for the most despicable larvae among North America members of the order Lepidoptera might be the native tent caterpillars that year-in, year-out infest wild cherry trees—but these larvae provide food for titmice and yellow-billed cuckoos, and their cocoons, with some persuasion, yield usable silk. Another contestant might be the larva of the gypsy moth, imported from Europe in the late 1860s as a silk moth; these caterpillars, polka-dotted with red and blue, wearing tufts of golden hair, produce no silk for commercial harvest—produce no good at all that I can see—but rather leave defoliated forests in their wake. As for hornworms, what good are they? Apart from the Hippocratic remedy, I know of only two benefits that they confer on humankind. The first involves only a single spike-tailed species, *Ceratomia catalpae*, the "hornworm of the catalpa tree." It is intentionally gathered up by anglers, turned inside out, and threaded onto a fishing hook as bait. The second is that entomologists find hornworms useful for research into matters like insect hormones and behavior; the caterpillars are large and easy to raise under laboratory conditions (nor has anyone ever complained about their maltreatment). But in the fashion of all successful creatures, hornworms care not a whit about being useful to anything other than themselves. No matter what the family, genus, and species, they and their creeping, crawling ilk are living proof of a primordial law: Whatever can be eaten, shall be. And yet, and yet—in the case of hornworms, ugliness and insatiability lead to something else, something that I gladly perceive as beauty. Metamorphosis works wonders in the hideous worm.

The family is that of the Sphingidae, the sphinxes. And the creature's caterpillar stage is responsible for the family's name, an ancient name that brings to mind pharaohs and pyramids, or Oedipus and riddles. The monumental Egyptian version—its lion-body recumbent and its regal head (thought to be that of the pharaoh Kafre) alert and upright—still guards the desert near Giza. But the Sphinx sent by the goddess Hera to punish the Greek city of Thebes has not been seen since the days of Oedipus. With a woman's head on a lion's body, the wings of an eagle, and a writhing, lashing snake for a tail, she crouched atop a mountain on the road to Thebes, barring the way to travelers. It was her habit to taunt all passersby with a riddle: What being with but one voice goes on four legs in the morning, two at noon, three in the evening, and is weakest when it has the most? She would most bloodily dispatch all those who failed to answer, and that was everyone until Oedipus, journeying toward Thebes to learn his fortunes from an oracle, was challenged but prevailed: Man, who crawls as a baby, walks upright as a youth, and leans upon a stick in his old age. Though it is not recorded, I think that the Sphinx must have screamed, must have shivered the air and set the mountain to trembling with her rage and disappointment. But it is known that she leaped from the mountain in defeat, and her broken corpse was found in the valley below.

It is, however, not her monstrosity (though perhaps it should have been) that led the classifiers of insects, men learned in classical studies as well as entomology, to take one gander at a typical horn-worm and promptly attach her name to the whole family. Rather, it is her posture, and that of her Egyptian counterpart. Head and thorax uplifted to appraise the world, abdomen stretched out flat behind, the sphinx caterpillar unwittingly mimics its namesake's pose. And, wonder of wonders, in those still moments it ceases chewing. But then the tireless jaws begin to scissor at the leaves again. Like cookie crumbs falling from a small child's mouth, bits of greenery drift to the ground. And if there are enough caterpillars at work, you can hear their droppings steadily bombarding the leaves below.

Yet the reason for this gross, obsessive appetite is transformation. Egg to larva, larva to pupa, pupa to adult, moth and butterfly retell an old tale: once again, the ugly duckling turns into a swan; once again, the frog becomes a prince. But the story of butterfly and moth occurs in real-world terms, which makes their onrolling transformations more marvelous than anything imagined in a fairy tale.

The adult sphinx moth is indeed a marvel. Ash sphinx, laurel sphinx, pawpaw sphinx—some have common names centered on their favorite food. Waved sphinx, one-eyed sphinx, clearwing, and five-spotted hawk moth (the tomato hornworm's apotheosis)—some are named for notable features of their appearance. Some names, like those of Carter's and Abbot's sphinxes, honor people, while places are mentioned in others—Canadian sphinx and Carolina sphinx (the elegant transfiguration of the lowly tobacco hornworm). Myth also figures—the Pluto sphinx, recalling the Roman god of the underworld, and the Nessus sphinx, named for the centaur whose burning blood was used to kill the hero Hercules. Still other names seem to consist of commentary—rustic, hermit, plebeian, and mournful sphinxes. All these moths, and the hundred others that I have not mentioned, are not just beautiful but in some measure astonishing. Their colors tend toward the conservative: blends of black, gray, and brown, often embellished with orange spots or patches. And the clearwings have no color at all in the center of their wings, which are scaleless and as transparent as a windowpane. A hummingbird clearwing visits the butterfly bush planted beside our front deck; true to its name, wings beating so fast that they're nearly invisible, it hovers above the flowers and sips the nectar with its long, long tongue. But in respects other than color, the sphinxes are spectacular indeed. Whatever their size—most are medium to large—they possess long, pointed wings that project from their plump, furry bodies like the elongated wings of a sailplane. With those wings, they sometimes attain a ground speed of thirty-five miles an hour, outflying most insects of any kind, including the other moths. Flight takes warmth; they contract their flight muscles until sufficient heat is generated to give their wings lift, and once airborne, they maintain a body temperature of 100 degrees Fahrenheit. Nor are their forays into the air fluttering and windblown; instead, maneuvering with exceeding strength and skill, they are able to rise vertically in the air and even to fly backward. The sphinxes may also be the least silent of moths. By inhaling and exhaling air, they can produce a chirp, which is the music of love, designed to attract a mate.

At the end of June, the green beans are done, with twenty-three pints put up and more given away. July brings fourteen pints of limas, which translate into eleven pounds of shelled beans. By the end of July, the tomato vines begin to give out, though not because of hornworm appetites; the falling yield is due rather to a month of fierce sun and little rain. But I have better than one hundred pints in the freezer and on the shelves—tomatoes stewed with onions and green peppers, tomatoes plus jalapeños, tomatoes with eggplant or in wine sauce, tomatoes canned plain, green tomato pickles, and sweet red tomato relish, along with juice, of course, and soup. The crop has been big enough to satisfy every comer, from orchard orioles to box turtles, from hornworms to woman. In mid-September, putting the garden to bed, the Chief begins with removal of tomato cages and stakes.

"Hon," he calls, "come look!" Inside fixing lunch, I'm galvanized by his urgent tone. He stands by the steps to the back deck, a tomato stake in his hand. On the stake rests a moth. The wings, folded over its body and patterned in the subtlest shades of brown, beige, and charcoal gray, look newly dry. The moth has emerged from the underworld only this morning. When at last it flies, the orange spots on its abdomen show clear: *M. sexta*, Carolina sphinx, scourge of the Chief's tobacco-cropping days. It lands three feet away on the trunk of a loblolly pine, where it is camouflaged by the barklike appearance of its wings.

That instant, my mind rewords the riddle of the Sphinx: What being sleeps at dawn, crawls at noon, then sleeps again but wakes in the evening to winged flight?

Carolina Sphinx (*Manduca sexta*)

# LEGS

Garden Centipede (*Scolopendrid sp.*)

## Centipedes

"THERE I was, sixteen years old, in the little basement apartment my dad had built for me. And there *it* was, a centipede, just a-gittin' it across the wall. I hit it with a phone book. It came in two—one teeny piece and the other with all those legs that kept moving and moving." Laura shudders. "Eew, that did it!"

Laura, wife of my elder son, is a down-to-earth woman, good at mothering and growing flowers. She's not usually fazed by anything that creeps or crawls, slithers, bites, or stings—except for centipedes. That apparition on her wall, that invasion of her space and her sensibilities, might well have been foreseen, however, had anyone been thinking at the outset of the preferred habitat of these small, swift, many-legged animals. The apartment, consisting of a tiny living room and bedroom, was ideally damp and dark. Laura's father had thought of pleasing her, while Laura, being the age she was, had mostly considered her newly acquired ease in sneaking out.

But those legs, those remarkable, multiple legs—she's brought up an interesting point. It occurs to me to wonder just how a centipede can move without treading on its own toes and tripping all over itself. As an old ditty has it:

> The centipede was happy quite
>     Until a toad in fun
> Said, "Pray, which leg comes after which?"
> That worked her mind to such a pitch,
> She lay distracted in a ditch
>     Considering how to run.

Surely, any other means of locomotion would be more efficient, from the snake's no legs to the spider's eight. More than eight legs seem a nonsensical abundance or, worse, a repulsive superfluity. Yet centipedes thrive, and so do millipedes with an even greater number of legs (where centipedes have one pair of legs per segment, millipedes have two and can also curl themselves into a tight little spiral). And there is an ever-scurrying host of other similarly overen-

Millipede (*Parajulus impressus*)

dowed, like horseshoe crabs and the tiny centipedelike pauropods with eight to ten pairs of stumpy legs.

All of these belong to the kingdom of the Animalia and to its most populous subdivision, the phylum Arthropoda, the "joint-legged" animals. Insects in their uncountable legions are arthropods. So are the crustaceans—"shell creatures"—like the barnacles, blue crabs, and shrimp found in my underwater front yard in North Carolina; the chelicerates, or "horny claws," which include both horseshoe crabs and the many arachnids—spiders, ticks, chiggers, mites, and scorpions; the trilobites—"three-lobed creatures"—now extinct, that swarmed in Cambrian seas 450 million years ago; and the uniramians—"one branchers"—a subphylum that gathers in several classes of terrestrial animals, among them the six-legged insects, or hexapods, and the mob of Myriapoda, which means "ten thousand–leggers." (Whoever named the myriapods—and this large group's "hundred-legger" centipedes, its "thousand-legger" millipedes—may easily be forgiven repeated exaggeration: the legs of these animals are often the first thing noticed. And there is a European species of centipede that has committed an anatomical exaggeration all on its own: it sports no fewer than 177 pairs of legs.)

"One brancher" may seem a curious term. But it means that the uniramians have one limb, and only one, extending from each possible location, while the biramous marine arthropods—the crustaceans, horseshoe crabs, long-lost trilobites, and others—have two, one of which is easily recognizable as a limb, while its partner is feathery and gill-like. The two-branched animals inhabit the fossil record almost as far back as it goes, but not the one-branched kind. So it is reasonably thought that, long ago, the arthropods evolved and diversified on land after a group of soft-bodied marine animals with somewhat hardened heads had left the sea behind forever, and taken up residence amid the heavy vegetation of coastal wetlands. They may have crept ashore on lobopods, or lobe legs, which are highly variable knoblets of flesh able to be extended or retracted; a set of lobopods near the head—future mandibles—would have been used to spoon food into the mouth. The first requirements of these creatures, the needs that shaped their evolution, were to shield themselves from both water and air, the former lest their bodies absorb a chemically imbalancing amount of freshwater, the latter lest they dry out completely in the seaside breezes. And they had to find food. Their legs developed sections, from the coxa, which is closest to the body, on through the trochanter, femur, and tibia to the often many-jointed tarsus; some joints swung like hinges and other like pivots. Their skins turned into armor—an exoskeleton of hardened cuticle that protected the animals' soft innards. The ancestors of flies and moths grew wings and flew; the aboriginal spiders and myriapods opted for a plenitude of legs and good ground speed, while the protograsshoppers and crickets developed a sturdy third pair of legs that gave great jumping power; still other land-going arthropods evolved an ability to sting or to curl into a defensive ball. Some reached adulthood though metamorphosis, while others hatched out as miniature versions of the grown-up arthropod and attained greater size through successive molts. But whatever routes they took, they thrived—the fossil record (for what it's worth) shows no massive extinctions—and became the present multitude of uniramous animals: the insects, the myriapods, and the minuscule pauropods and their close cousins the symphalids.

Of them all, the centipedes are indeed a mighty crew. Ranging the world, they tally in at about twenty-eight hundred species and warrant a myriapod class of their very own, the Chilopoda—the "jaw-leggers." That class is in turn divided into two subclasses, the

Epimorpha, "whole shapes," which hatch complete with every one of their many adult segments, and the Anamorpha, the "growing shapes," which hatch without the full complement but gain segments as they molt. These subclasses are each further divided into two orders with wonderfully descriptive names. The epimorphic kinds, both short-legged, are, first, the Geophilida, the reclusive "earth lovers," or soil centipedes with 31 to 177 pairs of legs, that burrow into the ground in the fashion of earthworms, and, second, the Scolopendrida, the "millipedes," so named because of their superficial resemblance to true millipedes—the dark dorsal surface and all those legs, legs, legs, from 21 to 33 pairs, depending on the species. The scolopendrids, sometimes called garden centipedes, are a primarily tropical group that in its largest incarnations reaches nearly a foot in length. The anamorphs, which never attain such a truly monstrous size, consist of the Lithobiida, the "stone" centipedes often found under rocks, and the Scutigerida, the "shield bearers," named for the hardened scutes, or plates, on their backs. The shield bearers, commonly known as house centipedes, are the critters with superlong, spidery legs that haunt sinks and bathtubs and run like speed demons across damp basement walls like Laura's. But no matter to which family they belong, centipedes are primarily nocturnal. All of them prefer to hang out in dim, dank places, like rotting logs, leaf mold, moist soil, or drains. All are predators; all are fiercely dedicated carnivores.

It is their method of catching prey that has awarded them the name of jaw-leggers. The pair of jointed limbs on the first body segment after the head have become modified into poison claws, sometimes called poison jaws (the technical term is toxicognaths) that inject a paralyzing venom into hapless victims, like worms or arthropods of other species. These claws are never used for locomotion but are held so closely under the mouthparts that they appear to be part of the animal's eating apparatus. Which indeed they are, for the poison they pump out is used for external digestion; it works not just to stun prey but helps to break down its soft tissues. Peculiarly, the choices of prey and eating habits are not the same with anamorphs and epimorphs. The latter, all short-legged, and some as short-sighted as Mr. Magoo but most of them completely blind, seek their dinners in confined, night-dark places like subterranean crevices and the undersides of decaying logs. They don't, however, need to see in the dark. Other senses and abilities help them out. While the scolopendrids skitter and wriggle into the merest crannies, the soil centipedes thicken segments of their bodies in earthworm fashion to exert a strong thrust forward and thus move through the earth in search of prey. It has been said that, blind as they are, both kinds of epimorphs must "stumble on food by accident." When it has been found (by accident or by intent), they swallow it whole if it is small or use their strong and hefty poison claws to paralyze and tear holes in larger prey, then insert their heads to suck up the soft parts. One large scolopendrid, *Craterostigmus tasmanianus*, which has some eyesight and is native to Tasmania and New Zealand, has been observed using its poison claws to pull termites out of cracks and to drill tunnels through compacted decaying wood. The long-legged anamorphs pursue another way of hunting. They look for prey with capture aforethought and are able to deal with tougher, harder food. The little stone centipede is particularly good at weaseling into crevices but without thrusting or pushing as the geophilids do. The house centipede, *Scutigera coleoptrata*, is surely the mightiest hunter of all the chilopods. (Its genus is "shield bearer," but the species name, bestowed by the master taxonomist Linnaeus himself, seems to mean "sheath wings," perhaps because its body, like a beetle's, appears to be sheathed. That's only a guess, for the name makes no good classical sense; nor can we wake Linnaeus to find out what he meant.) The house centipede is blessed with excellent vision; it works in more open places and can spot a fly or a spider before the victim is aware of impending doom. And in the chase, the creature runs like a greyhound, attaining top speeds in an instant. This kind of swiftness leads to great success in pouncing on and stunning normally wary food creatures like cockroaches and flies. The poison claws of the anamorphs are small and light compared to those of their epimorphic cousins, but their large, powerful mandibles can cut well-armored food like a beetle into pieces small enough for swallowing.

I hear the question rising: What hurt can a creature with poison claws put on a human being? The answer is, Not anything you'd notice much—unless, that is, you were seized and injected by one of the tropical models nearly a foot long. Even then, the bite, though uncomfortable, would hardly be fatal. Some scolopendrids can also use the hindmost pair of legs to pinch, but such an attack would produce more fright than pain. Our kind, however, is not irrationally averse to the very notion of pinches and venomous claws (not to mention the frisson raised by the skitter and scurry).

Centipedes do, however, have their human uses. We have managed to render the word *centipede* impotent by detaching the name from the animal and applying it both to a lawn grass, much favored in the South, that sends out fine-textured runners and grows so densely that it shuts out weeds, and to a computer game in which creepy-crawlies invade a mushroom patch. (That game is surely a way of transforming computer "bugs" from objects of frustration to subjects of fun.) And the people of Silver City, New Mexico, featured various contests as part of their annual Mining Days festival, which was discontinued, alas, in the mid-1990s. Prizes were awarded to the people fielding the fastest, heaviest, longest, and best-dressed hundred-leggers. Yes, like the star performers in a flea circus, some entrants wore costumes. Why centipedes? As one former resident puts it, "Little desert towns have to think of something to celebrate."

Nonetheless, an unexpected sight of the real thing can be off-putting, to say the least. I confess to being utterly unnerved, to flinching and yelping when I settled down on the sofa one chilly evening to read, reached for my favorite throw, and found a motionless centipede snuggled therein. Not the long-legged sink-and-tub kind but a dark, short-legged sort, which I immediately shook onto the floor—and almost as immediately regretted not having observed. The sharp eyes of my husband, the Chief, found it half an hour later. It had traveled thirty feet from the living room well into the hall. Becoming brave, I tweezered it, put it in a small jar with alcohol, and later sent it to the entomologists at North Carolina State University. Behold: a juvenile scolopendrid—*Hemiscolopendra punctiventris*, to be exact. (The name, a fine example of taxonomical, not classical, Greco-Latin, means "point-bellied half-millipede.") Along with the identification, the entomologists sent a detailed note: "This particular species is native to North Carolina and is commonly found in woodland areas. Our dry weather has sent many insects and other critters, such as millipedes, indoors in search of moisture, and it is likely that the centipede entered the house in search of prey." Then, with the assumption (natural but quite mistaken) that I had sent in the specimen because I wanted mainly to know how to commit acts of extermination, the note concludes: "Since only one specimen was found, you have solved the problem. Spraying is not really necessary, since these incidences are usually sporadic, and it is difficult (often impossible) to target specific areas to treat indoors." But within days of finding the first, I came across two more indoor point-bellies, one alive, the other completely dessicated. And though there's still an involuntary flinch when one makes a sudden appearance, I'm more interested in counting their legs. *H. punctiventris* boasts twenty pairs plus one—the concealed poison claws.

Flinch factor aside, a few things may be said in favor of centipedes. They do, for example, exhibit family feeling. The scolopendrids guard their eggs until they hatch; females curl around them and may even lick them on occasion, supposedly to keep them clean. There also seems to be some parent-offspring communication in all the chilopod orders. Centipedes may also serve as objects of pity, for they are not always the winners in the eat-or-be-eaten game. One species of ant, *Amblyopone pallipes*, for example, has a taste for hundred-leggers, but because this kind of prey is often far too large to lug home, *A. pallipes* moves its larvae to their dinners. Centipedes are not without defenses. The little stone centipedes, for example, throw out sticky threads from their posterior legs to entangle predators like ants and wolf spiders. And silken strands are spun out by the male centipedes of some species to guide the females to their spermatophores, or sperm packages. Sex among these centipedes is not a matter of penetration but rather one of pick up and insert.

What you really want to know, if you're still with me, is how they run. Question: What goes ninety-nine, thump, ninety-nine, thump? Answer: a centipede with a wooden leg. Only a human being could construct that joke—or this one: If the first fifty legs of a centipede are going forty miles an hour, what are the last fifty doing? Hauling ass. The centipede has no interest, of course, in such frivolity. It lives to run and runs to live.

Just how it runs is a matter that has received considerable study. Centipedes of every family have been kept in laboratories, where, alive, they are incited to thicken their segments or do high kicks for photographs and to scurry across smoked paper, leaving telltale patterns of footprints as they go; where, dead, they are dissected and subjected to microscopic examinations of their most intimate anatomical details, from their brains and digestive systems to the muscles in their bodies and legs. In that last department, it has been discovered that the animals are particularly well equipped with intrinsic (in-leg) and extrinsic (in-body) muscles affecting their locomotion. Each of these muscles tends to several tasks: flexing, lifting, lowering, or rotating the appendage to which it is attached. (The rotation is an action to which science has given the term "leg rocking"; it helps position a leg for the best push forward.) The geophilids are the pikers in the muscle department, with only thirteen extrinsic muscles for each leg, but multiply that thirteen by forty legs—the grand total is 520. The scolopendrids achieve a somewhat higher count at nineteen extrinsic muscles for each leg, and the stone centipedes boast twenty. The champs, however, are the scutigerids, with an amazing thirty-four each. The difference explains why the scutigerids get out of the starting gate so fast, while the epimorphic species, less amply endowed, need to start off slowly and shift up to speed. But the mere fact that each of the many legs is fitted out with a boggling number of muscles does not in itself explain how the animals manage not to trip themselves. As big centipede says to little centipede in a 1998 *Wall Street Journal* cartoon, "Strength and speed are useful, son, but coordination is *crucial.*"

Imagine a chorus line. Imagine the Rockettes or a ruffled flurry of cancan dancers or a troupe of high-kicking showgirls in Las Vegas. The main difference is that while the line is composed of separate entities acting as one, the centipede's parts are all irredeemably attached one to another. Then, while a chorus line may specialize in synchronous movements, its members stepping uniformly in the same direction at the same time, the centipede runs with undulating leg movements known as metachronal rhythms. The legs do not work in unison but rather in succession. At low speeds pairs of legs operate in the same fashion, but in an all-out dash, they each take on different roles, with some on the ground, others in the air. At any given moment, some legs perform the propulsive backstroke that pushes the animal forward, while others lift off the ground, stretch out, and swing in preparation for their turn at pushing. A centipede runs, says one researcher, "by a series of gaits in which the phase difference between successive legs and the relative duration of the forward and backward strokes by the leg change harmoniously." The gaits depend, that is, on which legs are moving and for how long they lift or push. These movements look something like grass rippling beneath a breeze. But all centipedes run with immense precision. The tracks recorded on smoked paper show distinctive patterns for various species, with successive footprints often being placed on the one same spot. In their fastest mode, the scolopendrids, which tend to weave a bit from side to side as they run, use only three legs at one time for propulsion, while the other thirty-seven move through the air. The stone and house centipedes, many of which which have only fourteen pairs of legs to run with, fly straight ahead like arrows. The longer the legs, the longer the stride of an animal, and the greater its speed. And some amazing speeds have been recorded in the laboratory, with *S. coleoptrata*, the house centipede, clocking in at more than sixteen inches per second, though it is thought to run much faster in the wild.

The leg movements that Laura saw, however, were chaotic, inspired by the involuntary muscular contractions that occur after death. I hope that I can convince her, as I've convinced myself by looking into their lives and habits, that centipedes, all unaware, are full of grace. Aside from which, they eat cockroaches.

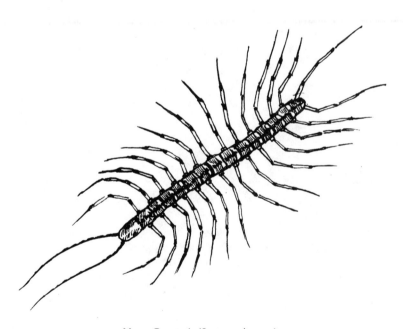

House Centipede (*Scutigera coleoptrata*)

# HERITAGE

Kudzu (*Pueraria montana*) on Cabin

## Kudzu

"Goats do a pretty good job," says the cooperative extension agent. He goes on to recommend a popular herbicide in case I don't have a goat.

I've just asked him how kudzu, that perennial and extraordinarily aggressive vine, is best kept within bounds. In a sense, my question is rhetorical, for none of the stuff grows on the riverside land that my husband, the Chief, and I call home during the balmier months of the year. But in another sense, it's not. Go a mile or two inland: there, with lush exuberance, the vine has seized many unfarmed portions of the landscape.

Throughout the South, kudzu creeps with stealthy swiftness over brushpiles and fences. It climbs trees and telephone poles and casts its soft but heavy net over thickets and hedgerows. It enshrouds abandoned houses, tumbledown tobacco barns, rusted appliances, and junked cars. It sneaks into gardens and plowed fields. Displacing innocent native vegetation, it twines, curls, shoots upward and outward with relentless green insistence. In its wake, power outages occur, and trains have been derailed. By the mid-1990s, kudzu had laid claim to more than 7,000,000 acres—almost 11,000 square miles—of the South. Monstrous roots thrust deep into the earth of at least sixteen states, ropey vines embrace the landscape, and leaves smother it in a big, soft, fuzzy, unbreakable hug. Once the vine invades any location, getting rid of it is well nigh impossible. And it seizes another 120,000 acres every year, a rate that can only increase with the increase in the plant's domain.

Only a decade-long drenching with herbicide could end such a massive takeover—if, that is, the anti-kudzu factor of the population, and that's almost everyone, could be rallied into a concerted effort. (A more important consideration for many of us might be the degree of comfort we feel about using manufactured herbicides in the first place.) As for goats, I doubt that the world has enough of them to eat their way to a solution of the problem.

Kudzu *is* a problem. Not only are fortunes in private and public money spent annually on its control and eradication, but the vine is rude enough to affront human sensibilities; unruly, running wild, operating only by genetically encoded instructions, it defeats our every attempt to assert our superiority. It's vegetation with a vengeance, a sprawling, mindless, inordinately grabby menace. Or, as my Southern-speaking friends and neighbors have put it, the damned stuff's a right sorry mess.

Yet, it has virtues. Without them, it wouldn't have spread so far so fast.

Kudzu came to America by invitation more than a century ago. Now it has moved in, the very model of a sorner—a guest, that is, who overstays the initial welcome, becomes entrenched, and places undue burdens on an unwilling but helpless host. Yet this native of the Far East was escorted here with much praise for its beauty and usefulness. And it is an attractive plant. The young vines are tender, the more mature vines, woody. The deciduous leaves, sometimes entire but usually lobed, grow in groups of three on softly hairy stems; each one is as big as my hand. The leaves, along with the young vines, provide certifiably nutritious fodder for livestock. The densely clustered flowers, purple as Concord grapes and just as fragrant, bloom in late summer. With the onset of cold weather, the leaves are shed and top growth dies back, leaving dry, sticklike tangles. The fuzzy brown seedpods look something like woolly bear caterpillars. The plant, however, rarely produces itself by seeds, which resist sprouting unless the hard seed coverings have been scarified—that is, softened or cut. The alacrity of its spread is accounted for partly by the fifty- to one-hundred-foot growth of the vines each season—more than a foot a day—and by its habit of sending down roots whenever a leaf node touches the ground. These root crowns, like those of ivy, easily produce new plants. But when a seed manages to sprout and the resulting vine does not lie on the ground but shoots aloft, climbing a fence, a tree, or a telephone pole, the resultant root develops massive proportions. Old roots have tapped as deeply as twelve feet into the ground and may weigh anywhere from two hundred to four hundred pounds.

The Japanese, who make an edible starch from the roots, are usually given full credit for kudzu's introduction. Its very name is an Americanization of *kuzu*, the Japanese word for the plant. It was they who exhibited this country's first living, growing specimen in their pavilion at the Philadelphia Centennial Exposition, which opened in May, 1876, and they showed it again in 1883 at the New Orleans Exposition. The Japanese, however, were not entirely responsible for kudzu's entry and all the ensuing complications. An American was involved, one Thomas Hogg (1820–1892), who had been sent to Japan in 1862 at the behest of none other than President Abraham Lincoln, and there, for the next eight years, he filled the post of United States marshal. Toward the end of his tenure, he worked closely with the Japanese customs service and, in 1872, returned at the Japanese government's request to continue this work for another two years.

I came, I saw, I picked it up and brought it home—this might be a guiding principle for many travelers, known and unknown. Their ranks include Marco Polo, Columbus, Cortés and de Soto, Captain John Smith, Captain James Cook, and Charles Darwin. As they sailed the seven seas or slogged through the New World, they collected everything that had ignited their admiration, beckoned to their greed, or simply tickled their curiosity. And what a hodgepodge the collections were—goods and curiosa that ranged from spices and gold to lizards, raccoons, and sassafras trees. Japan came late into the scrutiny of the Western world, but when the country was at last opened to foreigners in the mid-1800s, a good many of the Europeans and Americans who went there followed the grand, acquisitive example of earlier travelers. And how could the Honorable Mr. Hogg resist? The son of a noted horticulturist who ran a nursery business in New York, he had learned botany in his cradle days. Later, the decade of service in Japan provided him with ample time in which to indulge his passion for plants and also send home many specimens that had caught his fancy. He sent them on their transpacific journey in a recently designed container called a Wardian case, after its inventor Nathaniel B. Ward; it was a sealed terrarium with a glass dome set atop a sturdy wooden box filled with soil in which the plants had already taken root. Some of the plants that Thomas Hogg introduced into the United States are popular today, among them the katsura tree (*Cercidiphyllum japonicum*, "Japanese tree with leaves shaped like a weaver's shuttle") and, especially, Japanese stewartia (*Stewartia pseudocamellia*, "Stewart's false camellia"), with its exfoliating mauve and silver bark and its huge flowers comprising five snowy petals around an orange-gold central disk. But stewartia's popularity is deserved; it's a well-behaved plant, possessed of the good manners that kudzu sadly lacks.

Yet on its home ground in the Far East, kudzu has ever observed the proprieties. Climate and native pests help keep it in check, along with its regular harvest for a host of human uses. (I think it likely that its spread in densely populated Japan is also limited by the need to make the most efficient use of available land.) The plant is assigned to the genus *Pueraria*, which is divided, depending on the source used, into anywhere from ten to fifteen species, originating variously in China, Japan, Taiwan, or India. *Pueraria* seems a classically Latinate designation—except that it yields no meaning when I consult a lexicon. I consult a friend instead, the poet Jeffery Beam, who is most fortuitously employed in the botany section of a noted university's biology library. He finds that the name has nothing to do with Latin, nor is it connected with anything Oriental. Rather, Switzerland's Augustin Pyramus de Candolle (1778–1841), one of the truly great nineteenth-century botanists and a member of an important botanical family, named it for his compatriot and fellow plantsman Marc Nicolas Puerari (1776–1845), who had donated his private herbarium—his collection of dried and systematically arranged specimens—to the family. M. de Candolle bestowed the generic name no later than 1825, the year that *Pueraria*, according to the custom of the time, received a sumptuous description in Latin. Just how he and

Puerari came across kudzu in the first place remains a secret, but it's surely safe to think that some European traveler, impelled (like Marco Polo and Thomas Hogg) by collector's fever, saw it, picked it up, and brought it home.

The vine that has thrust its roots deep down into U.S. soil is *P. montana*, "mountain kudzu," which was known until very recently as *P. lobata*, "lobed kudzu" (botanical nomenclature, like that for all other living things, undergoes cyclic upheavals). But the American venture was by no means *P. montana*'s first move into a foreign land; native to China, it was brought into Japan no later than the sixth century A.D., and there it quickly, contentedly settled in, appearing not only as a feature of the landscape but also as an ingredient in recipes, an item in the natural pharmacopoeia, and a subject for poems.

Every kudzu, of whatever species, belongs to the Leguminosae, the pea family, which comprises a stunning diversity of flowers, shrubs, vines, and trees, all of which produce seeds contained in pods. The snap peas growing in our spring garden, the green beans we've yet to plant are both kin to kudzu, as is the soy that will be harvested, come September, in the fields of nearby farms. Also part of this huge family are the various clovers and the spurred butterfly pea, a delicate lavender-colored wildflower, that rejoice in our seldom-mowed yard. So are the tough-rooted plants called devil's shoestring that my brothers grubbed out of the pastures on my father's farm, the trefoils with seeds that stick tight as ticks to my socks, and the poisonous locoweeds of the Midwest, not to mention the gentler bluebonnet, pride of Texas. And so are shrubs like broom and woody, climbing vines like wisteria, along with ornamental trees—acacia, redbud, mimosa, honey locust, and a host of flowering, sweet-smelling others. Except for the frozen polar deserts, no part of the world lacks members of the family.

Kudzu's arrival in America may have been predestined. Even without Thomas Hogg and the showcasing of the plant at the Japanese exhibits in 1876 and 1883, it would have made its way to the New World for both aesthetic and culinary reasons, and also, it's likely, for its amazing, colossal, downright stupendous habits of growth (Americans have ever been enamored of wonders, freaks, and marvels, indeed of all manifestations of the biggest, the best, the least, the worst, and similar extremes). But its entry, though thoroughly triumphal, was accomplished with little fanfare, and the plant's initial popularity was based on its leafy beauty and the friendly shade it could cast on porches that would otherwise bake and swelter in the direct rays of the sun. For more than fifty years, from its introduction into the 1930s, kudzu seemed a benign, quite amiable alternative to the bee-rich grapevines, climbing red roses, twining blue morning glories, and other greenery often planted by porchside trellises to moderate the fierceness of the summer sun.

The story of kudzu's takeover is an oft-told tale that has been featured in many publications, from agricultural bulletins and forestry magazines to the front page of the *Wall Street Journal*. At least two books have been devoted to its history and real or putative uses. Suffice it here to summarize the events that led to the transformation of an ornamental plant known fondly as the "porch vine" into the apparently limitless, all-devouring vegetative sprawl that has entered folk parlance as the "vine that ate the South."

Although there were signs early on that kudzu had an appetite for escape and an unquenchable zest for proliferation, its partisans by far outnumbered the monitory skeptics until the early 1950s, a full seventy-five years after it had begun to snuggle itself firmly and probably forever into the South. Some who issued warnings were farmers who refused to let kudzu set one tendril, one rooting leaf node on their land. Today, their names have disappeared into the crevices of history. One name that survives, however, is that of David Fairchild, who worked before the turn of the century as a plant explorer, seeking useful specimens abroad, for the U.S. Department of Agriculture and in the early 1900s became the department's chief. In that capacity, he'd served in Japan, where his eye, like that of Thomas Hogg, had alighted on wild kudzu; his interest, however, was centered not on the plant's beauty but rather on its undeniable appeal to grazing and browsing animals and its high nutritive value. So he brought it along when he came back home. His particular misadventure happened in his own yard in Washington, D.C., where he'd set in seedlings. And there his vines grew quickly into luxuriant specimens that had a tendency to emulate Pinocchio: he'd planted kudzu, given it a good home, but it ran away. It covered his shrubs, climbed his trees, and made of itself "an awful tangled nuisance." This description, along with advice to exercise great caution, appears in his book *The World Was My Garden: Travels of a Plant Explorer*, which was written in 1902 but not published until 1938. By then, it was much too late for his warnings to bring the runaway back into even a semblance of domesticity.

But by the turn of the century, when David Fairchild first discovered kudzu's potential for trouble, others in both private and public life were promoting its virtues with the intensity and fervor of true believers. In their vanguard were Charles and Lillie Pleas, Quakers and dedicated conservationists, who had moved in the early 1900s to Chipley, Florida. There they not only promoted kudzu as a superior forage crop and the best-ever means to control erosion but also established a commercial nursery, which propagated the vine from seeds, cuttings, and root crowns and sold it well into the 1940s. On the outskirts of Chipley, a bronze historical marker, surmounted by the state seal of Florida, now honors their efforts: KUDZU DEVELOPED HERE.

But the Pleases' earnest and unceasing dedication centered on practical uses for the wonder plant. The work of Channing Cope took another form; it was a carnival of hype and hucksterism. From the late 1930s to the end of the 1940s, styling himself the Father of Kudzu and the Kudzu King, he extolled the virtues of the vine, writing often in the *Atlanta Journal-Constitution* and delivering a daily radio broadcast from the front porch of his house at Yellow River Farm. When he'd purchased the old farm near Atlanta, it had consisted of seven hundred eroded, unproductive acres, but kudzu, which he set in to improve the soil and serve as pasturage for cattle, worked its thickety green magic almost overnight. As possessed as a pearly-gates preacher by joyful certitude, he not only used his pulpits in the media to promote the vine's cause but also founded the Kudzu Club of America. With meetings, contests to see who could plant the most kudzu, and the election of kudzu queens, the club at its peak boasted twenty thousand members. But after Cope died in 1962, his dreams withered on the vine by the chill of disfavor, the kudzu he'd planted on Yellow River Farm overran the house from which he'd made his broadcasts and pulled it to the ground. To this day, however, the *Journal-Constitution* runs features on kudzu that grant the stuff a sort of quizzical praise, and towns in Georgia, Alabama, and Mississippi regularly hold kudzu festivals—all of which just goes to show that a bane serves as well as a blessing in giving folks an excuse to eat, drink, and make merry.

Channing Cope, however, leaped aboard the kudzu wagon well after it had picked up steam. Nor was it private efforts, like his and those of the Pleases, that caused a tidal surge of greenery to inundate the South. Credit—or blame, as the case may be—lies with the federal government and, more particularly, with Franklin Roosevelt's New Deal. In 1933, Congress brought into being the Soil Erosion Service, which two years later became the Soil Conservation Service (SCS), the name under which it operates today. In one tidy bundle, the work of the SCS addressed two of the major problems of the Depression era: the high unemployment rate and the rapidly failing productivity of Southern soil, which had been eroded and leached to exhaustion by two hundred years of farming methods that reaped crops but did not replace the nutrients needed to grow them. Enter the Civilian Conservation Corps (CCC), which aimed to keep young, unmarried men off the streets and out of bread lines by putting them to work at many soil-saving projects, from building flood-control levees to planting kudzu throughout the South. The U.S. Department of Agriculture published promotional literature on the vine, which could indeed keep worn-out farmland from washing away because it readily took hold in soil too poor to support anything else. The government not only established nurseries but also coaxed farmers into planting root-crowns and seedlings (some of the latter imported from Japan) by offering a cash incentive of up to eight dollars an acre—a fortune in those hard times. And once kudzu was on a roll, no one could stop its forward drive.

"We've got this concern that you're going to incriminate us or our agency," Bobby says in mock terror. Then he implores me, "Be kind to us, please."

Bobby is Robert Whitescarver, a slim, wiry man in his early forties who works for the SCS as a Natural Resources Team Leader for three large administrative units in the western part of Virginia. He's headquartered in Augusta County, my home territory, which lies in the center of the Shenandoah Valley; it's bounded by the Blue Ridge on the east and on the west by Great North Mountain, part of the long, lean Western Range of the Appalachians.

It's not kudzu country, or not in any worrisome way. The only place that I've seen it growing was along the back-yard fence of the house in which, two decades ago, I occupied the basement apartment. At that time, acquainted with the plant only by scowling rumor, I'd thought that the large-leafed vine winding itself around the fence wires was a watermelon, a volunteer that had sprouted, as they do, from a seed *thhhped* out by someone (perhaps me) on a hot summer day. Not a bad-looking plant. A rather decorative plant, in fact, and one that could perform a service by covering the naked fence wires. But a clue to its identity soon came along: the vine showed an extraordinary capacity for growth; indeed, it leaped forward several inches every night. Strip off the leaves, rip out the young root—I showed no mercy. It did not return the next year.

Bobby says that kudzu's not a problem in the valley—too cold—and that little of it was planted there in the 1930s and 1940s, the years that the SCS promoted it, propagated it, and set it ineradi-

cably into tens of thousands of acres in the South. With him, however, is the man he calls an "angel of the conservation movement."
This particular angel, gray-haired, somewhat rotund, and blessed
with a smile as ample as the whole outdoors, is retired District Conservationist Wayne Hypes, who admits to planting some kudzu in
the valley—"not much, though, not much." In his soft Southern
voice, the word comes out as "plainting."

"I know your mother, go to the same church," Wayne says,
introducing himself in Southern fashion, which aims to establish
connections wherever possible. And he tells me that it was he who
designed and supervised the construction of the small stream-fed
lake on my father's Augusta County farm forty years ago. (The lake
was stocked with bream and smallmouth bass; their descendants
may be caught today.) Only then does he embark on his tale of service with the SCS. Now seventy-eight years old, he well remembers
kudzu's glory days. In 1941, when he was twenty, he joined the SCS
and was sent forthwith as a soil conservationist to the CCC camp in
Mocksville, North Carolina, just southwest of Winston-Salem. He
describes the work done by the members of the corps: "The first
thing I saw them do was to construct terraces to cut down erosion
on cotton land and tobacco land, and the next thing I saw them do
was to plaint kudzu to heal up gullies." Then, as their supervisor, he
determined where they should terrace and where set in cuttings. In
November of 1941, he turned twenty-one; in February 1942, three
months after the Japanese attack on Pearl Harbor, he reported to the
Army and was subsequently sent to the South Pacific. Nor did he
leave kudzu behind. There he participated in the invasion of Moro-
tai, an island located just north of the equator about midway
between the Philippines and New Guinea. Once ashore, his company was assigned to dig anti-aircraft guns into the ground, build a
berm around them, and camouflage them from enemy planes.
Noticing wild kudzu everywhere and recalling its phenomenally
rapid rate of growth, he bade his men did up some plants and set
them on the berm of the gun emplacements. He remembers that "it

grew so fast our guns were completely camouflaged in just a few days." After the war, in 1946, he was back in North Carolina, where little had changed on the SCS front: "They were still plainting kudzu and building terraces."

Then I learn one reason that Bobby has granted Wayne Hypes the status of an angel. With some bemusement, as if he's still shaking his head at being tapped back then by serendipity, Wayne says, "I was one of the first—and I'm proud of this, though I don't know where I got the idea—I laid out contour strips in Richmond County in North Carolina, the town of Rockingham. They'd never heard of those before. They were all using terraces." He adds that he also laid out some contour strips on my father's farm. He's talking about the practice, now standard, of leaving the land as is and planting along with its natural elevations and declivities to maximize soil retention and minimize erosion, rather than bringing in manpower and machines to work a sometimes brutal reshaping. The use of contour strips may have saved more farmland than any other method. Nonetheless, terrace or contour strip, the mission of SCS was to cover every inch with kudzu. In 1947, Wayne was transferred to Virginia, where the vine had already become well entrenched east of the Blue Ridge. He comments on the kudzu that covers the roadsides along Route 29 from Culpepper down to Danville. It's truly abundant there—lush, rank, and powerful. Rearing up trees and poles, it turns them into long-necked, leafy-headed brontosauruses, browsing in the great fields of the sky. The sight is somehow beautiful.

Wayne and Bobby both comment on other takeover plants. Kudzu is not the only exotic that has made a successful landgrab. Bobby notes somewhat ruefully that the SCS also raised cuttings and gave enthusiastic support to the planting of the multiflora rose (*Rosa multiflora*, the "many-flowered rose") as a living fence. Like kudzu it was imported from Asia in the nineteenth century as an ornamental plant. It puts on a lush display of small, fragrant white blossoms in the spring and in the fall bears red hips that are immensely attractive to wildlife, especially birds. And this species lived a tame, unobtrusive life in yards and gardens until the 1930s, when some of its other gifts—a kudzulike ability to flourish in poor soil and stabilize it, a habit of dense growth that could form wind-breaks—came to federal attention. According to the wisdom of the time, multiflora rose would not spread on its own, for a seed fallen to earth could not sprout without prior scarification. But multiflora ran out of control, nor could anyone, not the SCS nor farmers who'd only wanted a living fence, call the renegade back. Birds, delighting in the fruit, performed the act of scarification in their digestive tracts: behold, multiflora throughout the eastern U.S., invading pastures and blanketing roadsides. I see it in the flatlands of Virginia and the Carolinas; I see it beside the switchbacks on West Virginia's steep mountain roads. Recapture is out of the question. But unlike kudzu, which left behind its natural controls when it crossed the ocean, the renegade rose is subject to insect predation and mite-borne disease. Therein may lie hope for some restraint.

In an effort to contain the aggressive spread of another one-time favorite of the SCS, the autumn olive (*Eleagnus umbellata*, "olive-chaste tree with umbels"), native to eastern Asia, restrictions have lately been placed on its planting. Wayne tells me something of its history with the SCS: "In the borrow areas of a dam—where you dig up dirt to build the dam—down in the clay and soil where plants don't grow well, we'd set out autumn olive. Wonderful wildlife food, berries about as big as a pencil eraser—a friend in the Soil Conservation Service made jelly out of them." Then he outlines the problem: wherever autumn olive is set in, it asserts its thickety dominion and elbows out all other greenery, from grasses to trees. Limits have since been put on its use, and its sale west of the Blue Ridge in Virginia is now prohibited. (But commercial mail-order nurseries advertise its appeal both to wildlife and to jelly makers, and you can order it, no questions asked.)

Their talk reminds me of other Pinocchio plants that were brought to the U.S. with entirely benign intent but, once established, quickly naturalized themselves, spread out, and created problems. Melaleuca (*Melaleuca quinquenervia*, "black-and-white tree with five-ribbed leaves"), native to Australia and sometimes called bottle-brush or honey myrtle, proceeded to help itself to large chunks of southern Florida's wetlands, including the Everglades, after its introduction in 1906. And where it takes hold, it crowds out native vegetation and wildlife and offers them no possible means of return. Fortunately, the rest of the United States is exempted from invasion by melaleuca's aversion to frost. Many parts of the Southwest, however, have been overrun by a Eurasian tree sometimes called the "kudzu of the West." It's the salt-cedar (*Tamarix chinensis*, "Chinese tamarisk"), named for its salt-retaining leaves, which was introduced in the late 1700s and planted along streams as a resource for pioneers seeking wood and flood control. It showed nothing but modesty during its first century here, but then it reached out and now occupies more than a million acres, no end in sight. Sopping up water from the rivers and dammed impoundments of a basically dry and thirsty land, it sends the vapor into the air, and when the leaves are shed, the salt they contain poisons the soil against other kinds of growth.

And these are only a few of the pest plants now on the rampage. In the matter of kudzu, Bobby again begs me to be kind. But there's no other choice. The SCS faced formidable challenges—a country out of work, the South turned into a wasteland—and did the best it could with the knowledge and means at hand. It's fair, I think, to say that kudzu helped in no small way to save the South. And it's angels like Wayne Hypes, like Bobby himself, who have done the rest, setting out and carefully cultivating modern methods of farming, soil restoration, and conservation.

In 1953, kudzu's bad manners—its utter lack of restraint, its reckless tendency to barge in where it wasn't wanted—led the U.S. Department of Agriculture to delist it as one of the cover plants allowed under the Agricultural Conservation Program. But only in 1962, nine years later, did the SCS stop recommending its use in developed areas (a too little, too late response for which Bobby Whitescarver and Wayne Hypes can hardly be castigated). Finally, in 1970, the USDA gave it official status as a common weed in the southern U.S.

How sluggish, bureaucracy! How swift and all-encompassing, the newfound weed! Kudzu has virtues, yes, some of which have been put to good use during the century-plus of its tenure here: it is ornamental; it provides good fodder, though the yield per acre is low; and it holds poor, washed-out soil in place. Nowadays, in these three areas, technology and horticulture provide far better choices. Yet kudzu is here to stay. As we slide toward the millennium, little has changed—except that there's more of the stuff than ever before. Nor can we really hope to grub it out.

It may be that the vine has put a lock not only on the Southern landscape but also on American ways of seeing it, with bright, well-meaning people taking a dark and hostile view of an alien species. Certainly, on its own green terms, kudzu has achieved a spilling-over, almost enviable success. Perhaps we're put off, even frightened, by its sheer abundance, and the silent speed with which it grows.

> Japan invades, Far Eastern vines,
> Run from the clay banks they are
>
> Supposed to keep from eroding,
> Up telephone poles,
> Which rear, half out of leafage,
> As though they would shriek,
> Like things smothered by their own
> Green, mindless, unkillable ghosts.
> In Georgia, the legend says
> That you must close your windows
>
> At night to keep it out of the house. . . .

So begins "Kudzu," the eerie poem composed by James Dickey (1923–1997), a Georgian who knew most intimately whereof he spoke. There, the leeriness, the outright fear that it provokes, take the aspect of a recurrent nightmare that, like the eponymous plant, cannot be extirpated. In the poem, as in life, a man making his way through a tangle of kudzu may plunge unaware into a pit concealed by the leaves, and the venomous snakes that shelter unseen beneath the dense cover will strike both man and the cattle that are grazing there; the hogs, sent into the kudzu afterward, catch and fling upward, killing the "living vine" of each snake that they root out. Nor is there an end to terror. Even with the windows shut tight at night, there is no keeping out this immane green power. The one mortal recourse against the threat is to take on the nature of kudzu, to let its blind energy surge through the veins till a man becomes just as strong as the vine itself, and just as concealing, as friendly to serpents. Only then might his life receive the awe accorded to the vine—that it "prospered, till rooted out." Almost wistfully, I think, the poem seeks to soothe a pervasive human uneasiness in the presence of any thrusting, seizing, unstoppable force.

But the insinuation of kudzu into Dickey's lines and its possession of his poem is not the victory for *P. montana* that it might seem. The poet does not take on the nature of the vine; instead, he has seized it and bent it to a purpose of his own devising. Along with ornament, forage, and erosion control, another good use for kudzu: to provide the controlling image for a poem—the soil, so to speak, in which the words and music put down their roots. Nor, by more than a millennium, was Dickey the first to put kudzu to this kind of work; it fills the same role in the *Manyoshu*, a collection of Japanese poems which was assembled around A.D. 600.

So the result of kudzu infestation is not necessarily rage, despair, or bad language. And many people have discovered virtue where popular perception, especially in the American South, has all too often seen only vice. To begin with, *P. montana* can satisfy the three most basic human needs, those for food, clothing, and shelter.

From soup and salad to noodles and dessert made from its starch, kudzu has immemorially satisfied appetites in the Far East.

Certainly the Chinese popped kudzu into their mouths long before the vine was introduced into Japan sometime in the earliest centuries A.D. (The Chinese character meaning *kudzu* was imported along with the plant and has been used ever since by the Japanese.) The oldest surviving Japanese record of kudzu as a food comes from the 800s and notes that the leaves of wild vines were gathered and eaten as a vegetable. And it is thought that by the 1200s, the roots were dug so that their starch could be extracted both for everyday cooking and for the preparation of medicines. Little-changed over the centuries, the process of starch extraction—digging the roots by hand, cleaning, cutting, and pulverizing them, washing them over and again, removing impurities, forming the resultant paste into blocks and slowly drying it—is a labor-intensive, time-consuming enterprise that may take as long as ninety days. For Japanese palates, at least, the work is worthwhile. Stirred into sauces, soups, and even lemonade, the starch acts as a thickener every bit as effective as arrowroot and cornstarch, and maybe more so. It gives a crisp, delicate coating to foods rolled in it before they are deep-fried. As a jelling agent—and one sure to please a hard-core vegetarian—it ranks up at the top with gelatin. The leaves of kudzu may be steamed, eaten raw, or dried for crumbling into dough. Bees make a delicious honey from the fragrant purple flowers, and the flowers themselves may be pickled in vinegar or steeped for tea, from which jelly may be made. Actively farmed in Japan, kudzu vines are set into land that might not otherwise be put to human use, like the steep cuts along roads and railways, where they not only hold the soil but also provide a green view for the traveler's eye. But though it's a staple in Japanese kitchens, it's seldom found as a food in the United States—unless, of course, one encounters it by rumor, or comes across it in a health-food store, or wanders into Atlanta's Kudzu Cafe, which features a kudzu quiche among the exotica on its menu.

And what does kudzu powder taste like? No one in the boondocks of coastal North Carolina carries it, nor can I find a supplier anywhere near the small Virginia town in which we spend our winters. But I find an 800 number appended to an article on kudzu cookery in the *Atlanta Journal-Constitution*. The starch is in stock—hurray!—at a natural-foods wholesaler in Eaton Rapids, Michigan, just south of Lansing. The French-Canadian clerk (I asked her what accent I was hearing) wants to know how I'll use it—dietary supplement or what? For cooking, I say. Less than a week later, two pounds of Japanese-made kudzu starch arrive in the mail. The bag is filled with small, flat-sided chunks of a substance that looks like bone-white chalk. No eagerness touches my palate. Nonetheless, expecting to bite into something crunchy-chewy, I do pop a tiny piece into my mouth: no taste at all, it melts instantly and slides on down with silken ease. Seeing that I neither wince nor sputter, my husband, the Chief, pops one into his mouth, too. He nods. I have his assent, then, to use the starch not just in my food but also in dishes that are cooked for him. Stirring it beforehand into a little water, I add it to canned tomato soup and homemade gravy, both of which gain pleasant body. Unlike cornstarch or flour, it does not lump. Ideal stuff, I think, for thickening a gumbo. As for quiche, oh, yes: the Atlanta paper also printed the recipe from the Kudzu Cafe.

Whereas roots, leaves, and flowers all provide food, the ropey vines may be coaxed into yielding fiber. The traditional work of turning kudzu into filaments ready for weaving commands a week of scrupulous attention. The vines must be gathered when they are still young and flexible. Then they are cooked, wound into coils, and soaked in cooling water for a full day. The next step is fermentation, which decomposes the outer skin and allows it to be slipped off, revealing the two layers of fiber that enclose a woody core. More soaking follows. At last the cores can be extracted, and the fibers freed. They are made into loose hanks from which single strands of kudzu fiber are carefully pulled. The short strands are then knotted by hand into filaments, as long as 250 feet, that can be woven on a loom. The fabrics produced range from light ones with a tight

weave and the shine of silk to loosely woven netting used to catch
fish. Every step in the process, including the weaving, is done by
hand. And the "grass" in the popular grass-cloth wallpaper is none
other than kudzu, woven for the most part in South Korea, which,
spotting a commercial winner, imported the vine from Japan. The
fibers are also used to make decorative papers.

As for shelter, that use of the vine may not be an Oriental phenomenon. It may, indeed be restricted to the United States. But
given the dense tangles formed by kudzu everywhere it grows, it's
easy to imagine the inviting recesses, the cool and shadowy grottoes
beneath the green leaves. I have, however, come across only a single
record, which places kudzu housing right where it might be
expected—in Georgia. The story, in the September 4, 1991, issue of
the *Atlanta Journal-Constitution*, told of the threat that the planned
construction of a new leg of the Presidential Parkway posed to the
dozen or so homeless people who had taken up residence in the
kudzu overrunning a vacant lot at Carter Center in Atlanta. I do not
know their fate. But parkway or no parkway, it's nonetheless sure
that city ordinances and social services (not to mention common
decency) would not permit human beings to keep on living in quarters with high snake appeal but no plumbing.

The modern uses of kudzu don't stop with providing for the
three basic human needs. The plant plays a not insignificant role in
the traditional pharmacopoeia of Japan, China, Tibet, and many
other countries of the Far East. There, herbal teas, brewed from dried
roots sometimes mixed with other roots, seeds, and spices like ginger
and cinnamon, have ever been imbibed to remedy a grand hodgepodge of ailments, including headaches, muscle stiffness, congested
lungs and sinuses, measles, hangovers, and constipation. The white
starch extracted from the roots may be eaten dry (take one chunk, let
dissolve on tongue) or mixed with liquids to make a cream. The
starch is considered every bit as potent as the teas against a host of
discomforts and diseases, from obesity and a flaccid libido to gonor-
rhea, dysentery, and anemia. These days, cubes of dried root, along

with starch for culinary or medicinal purposes, are available here and there in the United States at Oriental markets and stores devoted to natural foods or alternative healing methods. Some of these stores carry ready-made tea mixes and kudzu creams. Many do not offer the raw starch but sell instead nutritional supplements in the form of five-hundred milligram gelcaps, which contain the powder, often in combination with other herbs like Saint-John's-wort.

Kudzu might seem, at a superficial glance, to be a miraculous, all-healing remedy stirred up in the same kettle as the patent medicine of any snake-oil salesman. Recent research shows, however, that it may yet gain an honorable place in Western medicine. In 1992, the University of North Carolina at Chapel Hill issued a press release: David Overstreet and Amir Rezvani, both professors of psychiatry at the university's center for alcohol studies, had investigated an ancient Chinese application of a medicinal tea brewed from seven herbs, including kudzu. There, the tea has traditionally been prescribed not only as a relief from hangovers but also as a preventative for the overindulgence that causes hangovers in the first place. What real effects, if any, does the tea have on alcohol consumption? Working with David Lee, an organic chemist, Drs. Overstreet and Rezvani conducted experiments with three kinds of rats—Finnish, P-rat, and fawn-hooded—specially bred to prefer alcohol to water (like some people, some rodents not only relish booze but develop an addiction to it). Five of the seven herbs proved ineffective, but kudzu, refined to a chemical named puerarin by the researchers, served to reduce the impact of alcohol on the drinker and also to diminish intake. Similar work with a kudzu tea was announced in 1993 by Wing-Ming Keung and Bert L. Vallee of Harvard Medical School. Conducting their experiments with hamsters, the scientists discovered that animals injected with an extract from kudzu root reduced their alcohol intake by half, as opposed to the furry inebriates that had been injected with a substance containing no kudzu. These studies found that not one but two chemical compounds, which the researchers called daidzin and daidzein, were at work to reduce the desire to drink, as well as to make heads less tender and bellies less queasy the next day. It's a long way, yes, from rodents to *Homo sapiens*, but these experiments strongly reinforce the conclusions of Chinese folk practice.

And still no end to the uses of kudzu. Wherever it grows, its tough but flexible vines have furnished material for all manner of objects: tumplines, backpacks, and ropes for suspension bridges in Mongolia; birdhouses and rustic frames for pictures and mirrors in the American South, along with twisted wreaths, like those made of grapevines, for decorating front doors; and everywhere, baskets— dainty baskets the size of a small bird's nest, baskets big enough for a grown woman to sit in (if she so desires), and free-form and traditional baskets of every size between.

Then there's "Kudzu," the cartoon strip devised by Doug Marlette and now syndicated throughout the country. One of its main characters is the slightly befuddled Kudzu Dubose, an adolescent good old boy with high aspirations and acne-plagued skin. Another is the preacher, dressed circuit-rider style, who finds more trouble in technology than in the Devil. His church, of course, fields a youthful basketball team notable for falls and fumbles. Just as the plant has become the quintessential Southern vine, its cartoon namesake is a quintessential Southern way of pointing a finger at oneself and poking fun.

The almighty vine has also assumed an ineradicable position in the English language—that of an all-purpose pejorative epithet. Its application to salt-cedar—the kudzu of the West—has already been noted. The aquatic weed hydrilla (*Hydrilla verticillata*, "whorled water-plant"), another import from Asia, has been damned as "water kudzu" for the stealth and speed with which it covers ponds and dislodges native vegetation. Nor is condemnatory application of the word confined to the plant kingdom; like the vine, it shoots out and grabs whatever it can. An introduced Asian eel has been called "animal kudzu." The separatist members of a Georgia militia have been dubbed the "kudzu commandos."

And still no end to kudzu's deployment in the New World: although it's no longer the plant of choice for stabilizing soil and healing erosion, it is used to this day on precipitous embankments, and it continues to fill its two other early roles, those of forage and ornament. An American couple in the western part of North Carolina have devoted ten acres of their three-hundred-acre farm to the intentional cultivation of *P. montana;* after harvest, they either ensile or bale the leaves to serve as fodder for cattle and horses. And farmers in Alabama harvest wild kudzu for hay. In dry spells, especially, when other plants fare poorly, kudzu makes a crop because its root is able to tap deeply into sources of moisture. On a far grander scale, in 1990, the Japanese food-processing company Sakae Bio acquired 165 acres in Lee County, Alabama, strictly for the purpose of growing kudzu. Providing forage is not Sakae Bio's aim, however; this geographic outsourcing is meant to satisfy the human demands at home, where some fifteen hundred tons of the starch are consumed every year. In the realm of ornament, kudzu is still sold commercially to back-yard gardeners; my daughter-in-law Debra, who grows prairie plants in her small suburban yard in central Illinois (and also advises me on matters such as starlings), has found the vine in a local garden center, where it was labeled as an annual. Untrue, I thought—until Debra reminded me that killing temperatures of 25 degrees below zero are not uncommon in the great flatlands of the Midwest.

And who knows where the vine will pop up next?

Bane or blessing? Clearly *P. montana* is some of both, though the harm it does may well outweigh the benefits. Where the plant runs wild, it may destroy domestic species, especially young hardwoods, and it creates a self-sustaining monoculture that shuts out biodiversity not only in the vegetation but also in every other form of wildlife. But I'd like to quantify the economic damage wrought by kudzu, to learn its real costs to farming, forestry, and the taxpayer (to say nothing of the costs to utility companies and the budgets for railway maintenance). So I write to two government agencies, the department of agriculture and the department of transportation, in each of six Southern states: Georgia, Alabama, and Mississippi, the three that are most seriously overrun, plus North Carolina, South Carolina, and Tennessee. With the exception of Tennessee, which delivers only silence, the responses take one of two forms, depending on which agency replies. Agriculture, represented by the cooperative extension agencies, says, *Hmmm, interesting plant*, and sends informative pamphlets. The DOTs, responsible for keeping kudzu off the roads, speak with one voice: *Kill*. No one, however, provides any numbers. (Ironically, a USDA Forest Service publication, provided by the Mississippi Department of Agriculture and Commerce, plainly states that "lawmakers and public officials must be educated about the costs of noxious weeds" but gives no figures.)

I do gain other information. The plant is not troublesome enough in North Carolina for the state to declare it a weed. But few portions of this country are entirely safe from the deep roots and all-embracing vines; not even the Midwest with its arctic winters is free of the threat. While kudzu is happiest in the South's steamy heat, it will flourish almost anywhere, from the Atlantic to the Pacific, from the Canadian border to Mexico. The plant is capable of playing 'possum; the beholder is deceived by the fact that top growth is killed back by a hard frost. Beware: the root lurks underground unseen but quite alive. Once snuggled in, it becomes well-nigh unstoppable, absorbing nutrients from the soil and sending new leaves forth every spring to soak up necessary sustenance from the sun.

I learn, too, that for all its remarkable toughness, the plant is not immortal. It can be done in. South Carolina's DOT and its counterpart in Mississippi send lists of the herbicides that may be employed in the quest for annihilation. The names sound like a roll call of the Greek commanders who led their troops against Troy: amitrole, chlorsulfuron, clopyralid, dicamba, glyphosate, imazapyr, metasulfuron methyl, sulfometuron methyl, and paraquat. Some are selective, acting only on specific species; others kill every shrub, tree, vine, and blade of grass in sight. But herbicides alone won't conquer kudzu. There's one other ingredient without which nothing will be accomplished: persistence. The first few years of poison only discourage the vine; a full decade is required to do it in, a decade in which each sprout, each root crown must be relentlessly sought out and expensively sprayed. Even then, simply by sneaking under a neighboring fence or leaping from tree to tree, the villain may return, ready as ever to seize control. A spokesman for the Mississippi DOT says that herbicides merely keep kudzu at bay. (I've seen Mississippi kudzu along Interstate 55: for more than three hundred miles, from the Tennessee line straight on through Louisiana to the Gulf Coast, it swamps flat stretches of right-of-way with its huge green surge; its leaves, upholstering fences and brushpiles, turn them into overstuffed furniture for giants; its climbing vines convert trees and electric poles into long-necked and blowsy green dinosaurs.) And he offers the opinion that the only way to eliminate its dominion over the rights-of-way is to form anti-kudzu partnerships with adjacent landowners—together we stand, divided we give the vine permission to overrun and pull us down. His letter concludes with a wistful request that I let him know forthwith if I should come across any surefire methods for control.

(A goat, get a goat. Get any grazing or browsing animal. If the young shoots and leaves are regularly nibbled away and given no chance to regenerate, the roots will starve. But imagine Mississippi's highways, imagine any kudzu-threatened road, with goats, cows, horses, and sheep thronging the rights-of-way. Imagine the wrecked cars, the wrecked animals and people.)

Other methods of control include regular burning and cutting the foliage by hand. Several agencies are also trying to zero in on biological controls. Researchers at North Carolina State University have identified a native caterpillar, the soybean looper, as a creature with a usefully large appetite for kudzu; the trick, however, lies in connecting the caterpillars with kudzu leaves, which do not seem to lure egg-laying moths. Looking farther afield, the Forest Service studies the pathogens that infect the plant and the insects that feed upon it in China and Korea. To prevent new ecological disasters, perhaps greater in scale than that of kudzu, the biocontrols would not be transported here. Instead, a grand array of vegetation native to the American South would be taken to the Far East to see which, if any, species might be adversely affected. (Is it possible that one of these plants might be an American version of kudzu, a Pinocchio that runs away and grabs great chunks of the Orient?)

But kudzu can't be eradicated. Seven million acres and on a roll, it won't go away no matter what we do. Fire, chemicals, native predators, the backyard goat—all these amount to nothing more than temporary staying actions. What to do? Donald Ball, a professor and agronomist at Auburn University, sends me *Kudzu in Alabama: History, Uses, and Control*, an attractive pamphlet that he has cowritten with that state's cooperative extension service. In an accompanying letter, he writes: "Actually, I view kudzu not so much as being a despicable species as a multi-faceted species. . . . Furthermore, though not a native species, kudzu is so well adapted and widely established in the southern region that it has come to be associated with it and thus is part of our southern heritage."

Heritage, yes. And for all kudzu's tangible and widespread presence, its jack-in-the-box ability to pop up anywhere, he has also pointed out the invisible problem: How do we see it? American approaches typically range from disparagement to outright war—a war that the vine is preordained to win. Only an eccentric few of us find kudzu beautiful or worthy. And our perceptions both for and against depend, for the most part, on this criterion: Are we putting the plant to our own uses, or is it blindly, meanly appropriating something that we consider ours? In other words, who—or what—is in control?

Good, bad, or indifferent: no matter how we see it, kudzu has settled in and won't be budged. Roots in the earth, leaves to the sun, it will persist until the last trump. I've thought of comparing it to phenomena I find obnoxious, like the wild proliferation of pounding boom boxes or the unchecked spread of concrete lawn geese and decorative nylon banners, but no, there's nothing faddish about the plant. It's a force of nature, more on the order of azaleas and tobacco, country music, coon hunts, NASCAR races, and good old boys. It just plain *is*. Certainly, nothing obliges us to like it, but because we must live with it, the least painful way to come to terms with the doggone stuff may indeed be to see it as a heritage.

# THE WISDOM OF NATURE

Brown-Headed Cowbird (*Molothrus ater*)

## Brown-Headed Cowbird

IN early spring, I often hear the songs before I see the birds, little songs that tinkle lightly on the air like glass wind chimes. Then I'll spot a dozen of them, perched on a leafless mimosa or a black gum tree. The females wear a modest grayish brown, while the males are feathered in a glossy, iridescent black except for their heads, which are hooded in a rich golden-brown that is, not inappropriately, the color of a fresh cowpie. For these are brown-headed cowbirds, which hunt for their dinners in the vicinity of grazing cattle. Of all North American songbirds, the cowbirds alone are given to a most curious shiftlessness in matters of reproduction: dump and run, leaving not only their eggs but all parental responsibility to the harried care of a species other than their own. The brown-headed cowbird is known to parasitize at least 216 other kinds of birds, with the tally still mounting. And these days, this bird that puts the future of its species into a stranger's nest is found everywhere, from east to west, from north to south, clear across the country.

Once upon a time, not all that long ago, it was considerably more localized. In the days before Europeans discovered the North American continent and began barging across it, the brown-headed cowbird was found primarily in the central plains, with a range that more or less matched that of the plains bison. The bird followed the hooves of grazing bison or sometimes stationed itself right by an animal's head to batten on the insects stirred up as the great beasts moved and fed. Because of the close association, early observers sometimes called it the buffalo bird. Another fitting common name was cowpen bird, as in this description by Mark Catesby, the eighteenth-century British naturalist who wandered the colonies widely to collect botanical specimens and draw the birds, plants, insects, and animals of the New World:

> The bird is entirely brown, the back being darkest, and the breast and belly the lightest part of it. In winter they associate with the red-winged Starling and purple Jackdaw in flights. They delight much to feed in the pens of cattle, which has

given them their name. Not having seen any of them in summer, I believe they are birds of passage. They inhabit Virginia
and Carolina.

This description holds as true today as it did in the 1700s, except
that most observers—with one notable exception—have characterized the body of the male as black. Catesby's cowpen bird is often
seen abundantly amid the swarming winter flocks of the birds that
he called the starling and the jackdaw, birds known today, respectively, as the red-winged blackbird and the common grackle. It still
delights much to feed in the pens and pastures of cattle and horses.
And it is indeed a bird of passage, a migratory bird, although only
partially; its populations do not necessarily fly far from the breeding
grounds. Birds that summered and bred in northern parts may flock
slightly south for the winter, replacing birds that have themselves
flown farther south. The cowbird is also subject to what might be
considered a sudden whim—joining a fly-through flock of other
blackbirds. The species is certainly seen in Virginia and Carolina at
any time of year.

The oddball observer was John James Audubon, who received
a pair of dead brown-headed cowbirds in 1824 from a friend and
proceeded forthwith to paint them. He called the species by two
common names, cow-bird and cow bunting, the latter because of
the stout, nearly conical bills, constructed like those of the true
buntings, grosbeaks, and finches. In both the initial watercolor and
the subsequent plate, Audubon's female cowbird closely resembles
the real thing; it is the male that astonishes. His head is shown in
the proper brown, and his body the proper black—except for the
wing, which displays a vivid swash of Delft blue that extends into
the tail feathers. Was the male bird sent to him somehow damaged?
Had artistic license overtaken the sensory evidence? We'll never
know. But we do know that Audubon was boggled by the bird's
dump-and-run proclivities. "If we are fond of admiring the wisdom
of Nature," he wrote, "we ought to mingle reason with admiration."
Then he noted the bird's unadmirable habit of laying its eggs in the

nests of other species to the great disadvantage of the "foster parents" and their eggs, which contained chicks doomed to perish. But, bowing somewhat grudgingly to powers greater than himself, he remarked, "This is a mystery to me; nevertheless, my belief in the wisdom of Nature is not staggered by it."

Cowpen bird, cow bunting, brown-headed cowbird—whatever its common name, it is formally known as *Molothrus ater*, the "black parasite." It is a member of the Icteridae, the New World's indigenous, well-populated blackbird family, which includes not just the often showy blackbirds (like the red-wings with red and yellow epaulets, the yellow-heads with golden helmets) but also the meadowlarks, the flashy orioles of North and South America, the grackles clad in shimmering black, and two groups found only south of the border: the sturdy caciques and the colonial oropendolas, both of which weave basket nests. Nor is the black parasite alone in its genus but shares it with four others. Only one of those four is also found in the continental U.S., most often in the southern part of Texas and occasionally in southernmost California and New Mexico—the bronzed cowbird (also known as the red-eyed cowbird), *M. aeneus*, the "brassy parasite." The remaining trio consists of Central and South American species: the bay-winged cowbird, *M. badius*; the shiny cowbird, *M. bonariensis*; and the screaming cowbird, *M. rufoaxillaris*. One other bird, of a different genus, also belongs to this gang for reasons of anatomy and behavior—the giant cowbird, *Scaphidura oryzivora*, which is also of the dump-and-run persuasion. Its formal name means something like "hard-digging rice eater," but the word *Scaphidura*, which may refer to the bird's conical, finchlike bill, is hybrid Greco-Latin invented by a taxonomist; it makes no good classical sense.

The single exception among the five *Molothrus* species to the parasitic mode of living is the South American bay-winged cowbird, which appropriates a nest built the previous season by another species and then actually tends the eggs that it lays there. Curiously enough, it is this species—exclusively this species—that is parasitized by the screaming cowbird, its close cousin. The giant cowbird is almost as fussy, choosing primarily to parasitize members of its own blackbird family, the oropendolas and caciques. The other cowbirds are far more ecumenical, with 71 species recorded for the bronzed cowbird and 176 for the runner-up, the shiny cowbird. The brown-headed cowbird still leads the flock. The Plus-Minus style of symbiosis that cowbirds inflict on other species—and in their special, stealthy avian fashion—is known as brood parasitism. The females stow their offspring in another bird's brood by depositing their eggs one at a time in other birds' nests, usually after the nests are completely built and most often early in the morning when rightful proprietors are likely to be absent for a short stretch of time. When the egg is laid, the cowbirds fly away, shrugging off all further care. All that might be said in their favor is that the brown-headed, bay-winged, shiny, and screaming cowbirds are monogamous—no fooling around during the breeding season. But parasitism is without doubt a most peculiar way of life that leads me to ask why cowbirds behave as they do.

Gordon Orians, professor of zoology and environmental studies at the University of Washington, has investigated the cowbirds' seeming lack of parental feeling and responsibility. After all, most other birds not only recognize their own nestlings, sometimes amid a host of look-alikes, but also tend assiduously to the successful fledging of their young. Only one percent of all birds are brood parasites, most of them—mainly geese and ducks—intraspecifically amid their own kind and a few interspecifically with birds unlike themselves. Among the latter, the cowbird, along with the Eurasian cuckoo, is the prime example. With the exception of the moments of actual mating, it gives every single task of reproduction to others. And those others do all the hard work: build the nest, incubate the eggs, carry away eggshells after the hatch and fecal sacs after each feeding, and hunt from dawn till dusk to satisfy the bottomless appetites of insistent nestlings. Of those nestlings, the cowbird is usually the largest, able to seize every morsel of food brought in by the adult birds. To understand the cowbird's unbirdly behavior—how it came to abandon the usual parental role and how its habits affect both its own species and its hosts—Orians has asked five questions. How did brood parasitism start? Why did this habit spread? What changes, if any, has it caused in the behavior of parasite and host? How do changelings hatched in another bird's nest know what kind of bird they are? Is there communication between cowbirds and their hosts?

The answers to the first two questions must be speculative. No one was there to observe the aboriginal intrusion of a cowbird into the nest of a bird unlike itself or the subsequent behavior of that cowbird's duly hatched and fledged descendants. Though egg dumping—the occasional deposit of one bird's eggs in another's nest—is observed in many species, obligate brood parasitism is such a rare trait that its arrival on the avian scene is like a most peculiar and perverted accident. Its origin seems explicable, however, through looking at costs and benefits: at some point in the invisible reaches of the past, the hatch rate for the ancestral cowbird's dumped eggs and the fledging rate for her chicks must have been high enough so that it paid her and her offspring to continue taking advantage of the neighbors. As for the habit's spread, it may have to do with the fact that, with the single exception of one duck species, all brood parasites are altricial—born naked and blind, that is—and require several weeks of intensive feeding with high-protein food, mainly insects, before they can fledge. Though the adults are also granivorous, eating seeds, cowbirds have always relied to an almost perilous degree on the movements of large grazing animals—bison, horses, cows—to stir up the necessary insects as they move and munch. It is not always easy to find such conditions. "This suggests," Orians says, "that parasitizing the food-gathering abilities of host parents is more important than parasitizing their egg-covering abilities." Other inducements may have included fewer forays by predators on the hosts' nests than on the cowbirds', and an advantage in size—and thus in food-grabbing power—for cowbird hatchlings over their nest mates. Orians also posits a gradual beginning to brood parasitism, with eggs laid in foreign nests only now and then at first. But when successes began to outweigh failures, natural selection kicked in and set the cowbird on its future course.

How are the host birds to respond? The intruders' success often works to the hosts' detriment, drastically reducing the number of their young that fledge. Logic says that repeatedly parasitized species—red-eyed vireos, Eastern phoebes, lark sparrows, rare Kirtland's warblers, and the hundreds of others—would have evolved defenses over the millennia to protect their young. And indeed, a few have, but doing so must have been an almost fatally slow process. While birds recognize their own young, they rarely distinguish one egg from another. The basic syllogisms in a bird's brain may be stated this way: eggs in nest implies that the nest builder laid them; young bills agape stimulate the parent to invest in its own genetic line, the hatchlings in its nest, bringing them food and ignoring any other birds. Given these general rules of behavior, it's a matter of astonishment that any birds have been able to respond in a no-nonsense fashion to the cowbird threat. Yet a few species, like the gray catbird, the American robin, and Bullock's oriole, have become knowing rejectors, rather than unwitting acceptors. Experiments have shown that the female Bullock's oriole can indeed spot a foreign egg, and not just that of a cowbird but also of occasional egg dumpers like the loggerhead shrike and the house finch. She evidently learns to recognize her own eggs in a brief but crucial time that begins a few days before laying and ends when her first egg is placed in the nest. Just how she comes by this knowledge is still a mystery, but, as if she were picking up trash with a spike on the end of a long stick, she uses her bill to pierce the odd egg and cast it out. Some other songbirds also seem able to spot a ringer but are too small to use the oriole's tactic; they may either desert the nest and their own eggs or—as in the case of the yellow warbler, for one— reline the nest with a new floor that covers the old clutch, which will never hatch. There are no guarantees, however, that the rebuilt nest will not be parasitized. Yet the rejectors are few, and most species remain acceptors. Perhaps they've only recently been plagued by cowbirds, or, if the phenomenon of parasitism is not recent, circumstances were likely not conducive to the kinds of genetic change that would lead to the recognition and rejection of cowbird eggs.

How do young cowbirds know what they are? With many kinds of fowl, like geese, ducks, and chickens, imprinting is the secret: the first living creature the young bird sees gives it an identity. It will look to that creature for guidance in behavior and eventually seek a mate of that particular kind. This visual means of targeting a sense of self works perfectly when the creature seen is of the same species, but woe betide the gosling or chick led by first sight to think that it is something other than poultry, and in some cases that it is human. Other kinds of birds may also develop crippling attachments to humankind. My veterinarian daughter, who has worked to rehabilitate wild birds from various owls to a grown but featherless goldfinch, tells of an immature bald eagle returned to the greater world after months of mending in a cage: the bird had become so dependent on its keepers that after release, it would swoop down exuberantly and land on any handy human being just as if it were homing in on a parent. It could no longer fend for itself in the wild; recaptured, it will live out its days as a show-and-tell bird at a raptor rehab center. These birds with identity crises will fail at reproduction. But for cowbirds, despite the fact that the first creatures they see are birds totally unlike themselves, there seems to be no confusion, nor is there one jot of evidence that cowbirds have ever signally failed to perpetuate their kind. Researchers investigating the matter of how brown-headed cowbirds establish species identity have found that females raised in captivity respond with copulatory postures not to the songs of red-winged blackbirds, meadowlarks, or grackles but only to the lisping, tinkling wind-chime songs of males of their own kind. The knowledge of what they are is instinctive.

When it comes to communication between cowbirds and the species on which they batten, nothing could be more polite. The parasite will approach a potential host and offer something called a preening invitation display: stepping up close, bowing its head, and fluffing out its feathers. Often, the host will accept the invitation and proceed to groom the cowbird's feathers. And in a strange twist, some birds that do not engage in mutual preening among themselves will also respond in a positive fashion to the come-on of a cowbird. Observers think that such behavior may work both to calm aggressive tendencies in the host and to ease the way for the cowbird's entry in the carefully selected nest. The eggs, however, are laid when the host is absent.

And still no end to the questions. Orians also asks, Why did brood parasitism evolve in the cowbirds and in no other group of New World song birds? Why is there but a single species of cowbird over most of North America? And why, he wonders, do some kinds of cowbird parasitize many hosts, while others are limited to a few or, in the screaming cowbird's case, to only one? Here are puzzles not just for professional research but also for investigation by the educated amateur (if curiosity is not dampened altogether by despising the cowbird's habits).

One matter that is not a puzzle is the reason that the brown-headed cowbird has dramatically expanded its range in the last three centuries from the grassy, bison-friendly prairies to the country's entire sweep. The bird has simply followed our lead. We've facilitated an increase in its range and its numbers by clearing forests in favor of farms (more grain, more grazing livestock stirring up bugs) and also by establishing more forest-field edges (easier access to host nests) as we alter the landscape to develop living spaces for ourselves.

It may not be amiss to remember, too, that what cowbirds do is the birdly equivalent of leaving an infant on the doorstep of a church or a hospital: someone will accept the babe and care for it. The offspring's welfare is the prime consideration. The differences are two. First, the human mother practices parasitism on her own kind, while the brown-headed cowbird ventures forth and is parasitic only on species other than its own. Second, and more important, the human action is facultative, while that of the bird is obligate. In other words, the woman has options, but, in the wisdom of nature, the bird has been given no choice at all.

# THE DEW LOVERS

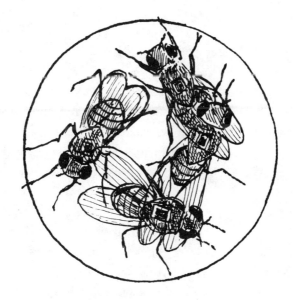

Fruit Fly (*Drosophila melanogaster*)

*Drosophila* Fruit Flies

THE sassy little fruit fly that is sometimes called a vinegar fly evokes a full range of human responses in the people I know, from outright loathing through amused tolerance to the most ardent boosterism.

My husband, the Chief, for example, hates them. He does not cotton to any common arachnid—spiders, ticks, and chiggers—nor is he fond of mosquitoes, Japanese beetles, and tobacco hornworms, but his attitude toward all of these is live and let live. The one form of life that he most truly and passionately despises is fruit flies. They insinuate themselves into our North Carolina mobile home through doors opened in passing or the mesh of screens. They hover thickly around the cantaloupes ripening on the kitchen counter and over the bananas the Chief uses in his peanut butter sandwiches. They cluster on the tomatoes he's brought inside to mature. They drown in his diet cola and his beer. And, oh, they are canny at sneaking into the freezer, where they become entombed in ice cubes. They are even bold enough to investigate his person, winging silently in front of his face, landing lightly on his T-shirt. He swats them, but as soon as one is gone, another takes its place. I do not understand this enmity. For fruit flies neither sting nor bite. They can be annoying, yes, but in a mild way, and when summer vanishes into fall, they also vanish (though I've known them to resurface in winter when we visit the trailer and turn on the heat.)

My daughter Elisabeth, the veterinarian, belongs, however, to the tolerant camp. She lives year-round with fruit flies that are a naturally occurring by-product of something that I invariably think of as The Habitat. Once upon a fairly recent time, when she and I were driving down a narrow street on the outskirts of my small Virginia town, she spotted an Eastern box turtle moving like an eight-inch tank along the side of the pavement. At her command, I stopped and backed up. She scooped up the turtle, which sensibly withdrew into its shell—a damaged shell with a great but well-healed gash three inches long near the crest. As soon as she saw that, the turtle received a name: Crash. Crash was then identified as a male because of his red eyes and concave plastron. She took him to her Midwestern home—husband, children, cats, dogs, and sometimes wild birds, like great horned owls, that she is rehabilitating. There, amid the mob, Crash was installed in a large aquarium fitted out with a floor of wood chips, leaves, and other dry material. And there he lives today, though his surroundings have changed somewhat. Earthworms now inhabit the lower levels of the flooring; when Crash is struck by a craving for meat, he roots around for them. The thick layer of wood chips on the surface has turned into a fine, feathery stuff like dried grass, and in it three American toads have made burrows for themselves. The surface is also covered in one corner with Crash's preferred food, vegetables and fruit (as box turtles age, their tastes become more vegetarian). And over the potato peelings, old carrot tops, brown banana skins, chunks of melon, and other rotting comestibles, the fruit flies swarm. They in turn serve as snacks for Crash and feasts for toads. With The Habitat, a miniature ecosystem has been achieved. Luckily, the fruit flies stay confined within its glass walls, and so does the somewhat sour and fusty stench.

It's our friend Jay Hirsh, a molecular biologist and professor at the University of Virginia, who stands firmly with the boosters and studies the little fly with something akin to passion. He is among its premier proponents, the breeders and researchers who devotedly rear these creatures in the interests of science. Indeed, since the beginning of the twentieth century, scientists—geneticists, in particular—have swarmed around the little fly, attracted by its huge chromosomes and easily identifiable genes. These virtues have led to its use in highly successful studies: providing proof of Gregor Mendel's theory that physical characteristics are heritable, parent to offspring; developing gene-mapping techniques; and establishing that chromosomal changes lead to heritable variations. Only recently, Jay's small but dedicated group of researchers has conducted a series of experiments that expose the little flies to crack cocaine, about which more shall be set forth shortly. First, it must be said that messing around with fruit flies has more than once provided entry to a Nobel Prize. In 1933, Thomas Hunt Morgan received the Nobel Prize in Physiology or Medicine for work with fruit flies that revealed the chromosome theory of heredity—that the linkage of genes on chromosomes accounts for distinct hereditary traits. Thirteen years later, in 1946, Hermann Joseph Muller won the same prize for demonstrating that X-rays can speed the rate at which genetic mutations occur in these tiny, winged insects. In 1995, that prize was shared by three developmental geneticists, the American Edward B. Lewis, Christiane Nüsslein-Volhard of Germany, and her American colleague Eric F. Wieschaus, who used the little flies to discover the genetic controls for the early development of the embryo.

What are these flies anyhow? To begin with, they belong to the order Diptera, "two-wingers," which first received its name from the Greek philosopher and naturalist Aristotle. The Diptera are far and away the most despised creatures in America, and they comprise a vast, multifarious tribe that includes formidable members like houseflies and hard-biting horseflies, blowflies, botflies, and the tiny blackflies that make a misery of summer nights. And some kinds of fruit flies are surely as pestiferous as any of these, wreaking havoc on both human pocketbooks and peace of mind. Science has sorted the insects commonly called fruit flies into two families, the Tephritidae, or "ash-colored" family, named for the generally gray to black bodies of its members, and the more colorful Drosophilidae, or "dew-loving" family, named for its most common source of sustenance. The ash-colored flies, sometimes known also as the Trypetidae, or "borer" family, are generally much larger than the dew lovers, two- or three-tenths of an inch to the smaller flies' seven- or eight-hundredths, and it is this family's larvae that are able to bore into every part of a plant, from root and stem to leaves and fruit, and so bring devastation to field crops and orchards. In their number are pests like the apple maggot, which also dines on pears, plums, and blueberries. The Tephritidae also include a great gaggle of species

that feast selectively but voraciously on such economically important plants as celery, cherries, and olives. The black-bodied Mediterranean fruit fly—the Med fly notorious for ravaging citrus groves—
is one of them. Some of these ashy fruit flies, however, are known to
content themselves with goldenrod, while others breed harmlessly
in the heads of composite flowers like daisies. The Swiss zoologist
Walter Linsenmaier, known especially for his meticulously rendered
paintings and drawings of insects, has this to say about the Tephritidae, noisome or not: "The little fruit flies . . . are striking because of
their particularly beautiful wing designs." And the wings of many
family members are indeed lovely, patterned with delicate geome-
tries of light and shadow. But it's not the ashy flies that drive the
Chief to desperation and bad language, not these that feed my
daughter's turtle and toads or pave the way to Nobel Prizes. The
dew lovers, though they're little things to look at and clearly
unaware of their own importance, fill these roles.

The dew lovers by whatever common name they're known—
fruit flies, vinegar flies, pomace flies—ignore the crispness of an
apple on the tree or the melon still growing on the vine. Vinegar
and pomace: these names point to their fondness for spoiling fruit,
for the ooze of a decaying peach, the mush and ferment of a tomato
going bad. And this is the "dew" that these tiny flies love, this ooze
and mush. More particularly, it is the yeast of the fermentation
process on which they feed. They also lay their eggs in rotting fruits
or fungi, which provide an instant yeast supply for the newly
hatched larvae. These maggots are so small, matching the adult flies
in length, that they hardly seem in the same league as the much
larger larvae—the soft, pale, squirming get of house- and horseflies—usually thought of as maggots. In their penultimate stage, the
larvae pupate and then emerge as adults by way of a curious mecha-
nism called a ptilinum, a sac on the front of the head that alternately
swells and shrinks and works like a piston to break open the hardened pupal casing. On emerging, the heads of the new adults are
huge with fluid, and their wings are tiny, flat, and intricately folded.

Flight is achieved after the flies do something that Jay Hirsh calls "blowing up their wings": they use the muscles in their heads to pump the fluid rearward to expand and straighten the wings. The adults, females slightly bigger than males, are well equipped with both small hairs and bigger hairs called sensory bristles—on their heads and bodies, their legs, the edges of their wings, and also on the surface of the wings—though it takes a magnifying glass or microscope to see them well. On their forelegs the males have special hairs called "sex combs" that are used to stimulate female interest. But, unmagnified, the flies merely look a little fuzzy. There are about one thousand species of these flies, which are found the world around, and every one of them belongs to the genus *Drosophila*, the "dew lovers." Their colors, anatomical arrangements, and mutations are many: classic red eyes or aberrant white eyes; wings long, short, vestigial, or curly; bodies colored in earth tones—russet, dark gold, warm cider brown, with sooty rings around the plump abdomens of the males. These, with other features, occur in a gallimaufry of combinations. But no matter what they look like, their lives are uniformly short: twelve to fourteen days from egg through pupal stage and, under laboratory conditions, a few weeks to a month or two of adulthood. The brief life cycle may be timed to the here-today, gone-tomorrow nature of decaying fruit.

In part, it's the short life span that has made the dew lovers attractive to researchers in the last century. The little flies have been around long before humankind came down the pike—and they were surely on hand for Eve's famous chomp into the apple, a fruit as irresistible to a vinegar fly as it was to the woman, although for different reasons. Nor have they been neglected in the realm of poem. John Milton, for one, called on them to provide a simile in *Paradise Regained*. Satan, vanquished, tries to fend off the troubles caused by his "bad success," but they come back unceasingly ". . . as a swarm of flies in vintage time,/About the wine-press where sweet moust is power'd,/Beat off, returns as oft with humming sound." It might be argued here that Milton was referring to some other sort of fly, for fruit flies are not given to humming in a way perceptible by humankind, but Milton's venue is entirely correct—the must, which is the pressed-out juice and the pulp and skins of crushed fruit, along with the yeasty ferment of wine making.

Scientific use of the dew lovers started in earnest as the nineteenth century rolled into the twentieth. And one species in particular, out of the thousand species available, has become the star of research projects: *Drosophila melanogaster*, the "black-bellied dew lover," so named for the dark striping of the gaster, or abdomen, in the male. It was in 1909 that the future Nobelist Thomas Hunt Morgan chose *D. melanogaster* as an experimental organism, first for the mapping of chromosomes and then for investigations of genetic inheritance. Use of this particular species then spread from Morgan's "flyroom" at Columbia University because he and his colleagues frequently contributed not only their stock of knowledge but their stock of experimental flies to other laboratories. Because these little flies live only briefly but breed prodigiously, many generations can be produced during the course of any given study, and changes from one generation to the next can be followed quickly. Also, the flies have only four chromosomes, which are ribbonlike in the salivary glands and one hundred times the length and thickness of the average chromosome. Only an ordinary microscope needs to be used to map the genes of such giants as objects lined up and linked in a series. Thus, researchers have been able to determine exactly the inheritance of traits and, recently with Nüsslein-Volhard, Wieschaus, and Lewis, to find and identify 140 of the genes that are crucial to embryonic development.

Why be concerned about the embryonic development of a *fly?* And why subject a fly to crack cocaine? I visit our friend Jay in his lab at the University of Virginia. En route, I think to myself that mad scientists come in two varieties, those who create Frankensteins and those who play with fruit flies. Jay, trained as a biochemist but converted to molecular biology, is decidedly of the latter persuasion. He's a slender man, now in his early fifties, with a lean El Greco face and a short, neatly trimmed dark beard that is beginning to grizzle. And he's a man with talents that I find even more arcane than his cutting-edge research with fruit flies and cocaine. Jay can and does perform astonishing feats like fixing balky plumbing, mending broken appliances, and repairing cars. In his lab, he is the consummate scientist, introducing me enthusiastically to both his students and his wild-type Oregon-R flies, one of the two strains of *D. melanogaster* most widely used by investigators. The other strain is the Canton-S. Both were captured near the beginning of the twentieth century, in Oregon and Ohio respectively. The strains are similar in most physical respects, but because there are what Jay calls "significant behavioral differences in learning and memory," the strain used at the start of an experiment is always precisely specified.

"We study behaviors here that are very robust, *not* finicky," Jay says. And the flies are to be seen, safely bottled, in every nook and cranny of his lab, which came into being in January 1991, when he transformed a large area used for storage into a research facility. There is the high-bench area where experiments are performed on waist-high counters, from which rise tiers of shelves, like library stacks, that are filled with books, beakers, vials, microscopes, computers, and television monitors. The low-bench fly room, with counters at desk height, is decked out with similar equipment, plus dissecting microscopes and a swarm of clipped-out comic strips taped to the walls. Then, around the corner, is the nameless L-shaped room in which crack cocaine is administered to small, carefully monitored batches of flies. There, one of Jay's graduate students, Rozi Andretic, a lithe redhead from Croatia, also studies the genetics of the flies' circadian, or daily, rhythms of activity and quiescence (these are the rhythms that, when they are disrupted, cause jet lag in human beings); she particularly investigates some mutants that do not respond at all to alternations of light and darkness. "Wine flies," she calls them, translating *vinske mušice*, their name in her native language. This room also contains two walk-in closets filled floor to ceiling with vials of flies. One is used for quarantining

new arrivals, dispatched by mail from various labs, including a fly
bank at Indiana University; the newcomers need to be isolated lest
they carry pests, like mites, that are inimical to the flies and, thus, to
the experiments conducted here. The other closet is a dark, chilly
place, kept at 18 degrees centigrade (64.4 degrees Fahrenheit) to
slow down the development of stock not yet needed for any project.
Though Jay calls these working quarters "low tech" because of the
inexpensive components found throughout—the low-cost TV sets,
for example, and the hair-dryer heating element that maintains a
constant temperature for the flies during the experiments with
cocaine—I find them to be an almost magical place in which tiny
creatures are magnified not just into an easily observable size but
into an amazing liveliness.

To introduce me to his tiny subjects, Jay first administers a
knock-out dose of bubbling carbon dioxide to immobilize half a
dozen black-bellied dew lovers. Peering at them through a microscope in the fly room, I see fruit flies as I've never seen them
before—as whole creatures, not hithering, thithering specks of protein. These are flies of the white-eyed sort; white, however, means
only that their large eyes are not the standard fruit-fly red but rather
the pale golden color of champagne. Their wings are elegant—
panes of transparent crystal with fine black venation, like the clear
leaded glass that one sees in churches and manor houses. Then I
watch unsedated flies scamper, flit about, and groom themselves
assiduously. Living as they do amid fermenting muck, they must
keep their wings clean if they're not to become enmired in their
food. Colleen McClung is also present in the fly room. She's
another of Jay's graduate students and co-author with him of a
recent paper on the lab's experiments with flies and cocaine. Young,
blond, well tanned, wearing shorts, and sporting a toe ring along
with toenails of iridescent frosted blue, she is busy putting instant
fly food into a batch of clean bottles. It's green stuff, made of oat-
meal and other grains and given its color by the addition of antibiotic and antifungal ingredients, plus a dash of food coloring. The

bottles, shaped like inverted funnels with a narrow neck and wide base, were originally manufactured for collecting urine samples, but they serve nicely indeed for housing fruit flies, with food at the bottom and flying space above. So that I can get a good, close look, the inhabitants of one bottle are magnified through a microscope, with their enlarged images projected onto a TV set. The adults flit busily about or groom themselves. I see one male lift a single wing and flutter it like a semaphore. He's doing something called "singing," or beating his wing so that it emits a low-frequency sound, inaudible to us but seductive to female flies. Meanwhile, the members of the up-and-coming generation—the larvae—have buried themselves head-first in the green glop that is their food, and their dark jaws work constantly, furiously, while their internal organs throb and their pale maggot bodies writhe with what seems a truly sensuous pleasure. The only stage of flykind not in motion is the pupal stage; the beige pupae, shaped something like tiny toboggans with squared-off ends, are attached to the bottle's walls.

And I learn from Colleen McClung that the scientists who play with fruit flies not only engage in robust experiments but do them with a sense of humor. They have a fine talent for coming up with memorably antic names for types of genes. Aside from fruit flies, one other creature is widely used for genetic research—the nematode *Caenorhabditis elegans*, or "elegant new-rod," for its rodlike shape—but the people who work with these worms assign them dour numbers rather than names. (It was announced in December 1998 that every last one of *C. elegans*'s 19,099 genes have now been mapped—a considerable feat.) Colleen tells me with great glee that fruit flies, on the other hand, may harbor genetic mutations that are named after vegetables, like *rutabaga*, or after human counterparts, like *dunce* and *couch potato*. There's also *fruitless* for a gene dictating interest in the same sex, *tinman* for a gene that instructs its possessors not to develop a heart, and *groucho* for a gene causing extra bristles that look like bushy eyebrows to grow above each eye. And there's *ether-a-go-go* for a mutated gene that inspires shaking in the legs of anesthetized flies. For purposes of easy reference, all fly genes have been given two- or three-letter acronyms; for example, *ether-a-go-go* is known as *eag*.

Then I visit the nameless room, where crack is volatilized in a sealed glass chamber inhabited by flies. The puff of smoke is generated by low-tech means—cocaine put on a coil with a Radio Shack connector. The flies do not snort the crack—they're not built to inhale—but rather absorb it through cuticle and trachea. Their subsequent behavior is recorded and also projected, a hundred times larger than life, by a video camera with a macro lens onto a TV screen. These experiments are blind. To keep them as objective as possible, the researchers do not know how much, if any, crack has been applied to each coil until the tests are concluded and the results analyzed.

Why do this? To begin with, how did the idea of experiments using crack ever arise? *How* had to do, Jay says, with "an evolution of thoughts" and, particularly, with trying to understand a type of fruit fly that spent an abnormally great amount of time in grooming behavior. As for *why*, he says this: "Fruit flies are a lot of fun. The exciting aspect is that you can do real genetics with them. The early genetics of flies was basically stamp collecting and categorizing genetic lesions [changes] that affected the external morphology of the fly. It was only in the 1960–1970s that it was generally realized that mutations affecting behavior and development of the fly could be easily isolated, and in the last fifteen years came the realization that the genes involved in these processes are highly related to genes involved in similar functions in vertebrates."

His investigations began in a definitely robust way: decapitating the little flies so that there was direct access to the nerve cord. How do they stay alive without heads? Their nervous system kicks in and directs them to follow normal routines in such matters as standing upright and grooming. Headless, they can live for days if they're kept moist and don't fly away. The beheading technique was developed elsewhere several decades ago; at the outset it was a tool for investigating mate preference. Till recently, Jay used it, "like pouring drugs down a hole," to deliver various substances to his tiny subjects to see if they would stimulate grooming. Many of them did. He puts on a videotape of headless bodies variously, soundlessly slurping in droplets of cocaine and neurotransmitters like dopamine and octopamine. But experimenting with decapitated flies proved problematic: though grooming behavior may certainly be observed in headless flies, it is impossible to use them to study genetics. Males that lack heads won't mate; there are no progeny to track for genetic traits. Worse, without heads, there are no brains. The fly's brain, however, is a main concern for Jay and his colleagues, who are particularly interested in the region in which dopamine and related compounds act directly on the brain's receptors to control motor activities. Cocaine, on the other hand, does not itself act directly on the receptors but rather seems to work in flies, as it does in vertebrates, to prevent the uptake and clearance of dopamine and related compounds.

What to do? How to revitalize investigations that were not only headless but seemed at a dead end? And a solution had to be found within only a few months, for the project's grant renewal was coming up with the speed of a runaway truck. Jay says, "The key that broke all this open was that, at a meeting, I ran into a clinician who deals with cocaine addicts." The subject of crack cocaine came up—freebase cocaine that can be smoked, that can be volatilized. With the proper papers, crack is available from the National Institutes of Health. It was obtained. In January 1997 the experimental chamber was set up—crack going up in a puff of smoke. And the little flies, able to take in the smoke through their skins, now kept their heads. As it happens, fruit flies don't really like smoke, but as captives they have no choice.

As Jay worked on writing the new grant proposal, Colleen McClung designed and monitored the new experiments. She found that when cocaine in volatile form is given to fruit flies, it induces behaviors astonishingly similar to those of rodents subjected to the drug. Depending upon the exposure to crack, the little flies exhibit seven progressively more severe responses, which occur within the short time frame of only 30 to 150 seconds, while recovery takes five to ten minutes. Their normal behavior involves the activities that so irritate my husband—skittering, flitting about, and pausing every so often to groom themselves. The seven stages of intoxication as recorded by Colleen McClung are these:

1. Intense nearly continuous grooming and reduced locomotion.
2. Stereotyped locomotion, extended proboscis. Some locomotion with simultaneous grooming. In this stage and those subsequent, the subjects lose their ability to fly and remain at the bottom of the container.
3. Low stereotypic locomotion in a circular pattern, extended proboscis.
4. Rapid twirling, sideways or backward locomotion sometimes accompanied by a front leg twitch.
5. Hyperkinetic behaviors including bouts of rapid rotation, wing buzzing, erratic activity with flies often bouncing off the walls of the container.
6. Severe whole body tremor, no locomotion, usually overturned with legs contracted to body.
7. Akinesia or death.

In the nameless room, I watch a fly in the throes of stage 3: slowly, aimlessly, out of control, it circles and circles.

The most important discovery here seems to be that when intermittent doses of cocaine are administered, *D. melanogaster* develops a behavioral sensitization. "The most exciting part of our fruit-fly research is this weird process called sensitization," Jay says. A first exposure to low doses of cocaine has little effect on the flies, but subsequent doses give rise to what Jay calls a "locomotory stir." A sort of frenzied bustle kicks in; that is, the fly cleans or flicks its wings in an almost obsessive fashion. For reasons not yet known, male flies are affected more severely than females. The flies do not become addicted to the crack, but that is because, with no say in the matter, they either take what is given or go without. (The dew lovers are capable of making choices, but how choice might affect their approach to cocaine is a subject for future study.) To observe effects, however, is only to explore the surface of sensitization.

"The vertebrate genome," Jay tells me, "is basically a quadruplication of the fly genome with a bit of elaboration." And here is the greatest boon that minuscule *D. melanogaster* provides to the serious inquirer: though the fly is certainly small and spineless, it is nonetheless endowed with biochemical pathways remarkably similar to those of higher vertebrates, including rats and human beings. The phenomenon is one known as the evolutionary conservation of genetic pathways. When a biological arrangement works, there's no need for evolution to tamper with the basic model. Elaborate on, perhaps, but not redesign. And not only developmental but neural and behavioral genetics may be traced in fruit flies, along with those affecting learning and memory. Thus, if the genetic pathways of things like embryonic development, circadian rhythms, neural function, and sensitization to cocaine can be charted in fruit flies, then the findings may well be significant for us. One researcher with a special interest in courtship has written, "Human counterparts have already been discovered for a number of genes originally identified in the fly. . . . These findings should provide insights into the molecular interactions that enable the central nervous system to produce behavior." Thus, flies may tell us how a fertilized egg receives its orders to grow. They may also indicate how alcoholism, for example, and sensitization to narcotic substances occur in our kind.

The crucial part of Jay's experiments will be to chart the fundamental biological pathways involved in the flies' responses to cocaine. So far, given the evolutionary conservation of pathways, his research indicates that things happening within the microcosm of a fly—the stereotyped behavior, the hyperactivity, the tremors—can, and do, happen within a rodent, an ape, a human being. And if the mechanisms of response to cocaine can be understood, then it may be possible to find methods of preventing or blocking addiction.

Fortunately, there are enough dew lovers in this world so that the Chief in his swatting, spraying mode cannot conceivably cause any diminishment in the world's population. I doubt, though, that I'll convince him to stay his hand or lay down the can of repellent. Nonetheless, I'm on the side of the proponents. There's much to be said in favor of the little fly. It may be infinitesimal, but it holds infinite promise.

# UNFINISHED BUSINESS

Snow geese (*Chen caerulescens*)

*Homo sapiens*

*Man was nature's mistake—she neglected to finish him—and she has never ceased paying for her mistake.*
                                    —Eric Hoffer, "On Nature and Human Nature"

WHICH species do you most despise? Which plant or animal do you think most pestiferous? Which would you like to see gone, and not just gone but gone forever? I solicit opinions from anyone who wants to answer. One Saturday afternoon, I posed the questions in Columbia, South Carolina, on the occasion of a lively book festival. I participated by telling a story from one of my books—telling because the six minutes allotted were enough for talk but not for reading the text except with unsuitably breathless haste. The final minute, however, was devoted to stating the questions about despicable species and requesting members of the audience to stop by afterward with answers.

And people did stop by to complain about and list their least favorite critters and weeds. Cockroaches led the pack with three votes. One vote each went to this hodgepodge of bugbears, harpies, and swatlings: pigeon, bat, mouse, opossum, squirrel, flea, gnat, mosquito, tick, chigger, cocklebur, and pines, not a particular species, like the loblolly, but all the durn pine trees in the world. One other animal may have received a vote, though it is hard to tell if the paragraph jotted on my notepad was meant to select a species or to deliver a short, emphatic lecture to the woman—me—who was asking silly questions. Signed "From a Zookeeper," it reads: "There are no despicable species—there is, however, ONE species which demonstrates great despicableness—the *only* species on earth that destroys the world in which it lives—the *human* species."

Zookeeper came, wrote quickly, and departed before I could note any details of age and gender. Conversation would have been welcome, along with Zookeeper's suggestions for ways in which our kind might learn to pick up the clothes, toys, fast-food wrappers, beer cans, and other detritus of our consumer culture that it strews in its own room.

But not everyone lives in middle-class comfort. Some of us live in mountain hollows or equatorial villages, on arctic islands or Asiatic steppes. Some people in this country live in cardboard boxes, on grates in urban sidewalks, or under the kudzu vines overrunning vacant lots. These rooms are strewn with hand-to-mouth necessity. The raw human search for food, fuel, and shelter leads to destruction. Birds lose the trees and grasses that they need for cover. Rivers run dry. The sands of the Sahel inch south a little farther every year. But who are we who live amid the comforts and luxuries—the delusive armor of technology—to tell those who lead hardscrabble lives that they ought to pick up their rooms and, while they're at it, limit their populations lest they starve? Control over reproductive life means individual freedom to some of us, but tyranny, mandated by the state or a privileged elite, to many others.

In 1711, Jonathan Swift's hero Captain Lemuel Gulliver was marooned on an uncharted shore by his mutinous crew. No stranger to fantastical adventure, he immediately set about seeking out the inhabitants of this land of pastures and tidy oat fields. He was sure that he could win the favor of any savages he might encounter, for he was well supplied with trinkets. The first living animals that he met were wild indeed—naked brown-buff creatures with no use for Gulliver's bracelets and glass rings. Though they bore a striking resemblance to humankind, they were endowed, male and female, with hairy chests and a ridge of fur down the center of their backs. Sometimes they stood on their hind legs, sometimes they went about on all fours, and all of them stank. Worse, when Gulliver took shelter under a tree, they climbed it most nimbly and shat upon his helpless head. "I never beheld in all my travels so disagreeable an animal," said Gulliver, "or one against which I naturally conceived so strong antipathy." He was rescued by two horses, one gray, the other bay, who turned out to be members of the land's dominant species, the Houyhnhnms. And these noble horses, dedicated to the use of reason in every aspect of their lives, kept the wild, stinking animals, which they called Yahoos, as beasts of burden and pullers of carts. The hides of the Yahoos were also turned into leather goods. Gulliver himself came to see the creatures as "the most unteachable of all animals, their capacities never reaching higher than to draw or carry burthens. Yet I am of opinion this defect ariseth chiefly from a perverse, restive disposition. For they are cunning, malicious, treacherous, and revengeful. They are strong and hardy, but of a cowardly spirit, and by consequence insolent, abject, and cruel."

From the beginning of his untoward adventure, Gulliver applied himself to learning the language of the Houyhnhnms and was soon able to converse at length with the horse he refers to as his master. Much of their talk consisted of questions and answers, with the horse earnestly interrogating Gulliver in an effort to understand his sudden apparition on these shores and to learn about his two-legged kind, and Gulliver responding at voluble and detailed length. As the years passed, one, two, three, Gulliver came so much to admire the horses' rule of reason that he wished to spend the rest of his days with these paragons of rational thinking and conduct. but after hearing descriptions of humankind as creatures who often sought their livelihoods by "begging, robbing, stealing, cheating, pimping, forswearing, flattering, suborning, forging, gaming, lying, fawning, hectoring, voting, star-gazing, poisoning, whoring, canting, libelling, free-thinking, and the like occupations," the Houyhnhnms declared Gulliver a Yahoo, albeit one with a semblance of reason, and bade him go. In the end, Gulliver himself came to think that "my family, my friends, my countrymen, or human race in general, when I considered them as they really were, Yahoos in shape and disposition, only a little more civilized, and qualified with the gift of speech." More terrible, the species of which Gulliver was a member employed their reason to multiply their vices and so exacerbate their degradation, while the Yahoos of Houyhnhnmland were only as barbaric as nature allowed.

Zookeeper would agree, I'm sure, that fiction echoes fact and that nothing has changed much in the last three hundred years.

How did we come to this not so pretty pass? An antic vision flits through my imagination:

There they are, the naked two of them, Adam and Eve, out there in that blasted thorn field. There's no going back to Eden, not with flaming swords barring the way. The two of them may be shivering. They're certainly wondering what happened. And here comes a snake, which seizes a frog and gulps it down whole. (Adam shudders.) Overhead flies a bird with grass for nest building in its bill. A cat tracks the flight with patient eyes, while its kittens play amid weeds and wildflowers. Over yonder, tiny flies swarm like motes of sunlit dust above the creek; a fish leaps open-mouthed. A mosquito lands on Adam's arm, another on Eve's right shoulder. Next thing they know, they're scratching and slapping and inventing the world's first invective. They also learn to pluck a leafy branch from the nearest tree, not to cover their pudenda—the parts, that is, they're supposed to be ashamed of—but to serve as a whisk and a surefire swatter. The problem is greater, however, than that of being preyed upon by a puny, six-legged insect with an unconscionably long proboscis. It's that they alone, out of all creation, don't quite know what they're doing. Everything else is perfectly at home, possessed of a niche and an inborn knowledge of how to live. But man and woman are restless, incomplete, unnatural, and probing always for something to ameliorate their psychic discomfort, which is as chronic and nagging as hay fever or a bad back.

The quest for remedies has come a long way—from the wheel to the Pentium chip, from cave paintings to computer imaging, projectile points to atomic weaponry, charms and healing spells to organ transplants, from the Ten Commandments and the Gospels to Einstein and to chaos theory. Man, according to Eric Hoffer

(1902–1983), the curmudgeonly longshoreman-philosopher whose aphoristic work was popular in the 1950s and 1960s, "has to finish himself by technology. . . . Lacking organic adaptations to a particular environment, he must adapt the environment to himself and re-create the world." And we, Adam and Eve's descendants, have truly, as Hoffer says, broken free from nature in many respects and removed ourselves from domination by her "iron laws."

But humankind once did occupy a niche. With an entirely appropriate metaphor, the ecologist Paul Colinvaux has defined "niche" as "more than just a physical place: it is a place in the grand scheme of things. The niche is an animal's (or a plant's) profession." Once upon a time, our profession was that of hunter-gatherers, seeking live animals for meat and collecting stay-put edibles like nuts and berries, insects, snails, and birds' eggs. It is part of today's conventional wisdom to ascribe to our tribal forebears a deep and healthy reverence for the natural world that provided them with meat and vegetables. I accept the existence of such reverence—it must have been as real as sun and rain, wind and the good earth— but I suspect that it sprang not from any phenomena outside the confines of the human body but rather arose out of that body's blind, unthinking, needy push to survive. Oak and bear and snake were worshiped not for their own sakes but in a cautionary way, for the benefits or harms that they could most certainly bring to our frail flesh. From seasons of plenty through seasons of dearth, the calories we earned were spent mainly on finding more calories. Our numbers were limited by the difficulties inherent in just getting by. It behooved us to revere the natural things whence came our strength, to appease the nonhuman powers arrayed against us, and to make our small magic lest we lose our lives.

For creatures such as we, capable of memory and forethought, the niche may have seemed as constrictive as outgrown clothes. But alterations were made some twelve to fourteen millennia ago when we discovered that animals could be tamed and plants made to grow in cultivated plots. Agriculture and animal husbandry have let us stay home, devoting less energy to the quest for brute survival and more to things like art, politics, and concocting remedies for everything, great or picayune, that bothers us. To this day, we struggle for completeness, if not perfection, and the effort has not been without success, at least in the realm of creature comforts. Although most of the planet's people still scratch out their livings in the thorn field, we who inhabit the industrialized West's neoparadise actively enjoy the apparent freedom from natural constraints that technology grants us. Nor are we mean enough to keep all the wealth to ourselves but parcel it out, sometimes efficiently, sometimes not. And all of us, those still caught amid thorns and those growing fat in a man-made Eden, increase our numbers, jostling and tripping one another as we crowd the earth. Human populations no longer depend on the cycles of nature for flourishing or failure. But these days, the only people who can afford reverence for earth, for plants and animals and stones and stars, may be those rich enough not to wonder where the next meal is coming from. We're still incomplete.

Yet how can better swatters and bigger mousetraps be enough to make us whole? How can computers and spacecraft and genetically engineered corn round off our rough corners, patch the cracks, and mend the recurrent breaks? Technology springs from human ideation. Then it's put together by human hands. A seized-up engine, however, is hardly the same as a myocardial infarction. Technology exists independently, not as part and parcel of our flesh and bones. It's armor, protecting the poor soft organism inside. Nor is it in any way finished. It can't be, not till we finish ourselves.

As the twentieth century wheels pell-mell into the twenty-first, the human situation is closely akin to that of the snow geese. What dramatically beautiful birds they are, entirely white except for their wingtips, which look as though dipped in midnight ink! I have seen the snows flying over the Neuse, their strong white bodies gathering up the last light and glowing against the gray of an evening sky. They fly far indeed, wintering on the mid-Atlantic seaboard, breeding in the Arctic tundra. But only a hundred years ago, at the turn of the last century, their wings were almost clipped forever; snow geese came within a feather of joining the dodo and the moa, the passenger pigeon and the Carolina parakeet in the dark realm of extinction. The cannonades of go-for-broke market hunting, along with wholesale slaughter on the nesting grounds, had diminished their number to a fragile few thousand. Thanks to protective legislation, however, and the creation of wildlife refuges, populations of snow geese have undergone an explosive recovery, with the species now totalling in the millions. Winter flocks more than half a million strong congregate these days in the salt marshes of the Delmarva peninsula. It might seem that they're out of danger, but no: they represent too much of a good thing. The problem stems from their not so dainty feeding habits. Not only do they relish the bird-attracting grains planted in refuges and nearby farms but they also instinctively grub out cordgrass (*Spartina alterniflora*), an important salt-marsh plant, then swallow its roots with great gusto, and go back instanter for more. In a phenomenon that wildlife managers call an "eatout," the living heart of the tidal marshes is devoured; where once acre on acre of tall cordgrass flourished, dry stalks rustling in the winter wind and roots waiting beneath the water for spring, mudflats remain. And the tundra of their breeding range—the Northwest Territories and Hudson Bay for the lesser snow geese, Baffin Island and even Greenland for the greater snows—is also being eaten alive. Nor can they be dissuaded from returning year-in, year-out to devastation. Waterfowl are often notoriously loyal to the same tiny patch of muskeg and moss where they themselves were hatched out, and they zero in on that one small spot to rear their own young. But more and monstrously more of them are crowding into the range with each passing year. In a book published in 1999, the naturalist and writer Scott Weidensaul, who has made an extensive popular study of migration, offers the inevitable comparison: "At last count there were more than half a million greater snow geese, and several million lesser snows—and both were continuing to increase at about 5 percent a year. Plot that on a chart and you get the sort of swooping, upward curve usually seen with articles on human overpopulation. And like humans, snow geese are consuming the very environment on which they depend."

What to do? Hunters' bag limits have been raised in some refuges, but the actual take of these birds has not made any dent at all in their gleaming hungry multitudes. Sooner rather than later, overpopulation will lead to starvation and disease. There is a gulf of difference, though, between the geese and us: the birds have no way to mend the ravages that they—and we in our zeal to protect them—have caused, but our kind holds its future in its own hands.

My friend Paul Cabe, the scientist who has provided me with much information about the swart and raucous lives of starlings, conducts genetic research these days on duckweed and also teaches environmental studies. He says, "Communities and ecosystems are not like engines—remove one part and nothing works, everything fails." And to their great aghastment, he tells his students, "For most species, extinction will not have catastrophic results. Other species will probably shift their abundance, but the world they live in will not disappear or fall apart."

To this it can be added that more often than not, extinctions are not noticed, not by ecosystem or community or ever-investigating, ever-nosy humankind. We are often not aware in the first place that a particular plant or animal, fungus, bacterium, or dinoflagellate had once had a lease on life but now is vanished from the earth. Nor has the absolute disappearance of these species impinged in any noticeable way—or in any way at all—on our two-legged, big-brained kind.

So what's the fuss about? Why work to save black-footed ferrets, torreya trees, and Wyoming toads, not to mention snail darters in Tennessee and California's minuscule Delhi Sands flower-loving fly? Are they necessary to the planet's spin? Ecological correctness shouts an immediate *Yes!* Reason takes a deep breath and says, *We don't know.* Reason also posits that they're highly localized species, all of them, and if they were gone, not just people but hardly any other living thing would miss them. Extinctions are a fact of natural life. Evolution, nonetheless, does not come to a screeching halt.

I think that one reason for the fuss—the often sentimental, sometimes woefully uninformed push to preserve, conserve, protect at all costs—has to do not with a sacredness supposedly intrinsic to all and any life but rather with a dimming of the rainbow's colors. And perhaps with the extreme endangerment of a peculiar, and peculiarly human—and hardly altruistic—sort of hope: if our behavior dooms species to sure vanishment, we may lose the magic plant or animal products that could cure us of mortality. There's something in the rage to preserve that speaks to an American fear of death.

The rage, however, is oddly selective. Nor do we have the courage of our common sense as often as we need it. If all humankind adhered to Jainism, an ancient and peaceful religion of India, everyone would follow the doctrine of *ahimsa* that prohibits doing violence to any living thing. Instead, we deem some lives worthy of defense but ignore, excoriate, or persecute others. Imagine a group calling itself People for the Ethical Treatment of Plants, and another called Save the Crustaceans. Trouble is, plants don't wag their tails or have big, love-me eyes; blue crabs and shrimp respond to us mainly by allowing themselves to be caught. If the life at hand is good to eat, it's fair game. If it annoys, stings, or sickens us, then we assume that we have leave to swat it, squash it, cut it down, or go all out to exterminate it.

Life on earth may well be compared to a rainbow. Red coexists here with purple and puce, along with the blooming or saddened rest of the spectrum. Blooming and saddening—the terms come from the dyer's trade and mean, respectively, brightened or dulled. In the best of all worlds there's room for both, as well as every tint and hue between. The scientists' term for the rainbow's colors is *biodiversity*. A ponderous catchword, but it means simply "variety of lifeforms." Loss of diversity means proliferating shades of gray—more starlings, more kudzu, more molds, although (who knows?) evolution, irrepressible as always, might ring some startling changes on these all too familiar themes.

But variety of life-forms has nothing to do with quantity. Only theology asks how many angels can dance on the head of a pin. With biodiversity, it's quality and coexistence that count. Do cherubim and seraphim, powers and dominions leap and spin along with the angels? In more mundane terms, what array of life-forms—what wings, roots, and fins, what creeping, dividing cells—exists in a given place? Are the niches filled? Who comes to visit or stay? Who leaves never to return? And, with the bustling arrivals and departures, what changes for better or worse are wrought in the community?

Why be concerned in the first place? Because, I am sure of it, the human psyche needs a sense of wonder as much as the body needs food, sleep, and breath. Because living with shades of gray, with concrete, steel, and oil, casts a grayness into heart and mind. But with our burgeoning numbers, grayness is all but assured. The natural philosopher and ecologist Paul Shepard has put it plainly: "Large human populations tend to degrade the environment, both as a source of food and as a medium of biological richness. Complex brains are an adaptation to biological richness and therefore require interesting surroundings." Then, speaking of our multifariousness, he has said, "Environmental requirements are greater and more exacting for humans than for most other species. . . . A wide flexibility of behavior requires suitable environmental diversity." And humankind, so studies of environmental planning say, also requires several features in its surroundings in order to find the greatest ease: natural light (no windowless factories), greenery (the African savanna where man was born, our present-day lawns), water (ocean, creek, deep well, or artificial pond), and mysteries (unexpected turns in the path that reveal new vistas). These mysteries have been defined by one set of researchers as surroundings that "give the impression that one could acquire new information if one were to travel deeper into the scene." The mysteries have to do with expectations then, and with the promises that every environment, urban or rural, should make to those who live and travel there.

We cannot always predict how and when the promises will be fulfilled, but living amid a rainbow of other lives ensures the keeping of promises—a flower around this corner, a snake around the next.

We've made things hard for ourselves. Animalcules and tiny plants exist in uncountable trillions, but most large, predatory animals are scarce on the ground, for their numbers are set by the calories available to support their way of life. Famine and feast, the energy contained in the food supply, determines the size of a population. With the taming of plants and animals some twelve to fourteen thousand years ago, however, we exempted ourselves from this stern mandate. We hoisted ourselves our of our hunting-gathering niche, the only animals ever to break free of their biologically ordained profession. Now the last kind of large-scale hunting in which we still engage is that for seafood. We have allowed ourselves to stay put in one place, a way of life that led to social pathologies like bloody, knock-down, drag-out wars over disputed lands and resources; any war, for all that it may seem just, is more truly a counterproductive survival strategy, for the fittest are sent to kill and be killed while the least fit stay home to perpetuate the species. Another pathology is the creation of hierarchical, status-based communities, with a few kings and priests at the apex, a mort of pariahs and slaves at the nadir. Worst is the despoliation of habitat, the fouling of our own nests, that Paul Shepard deplores.

Eric Hoffer is right: we are unfinished. But will it matter if we never manage to finish ourselves? Upright humankind—as bipedal creatures, we suffer from flat feet, creaky knees, and crooked spines. The major discomfort, however, lies in our intrinsic Yahoodom and in the dislocations caused by clambering out of our preordained niche. There's no going back, of course. But there may be a remedy in firming up a vital part of our psychic anatomy, namely the backbone, which is another name for moral intelligence.

Firming up to finish ourselves might not be a bad thing. In search of a means, I ramble through the classical underbrush in my head—

Io driven flystung across the continent of Asia, the riddle of the Sphinx, and Proteus, the prophetic Old Man of the Sea, transforming himself from animal to flame to water to tree and back to god. I envision the modern incarnations of these myths—May's blood-thirsty deerflies, the sphinx moths with their horn-tailed larvae, and the dangerous shape-shifting dinoflagellates in coastal sounds and rivers. And I home in, as always, on the ancient Greek concept of *kairos*—balance, moderation—and on the classical certainty that, as Paul Shepard puts it, there is "livingness in the inanimate," that the world in all its manifestations is sentient and holy—from mountains, rivers, and constellations, from alpine flowers, minnows, and cottontails to biting flies and fish-killing protists. As long as we have no Proteus to prophesy the consequences of our actions, it's meet to remember that there is no stasis, that the balance always teeters. So, it is meet also that we behave ourselves and that we step with care between the vines—wherever they may be—and not on them.

Late March, and the Virginia Festival of the Book is under way— hundreds of programs at dozens of locations all over Charlottesville, Virginia. Downtown is a mob scene, with parking catch as catch can. I'm out in the boonies, however, at Ivy Creek Natural Area with two other writers who have been invited by The Nature Conservancy to discuss the importance of place in our work and lives. Rob Riordan, a black Irishman who is the Virginia chapter's director of communications, acts as moderator. One of my copanelists is Dick Austin, a Presbyterian minister, who preaches and writes about a theology-based environmentalism, then practices what he preaches on his homeplace farm in southwest Virginia; the other is Christopher Camuto, a hunter, hiker, and astute observer who has written eloquently about the attempt to restore red wolves to the Cherokee Mountains, as the Great Smokies were once called. More important, Chris tries not unsuccessfully to recover the Cherokees' perception of that rugged land as a place that is sentient and holy. I speak, of course, about my three countries, the mountain-walled Shenandoah Valley, the flat-as-a-flounder Carolina coast, and the invisible but equally real territory that is archaic Greece. Its principle of *kairos* is no less valid today than it was then. We talk, the three of us, about our passionate connections to specific places and about how caring for and respecting these places can inform our behavior toward the rest of the world. And every person in this room has this same splendid chance to let a particular attachment to one back yard, one town, one farm, one valley, one mountain range provide a model for the ways in which we might comport ourselves in regard to the greater whole.

The large community room in Ivy Creek's recreation building is nicely filled. I think, though, that the people who have come to hear us are among the converted—those, that is, who are well aware that humanity must somehow pull back lest we, like the snow geese, damage our living spaces beyond their ability to sustain us. The questions and comments flow readily. They are variously earnest, fearful, moralistic, and filled with silver-lining hope. As our time is almost up, Rob recognizes one last member of the audience. A middle-aged woman stands and says, "We really have to be careful. Mother Earth needs us."

I respond. "Careful, yes, but think about it. Earth needs us? Isn't it really the other way around?"

To arrive at a modus vivendi, to realize comfort at last, it might not be amiss to ask if *H. sapiens* is truly necessary We're not localized but pop up all over the globe. What good are Yahoos to the workings of the world? Of what possible use to eagle, shark, and flower-loving fly? What would fail and fall apart if an asteroid sent us the way of the dinosaurs or if we used our rightly prized technology to do ourselves in? It's quite safe to predict some effects of our withdrawal from the scene. There'd be fewer starlings, for lack of cleared land, and fewer brown-headed cowbirds, for lack of new edges and, thus, access to host nests where woods border building lots. Now-toxic, fish-killing dinoflagellates would be largely restored to primal quiescence because they would no longer be chemically overfed. The deepest South might well vanish beneath a virid smother of kudzu. And everywhere else, forests would regenerate, and fish stocks be replenished. But all life would still be subject to natural cycles of feast and famine, of increase and diminishment; all subject to natural forces that our kind has nothing to do with, like El Niño, typhoons and hurricanes, volcanic explosions, earthquakes, and planetwide warming or cooling (we do contribute to warming, but how much we contribute is a matter on which no definitive pronouncement has yet been made).

Our self-taught tendency toward despicability is not in doubt. We're also dispensable. Indeed, the world can do very well without us. It doesn't need us to keep it spinning around any more than it needs deer flies and screwworms, starlings, sandburs, and poison ivy.

Rather, it's we who need the world.

Yahoo

# APPENDIX

House Mouse (*Mus musculus*)

# The Despicables Ratings

*What is your least favorite animal or plant? The one you despise most, the one you find totally useless, the one you'd love to see gone, and not just gone but gone forever?*

That is the completely unscientific question that I've asked friends and strangers encountered in my peregrinations during the last few years. The list leans toward the Southern experience of despicable plants and creatures, for most of my time is spent in the American South. But answers have come from as far afield as Falmouth, Massachusetts; Hermosa, South Dakota; and Jerez, Mexico.

As anyone might have predicted, the living things most intensely disliked are those that impinge unpleasantly or painfully on human existence. They appear without warning. Some bite and sting, while others raise blisters. Some are invaders, crowding out native species. Some merely annoy, but others may kill. All of them, be they exotic or homegrown, are exceedingly abundant. Swat, spray, and smash as we will, there is no getting rid of them. And here's the list.

It begins with the hodgepodge of despicables that come in last, only one mention each—but look at their exuberant diversity. Each is identified here as closely as possible with its formal name; in many cases, however, it is the generic creature that is despised, and only a loose, general identification can be made. Comments in parentheses are those of sworn enemies.

molds, fungi of various families, genera, and species

anthrax bacterium, *Bacillus anthracis*

slime molds, protists

*Giardia lamblia,* a parasitic protozoan that causes severe intestinal distress
*Pfiesteria piscicida,* a dinoflagellate

bamboo, *Phyllostachys aureosulcata* and others, all of them ornamental garden plants that tend to go invasively wild beets, *Beta vulgaris* ("I dislike their texture, their taste, their smell.")

common chickweed, *Stellaria media* ("It doesn't know when to stop.")

Eurasian milfoil, *Myriophyllum* spp., imported takeover water plants

ferns, order Filicales ("plants without character invariably used in restaurants without character")

greenbriar, *Smilax* spp.

Japanese honeysuckle, *Lonicera japonica* ("a curse on all")

jimsonweed, *Datura stramonium* ("offending prickers")

melaleuca, *Melaleuca quinquenervia*, an Australian tree that has become a pest species in Florida

nut grass, *Cyperus esculentus*

pennywort, *Hydrocotyle bonariensis*

phragmites, *Phragmites communis*, a nonnative wetlands plant, the presence of which indicates environmental degradation

pines, *Pinus* spp., any and all of them

rabbit tobacco, *Gnaphalium obtusifolium*

red-hot poker, *Kniphofia* spp. ("Summer is hot enough without having to look at those blazing spires. And the color does not blend well with *any* other color.")

Russian thistle, *Salsola kali* var. *tenuifolia*, an accidentally introduced Russian plant that is not a thistle but rather a member of the goosefoot family

salt-cedar, also known as tamarisk, *Tamarix chinensis*, an invasive exotic of the West

wire grass, also called Bermuda grass, *Cynodon dactylon* ("the plant that fights back unfairly")

yucca, *Yucca filamentosa* ("Those leaves remind me of swords and the stringy bits, of spider webs.")

ants in general, family Formicidae

earwigs, order Dermaptera, many families

louse, both *Pediculus humanus*, the head louse, and *Phthirus pubis*, the pubic louse termites, order Isoptera, many families

wasps and hornets, order Hymenoptera, many families

scorpions, Arachnida, order Scorpionida

zebra mussel, *Dreissena polymorpha*

sharks, class Chondrichthyes, many families

American eel, *Anguilla rostrata*

cat, *Felis catus*

chipmunk, *Tamias* spp.

coyote, *Canis latrans* ("They killed my cats!")

hyena, family Hyaenidae ("I've never seen one, but something about their looks and their scavenger lifestyle is tremendously unpleasant.")

mole, *Scalopus aquaticus*

opossum, *Didelphis marsupialis*

woodchuck, also known as groundhog, *Marmota monax*

Some of these votes may have been influenced by the environmental press. Two of the one-vote despicables, for example, appear on The Nature Conservancy's "dirty dozen" list of the least wanted species in America, and two others from that list show up later in my ratings. According to *Nature Conservancy* magazine (November-December 1996), the least wanted include the freshwater zebra mussel and the salt-cedar tree, each one an introduced species and both mentioned on my list above. Two others, which pulled in more than a single vote from my respondents, figure below: the purple loosestrife plant and the brown tree snake. Others of The Nature Conservancy's dirty dozen are the plants hydrilla, *Hydrilla verticillata;* leafy spurge, *Euphorbia esula;* miconia, *Miconia calvescens;* and Chinese tallow, *Sapium sebiferum.* One insect shows up, the balsam woolly adelgid, *Adelges piceae,* which attacks hemlocks; one terrestrial mollusk, the rosy wolfsnail, *Euglandina rosea;* and two marine species, the flathead catfish, *Pylodictis olivaris,* and the green crab, *Carcinus maenas.*

In my survey, the living things rating two votes each are: cocklebur, *Xanthium pensylvanicum;* crabgrass, *Digitaria sanguinalis* ("almost malignant"); tapeworm, class Cestoda; leech, class Hirudinea; house centipede, *Scutigera coleoptrata;* crickets in the house, subfamily Gryllinae (the qualifier "in the house" was specified both times); rock dove, *Columba livia* ("rats with wings"); starling, *Sturnus vulgaris;* bats, order Chiroptera; and *Homo sapiens.*

The following life-forms received three votes each: purple loosestrife, *Lythrum salicaria,* an aggressive imported plant; slugs, order Stylommatophora; sea nettle jellyfish, *Chrysaora quinquecirrha;* rodentkind, specifically mice, rats, and squirrels, *Mus, Rattus,* and *Sciurus* spp. respectively; and viruses, with special mention going to those for AIDS and the common cold. Of the rat, one sworn enemy writes, "No matter how rural and wholesome its habitat, it always looks *sordid,* as if it had come out of a sewer. The bristles on its otherwise naked tail are almost supernaturally repulsive."

Pulling in considerably more votes, with each earning about a third of the total earned by the most despicable species on the list, are two insects, fire ants, *Solenopsis* spp., and fleas, order Siphonaptera; and one grass, the sandbur, sometimes called sandspur, *Cenchrus* spp.

Second place is clogged by scourges that have individually racked up close to half the votes awarded to the winners of the despicable ratings. The occupants of the second-place niche are another insect, cockroaches, order Blattodea, family Blattidae; three kinds of arachnid—ticks, chiggers, and spiders; snakes; and one plant—you guessed it—poison ivy, *Rhus radicans.* With spiders, the generic kind figures most often in the tally, but several people particularly nominated the brown recluse, *Loxosceles reclusa,* and the black widow, *Latrodectus mactans.* Snakes, too, are mainly a generic category; to many of us, a snake is a snake is a snake, and has been so ever since Eden. One person, however, appended the word "poisonous," while a few people were definite, naming rattlesnake, *Crotalus* and *Sistrurus* spp.; copperhead, *Agkistrodon contortrix;* cottonmouth moccasin, *A. piscivorus;* and brown tree snake, *Boigo irregularis.*

(The last is a Far Eastern species introduced to islands in the Pacific; it has caused notable devastation to Guam's native wildlife, and there are fears that it may have reached Hawaii.)

And the worst of all, the greatest villains, the nastiest, the most despised? No contest here: the Diptera, the two-winged flies, swarm far ahead of the rest of the pack. They are despised by twice as many people as any other plant or creature on my list. Just plain flies, without any further qualification, were mentioned more often than not, but specific kinds were also damned: fruit flies, *Drosophila* spp.; houseflies, *Musca domestica*; deerflies, *Chrysops* spp.; the rabbit botfly, *Cuterebra cuniculi*; mosquitoes, family Culicidae; true gnats, members of several Dipteran families (which suffer from a bad press but are generally harmless); and no-see-ums, members of the family Ceratopogonidae (which most emphatically are not). The last have been characterized by one hapless victim as "jaws with wings."

Here the list ends, though any number of other species might be added. It will quickly be noted that some of the plants and creatures, like kudzu and hornworms, that I have written about make no appearance whatsoever on the list above, but they earn their own stories because of this conviction: they are so ubiquitous and so insufferably part of our lives that someone would have nominated them eventually. It will also be noted that the stories in this book do not include some of our most egregiously awful animal and vegetable pests. Some of the omissions in this book are due simply to the dizzying difficulties of having to focus on a particular one or two species out of the thousands possible, as with mosquitoes (though I could have chosen a monstrously large model with black-striped legs, the gallinipper, more formally known as *Psorophora ciliata*, which rises with a lusty and well-nigh unquenchable thirst during Great Neck Point's wetter summers.) Other omissions have received full discussion in other publications. Poison ivy, slugs, and cockroaches, for example, have been exhaustively examined in recent books, each scrupulously informed but nonacademic, and each devoted obsessively to the particular despicable species. I myself have treated elsewhere of such things as ticks and chiggers, jellyfish and snakes. But for anyone whose curiosity is still unslaked, an unapologetically selective booklist is included after the Notes section to help appease any lingering fancy for further investigation.

# NOTES

Full citations appear in the bibliography.

page xi    epigraph: Fabre, *The Passionate Observer*, 91.

*Living Together, Like It or Not*

10    "are not only cold-blooded": Dillard, *Pilgrim at Tinker Creek*, 64.
10    "No monkey makes light of shapes": Shepard, *The Tender Carnivore and the Sacred Game*, 47.
11    "embedded in our genetic memory": Conniff, *Spineless Wonders*, 108.
13    "What is not useful": Mather quoted by Matthiessen, *Wildlife in America*, 57.
14    "Swatting a mosquito": Evans, *The Pleasures of Entomology*, 11.
14    "if we are justified": Hubbell, *Broadsides from the Other Orders*, 85.
15    "What problems and characteristics": Shepard, 41.

*Prospect and Refuge: Sandburs*

23    "Spikelets 1-flowered": *Gray's New Manual of Botany*, 119.
23    "Unpleasant": Kraus, *A Guide to Ocean Dune Plants*, 59.
23    "Horribly spiny flower clusters": Brown, *Grasses*, 99.
24    It has been noted: Martin, *Weeds*, 27.

*The Barkings of a Joyful Squirrel: Gray Squirrel*

30    "He has not Wings": Lawson, *A New Voyage to Carolina*, 130.
32    "to a local officer": Kalm, *Travels in North America*, 52.
32    "Squirrels are the chief food": Kalm, 52.
33    "when the corn is juicy": Audubon, *Journals*, Vol. II, 496.
33    "the barkings of a joyful Squirrel": Audubon, *Journals*, 404.
33    *"Barking off Squirrels"*: Audubon, *Journals*, 460.

*Murmurations: European Starling*

48    ". . . said he would not ransom Mortimer": Shakespeare, *King Henry IV, Part I*, I.iii.219–26.

53 "That was lovely": Mozart quoted by Biancolli, *The Mozart Handbook*, 396.

53 "Full of wrong notes": Biancolli, 511.

53 "The Starlings are found": Cassell, *Natural History*, quoted in the *Oxford English Dictionary*.

61 "As if out of the Bible": Updike, "The Great Scarf of Birds," *Collected Poems*, 37–38.

62 "Since we never consider as a nuisance": Corbo, *Arnie, the Darling Starling*, 79.

## The Natural History of Proteus: Pfiesteria piscicida

67 "immortal Old Man": Homer, *The Odyssey*, Fagles translation, 4.431.

67 "swarming sea": Homer, 4.436, 4.477; other places.

67 "muster your heart": Homer, 4.466–71.

75 Yet a third element in fish killer's transformation: Shoemaker, "Fish Kills, Facts, and Pfiesteria."

## A Foot in the Door: The Fungi

91 "This is the foul fiend": Shakespeare, *King Lear*, III.iv.118.

92 One notable mycologist: Schaecter, *In the Company of Mushrooms*, 34.

93 "*Our kind multiplies*": Plath, *Collected Poems*, 140.

## The Creature with Nineteen Lives: Common Opossum

98 "An Opassum hath a head": Captain John Smith, *A map of Virginia with a description of the country*, 1612, quoted in the *Oxford English Dictionary*.

98 "The Weather was very cold": Lawson, *A New Voyage to Carolina*, 31.

98 "The *Possum* is found": Lawson, 125–26.

## Legs: Centipedes

118 "The centipede was happy quite": Mrs. Edward Craster, 1871, from *Bartlett's Familiar Quotations*, 13th ed., 750.

120  So it is reasonably thought . . . they had to find food: Manton,
     *Arthropoda*, 283–85.
122  "stumble on food by accident": Manton, 395.
122  One large scolopendrid: Manton, 391.
126  "by a series of gaits": Manton, 282.

## Heritage: Kudzu

136  "an awful tangled nuisance": Fairchild quoted by Shurtleff and
     Aoyagi, *The Book of Kudzu*, 12.
143  "Japan invades": Dickey, *Helmets*, 38.

## The Wisdom of Nature: Brown-Headed Cowbird

156  "The bird is entirely brown": Catesby quoted in Feduccia,
     *Catesby's Birds of Colonial America*, 124.
158  "If we are fond"; "This is a mystery": Audubon, *Original Water-
     Color Paintings*, Plate 59.
160  Orians has asked five questions: Orians, *Blackbirds of the Americas*,
     77.
160  "This suggests": Orians, 79.
163  Why did brood parasitism evolve: Orians, 87.

## The Dew Lovers: Drosophila Fruit Flies

169  "The little fruit flies": Linsenmaier, *Insects of the World*, 258.
170  "as a swarm of flies": Milton, *Paradise Regained*, IV.15–17.
177  "1. Intense nearly continuous grooming": McClung and Hirsh,
     "Stereotypic behavioral responses to free-base cocaine and
     the development of behavioral sensitization in *Drosophila*,"
     109.
178  "Human counterparts": Greenspan, "Understanding the
     Genetic Construction of Behavior," 78.

## Unfinished Business: Homo sapiens

182  epigraph: Hoffer, *Between the Devil and the Dragon*, 14.
183  "I never beheld": Swift, *Gulliver's Travels*, 215.
184  "the most unteachable of all animals": Swift, 251.

184   "begging, robbing, stealing": Swift, 239.

184   "my family, my friends": Swift, 261–62.

186   "has to finish himself": Hoffer, 22.

186   "more than just a physical place": Colinvaux, *Why Big Fierce Animals Are Rare*, 11.

188   "At last count": Weidensaul, *Living on the Wind*, 223.

191   "Large human populations": Shepard, *The Tender Carnivore*, 104.

191   "Environmental requirements are greater": Shepard, 121.

191   And humankind . . . "deeper into the scene": Falk, Kaplan, and Kaplan quoted in Hiss, *The Experience of Place*, 36–41.

193   "livingness in the inanimate": Shepard, 167.

# FOR THE BOOKWORM

A Reading Guide

## DESPICABLES ONE AT A TIME

Adler, Bill, Jr. *Outwitting Squirrels: 101 Cunning Stratagems to Reduce Dramatically the Egregious Misappropriation of Seed from Your Birdfeeder by Squirrels.* Chicago: Chicago Review Press, 1988.

> A compendium of maneuvers, some clever, some downright silly, to foil the furry marauders. Failure is assured.

Anderson, Thomas E. *The Poison Ivy, Oak and Sumac Book: A Short Natural History and Cautionary Account.* Ukiah, California: Acton Circle, 1995.

> A lively guide to *Toxicodendron,* the "poison-tree," including sections on how its oils affect humankind (a fulminating allergic reaction), how to cope with the allergy—and the plants themselves, and how to enter the annual Poison Oak Show held as part of a fall festival in Columbia, California. Many illustrations, from reproductions of old botanical prints to range maps and color photographs.

Cabe, Paul R. *European Starling,* No. 48 in The Birds of North America. Philadelphia: The Academy of Natural Sciences; Washington, D.C.: American Ornithologists' Union, 1993.

> Twenty-four pages that lucidly present the scientific facts about the bird and its rampage through the New World.

Gordon, David George. *The Compleat Cockroach.* Berkeley, California: Ten Speed Press, 1996.

_____. *Field Guide to the Slug.* Western Society of Malacologists. Seattle: Sasquatch Books, 1994.

> Short books both, with ample illustrations that help much to distinguish the German cockroach from, say, the wood cockroach and the banana slug from all others, large and small.

Hauser, Susan Carol. *Nature's Revenge: The Secrets of Poison Ivy, Poison Oak, Poison Sumac, and Their Remedies.* New York: Lyons & Burford, 1996.

> A slender volume that focuses on urushiol, the plants' toxic oil, and the "dastardly itching" that it almost inevitably causes. Learn here how it's possible to contract a raging case of poison ivy without even getting near the stuff. Learn, too, the soothing secrets of rubbing alcohol, hot water, and corticosteroids.

Hilyard, Paul. *The Book of the Spider: From Arachnophobia to the Love of Spiders.* New York: Random House, 1994.

> From page 80: "When Professor John Henry Comstock of Cornell

University was asked by a visitor, 'What good are spiders?' he replied, 'What good are they? They are damned interesting.'" Spider flight, spider silk-making, spider venom, spider myths, and much else are explained here in easily readable terms by a true arachnophile.

Lehane, Brendan. *The Compleat Flea.* New York: The Viking Press, 1969.
A circus of flea facts and lore, from physiology and curious adaptations to specific hosts to the creature's role as muse to both Mussorgsky and sundry pornographers.

Shurtleff, William, and Akiko Aoyagi. *The Book of Kudzu: A Culinary and Healing Guide.* Wayne, New Jersey: Avery Publishing Group, Inc., 1985.
A slim, elegantly illustrated volume on the botany, history, and human uses of the vine, along with a grand soup-to-cough-syrup compendium of recipes.

## ECLECTIC COLLECTIONS

Berenbaum, May. *Ninety-nine Gnats, Nits, and Nibblers.* Urbana and Chicago: University of Illinois Press, 1989.

———. *Ninety-nine More Maggots, Mites, and Munchers.* Urbana and Chicago: University of Illinois Press, 1993.
True tales about a wriggle of bugs, a writhing of worms, a loop of caterpillars, and a buzzing of flies. But sublimities like luna moths and fireflies are included, along with a multitude of small horrors like tent caterpillars, bagworms, bed bugs, and chicken lice.

Colinvaux, Paul. *Why Big Fierce Animals Are Rare: An Ecologist's Perspective.* Princeton, New Jersey· 1978.
Answers to questions like, Why are there so many species of living things and why far more of some than others? Chapters include "Why the Sea Is Blue," "The Nation States of Trees," "The Social Lives of Plants," and, not least, "The People's Place."

Conniff, Richard. *Spineless Wonders: Strange Tales from the Invertebrate World.* New York: Henry Holt and Company, Inc., 1996.
Visits with creatures, here held safely captive in print, that often inspire fear or loathing. One is monstrously larger than man, the others far smaller: giant squid, flies, leeches, fire ants, dragonflies,

tarantulas, fleas, ground beetles, earthworms, mosquitoes, moths, and hagfish, also known as slime eels. The essays entertain as they inform. Conniff has done the same thing for animals with backbones—sharks, moles, bats, bloodhounds "in full slobber," and many others—in *Every Creeping Thing: True Tales of Faintly Repulsive Wildlife*, Holt, 1998.

Lembke, Janet. *Looking for Eagles: Reflections of a Classical Naturalist*. Lyons & Burford, 1990.

Essays on earthbound creatures encountered while keeping eyes on the sky. Among those that may be considered irredeemably horrid are ticks, snakes, toads, jellyfish, and a parasitic isopod.

## THE HUMAN CONDITION

Hoffer, Eric. "On Nature and Human Nature," in the anthology *Between the Devil and the Dragon*. New York: Harper & Row, Publishers, 1982.

The views of an insightful and unregenerate curmudgeon, who writes with the passion of true belief. From page 37 "Man's being an unfinished, defective animal has been the root of his uniqueness and creativeness. He is the only animal not satisfied with being what he is."

Matthiessen, Peter. *Wildlife in America*. New York: The Viking Press, 1959.

A detailed account of the ways in which human alteration of the land—through activities like farming, logging, and hunting—have affected indigenous North American plants and animals since the days of European exploration and colonization. Citing observers from Captain John Smith and Cotton Mather to John Muir and more contemporary naturalists, Matthiessen makes extensive use of the historical record to document the changes.

Shepard, Paul. *The Tender Carnivore and the Sacred Game*. Athens and London: The University of Georgia Press, 1998. Originally published by Scribner in 1973.

An examination of human ecology, from our emergence as hunter-gatherers, with bodies specifically evolved to fit this niche, to our abandonment of the niche and the subsequent disjuncture between what we were biologically designed to do and what we have invented for ourselves to do, like binding ourselves to farms and cities. Shepard explains our evolution both physiologically and spiritually, with keen attention to the natural world around us as the means by which we form a compact with and a necessary reverence for other life. But even at this late date, and in these parlous times, he holds out hope that humankind may reconnect, in ways appropriate to this era of technology, with our true role.

## GENERAL BIBLIOGRAPHY

Audubon, John James. *The Original Water-Color Paintings for* The Birds of America. New York: American Heritage Publishing Co., Inc., 1966.

Audubon, Marie R. *Audubon and His Journals.* Vol. I. New York: Dover Publications, Inc., 1994.

_____. *Audubon and His Journals.* Vol. II. New York: Dover Publications, Inc., 1986.

Biancolli, Lewis, ed. *The Mozart Handbook.* Cleveland and New York: The World Publishing Company, 1954.

Borror, Donald, and Richard E. White. *A Field Guide to the Insects of America North of Mexico.* The Peterson Field Guide Series. Boston: Houghton Mifflin Company, 1970.

Brown, Lauren. *Grasses: An Identification Guide.* Boston: Houghton Mifflin Company, 1979.

Burkholder, JoAnn M., and Howard B. Glasgow, Jr. "Interactions of a Toxic Estuarine Dinoflagellate with Microbial Predators and Prey," *Archiv für Protisten Kunde,* Vol. 145 (1995), pp. 177–88.

Burkholder, JoAnn M., Howard B. Glasgow, Jr., and Cecil W. Hobbs. "Fish Kills Linked to a Toxic Ambush-Predator Dinoflagellate: Distribution and Environmental Conditions," *Marine Ecology Progress Series,* Vol. 124 (1995), pp. 43–61.

Burt, William Henry, and Richard Philip Grossenheider. *A Field Guide to the Mammals.* Boston: Houghton Mifflin Company, 1961.

Christensen, Clyde M. *The Molds and Man: An Introduction to the Fungi.* 2nd ed. Minneapolis: University of Minnesota Press, 1961.

Corbo, Margarete Sigl, and Diane Marie Barras. *Arnie, the Darling Starling*. Boston: Houghton Mifflin Company, 1983.

Covell, Charles V., Jr. *A Field Guide to the Moths: Eastern North America*. The Peterson Field Guide Series. Boston: Houghton Mifflin Company, 1984.

Dickey, James. *Helmets*. Middletown, Connecticut: Wesleyan University Press, 1964.

Dillard, Annie. *Pilgrim at Tinker Creek*. New York: Harper & Row, Publishers, Inc., 1974.

Evans, Howard Ensign. *The Pleasures of Entomology: Portraits of Insects and the People Who Study Them*. Washington, D.C.: Smithsonian Institution Press, 1985.

Everest, John W., James H. Miller, Donald M. Ball, and Michael G. Patterson. "Kudzu in Alabama." Alabama Cooperative Extension Service, Auburn University, Alabama. Circular ANR-65.

Fabre, Jean Henri. *The Passionate Observer: Writings from the World of Nature*. San Francisco: Chronicle Books, 1998.

Feare, Christopher. *The Starling*. Oxford and New York: Oxford University Press, 1984.

Feduccia, Alan, ed. *Catesby's Birds of Colonial America*. Chapel Hill and London: The University of North Carolina Press, 1985.

Fichter, George S. *Insect Pests*. New York: Golden Press, 1966.

*Gray's New Manual of Botany: A Handbook of the Flowering Plants and Trees of the Central and Northeastern United States and Adjacent Canada*. Rearranged and revised by Benjamin Lincoln Robinson and Merritt Lyndon Fernald. New York: American Book Company, 1908.

Greenspan, Ralph J. "Understanding the Genetic Construction of Behavior." *Scientific American*, No. 272 (April 1995), pp. 72–78.

Hiss, Tony. *The Experience of Place*. New York: Alfred A. Knopf, 1990.

Homer. *The Odyssey*. Robert Fagles, trans. New York: Viking Penguin, 1996.

Hoots, Diane, and Juanitta Baldwin. *Kudzu; The Vine to Love or Hate*. Kodak, Tennessee: Suntop Press, 1996.

Hubbell, Sue. *Broadsides from the Other Orders: A Book of Bugs*. New York: Random House, 1993.

Kalm, Peter. *Travels in North America: The English Version of 1770.* New York: Dover Publications, Inc., 1964.

Kraus, E. Jean Wilson. *A Guide to Ocean Dune Plants Common to North Carolina.* Chapel Hill: The University of North Carolina Press, 1988.

Lawson, John. *A New Voyage to Carolina.* Hugh Talmage Lefler, ed. Chapel Hill: The University of North Carolina Press, 1967.

Leahy, Christopher. *The Birdwatcher's Companion; An Encyclopedic Handbook of North American Birdlife.* New York: Bonanza Books, 1952.

Levi, Herbert W., and Lorna R. Levi. *A Guide to Spiders and Their Kin.* New York: Golden Press, 1968.

Lincoff, Gary H. *The Audubon Society Field Guide to North American Mushrooms.* New York: Alfred A. Knopf, 1981.

Linsenmaier, Walter. *Insects of the World.* Translated from the German by Leigh E. Chadwick. New York: McGraw-Hill Book Company, 1972.

Manton, S. M. *The Arthropoda: Habits, Functional Morphology, and Evolution.* Oxford: Clarendon Press, 1977.

Martin, Alexander. *Weeds.* Racine, Wisconsin: Western Publishing Company, Inc., 1972.

McClung, Colleen, and Jay Hirsh. "Stereotypic behavioral responses to free-base cocaine and the development of behavioral sensitization in *Drosophila.*" *Current Biology*, Vol. 8, No. 2 (15 January 1998), pp. 109–12.

Milne, Lorus, and Margery Milne. *National Audubon Society Field Guide to North American Insects and Spiders.* New York: Alfred A. Knopf, 1980.

Orians, Gordon. *Blackbirds of the Americas.* Seattle and London: University of Washington Press, 1985.

Plath, Sylvia. *The Collected Poems.* Ted Hughes, ed. New York: Harper & Row, Publishers, 1981.

Schaecter, Elio. *In the Company of Mushrooms: A Biologist's Tale.* Cambridge, Massachusetts: Harvard University Press, 1997.

Shoemaker, Ritchie C. "Fish Kills, Facts and Pfiesteria: My Patients and the River Told Me What I Had to Know." *The Washington Post*, September 21, 1997, p. c1.

Stokes, Donald. *A Guide to Observing Insect Lives.* Boston: Little, Brown and Company, 1983.

Swift, Jonathan. *Gulliver's Travels and Other Writings.* New York: Bantam Books, 1962.

Updike, John. *Collected Poems, 1953–1993.* New York: Alfred A. Knopf, 1995.

Waldbauer, Gilbert. *The Birder's Bug Book.* Cambridge, Massachusetts: Harvard University Press, 1998.

Webster, William David, James E. Parnell, and Walter C. Biggs, Jr. *Mammals of the Carolinas, Virginia, and Maryland.* Chapel Hill and London: The University of North Carolina Press, 1985.

Weidensaul, Scott. *Living on the Wind: Across the Globe with Migratory Birds.* New York: North Point Press, 1999.

Wilson, Edward O. *The Insect Societies.* Cambridge, Massachusetts: Harvard University Press, 1971.

Zim, Herbert S., and Robert T. Mitchell. *Butterflies and Moths.* New York: Golden Press, Inc., 1964.

# The Image of Guadalupe

# THE IMAGE
# of GUADALUPE

Second and revised edition

by

## Jody Brant Smith

✳

**MERCER UNIVERSITY PRESS**
*in association with*

*Gracewing.*

**FOWLER WRIGHT BOOKS LTD.**

1994

54-421-2 (USA)
44-325-0 (UK)

MUP/P121

*The Image of Guadalupe*
Second and revised edition
Copyright ©1994
Mercer University Press
6316 Peake Road, Macon, Georgia 31210-3960 USA
in association with
Gracewing/Fowler Wright Books Ltd.
#2 Southern Avenue, Leominster, Herefordshire HR6 0QF England

✳

The paper used in this publication meets
the minimum requirements of American National Standard
for Information Sciences—Permanence of Paper
for Printed Library Materials, ANSI Z39.48–1984.

✳

*Library of Congress Cataloging-in-Publication Data*
Smith, Jody Brant, 1943–    .
The image of Guadalupe / by Jody Brant Smith — 2nd rev. ed.
xvii+132 (+16 [illustrations]) pp.   6x9" (15x23 cm.).
Rev. ed. of: The image of Guadalupe. 1983.
Includes bibliographical references and index.
ISBN 0-86554-421-2
1. Guadalupe, Our Lady of.  I. Smith, Jody Brant, 1943–    .
Image of Guadalupe. II. Title.
BT660.G8S44    1994
232.91'7097253—dc20                        94-44152
                                                     CIP

# CONTENTS

# Illustrations

To the memory of
*Jody Bass and Gladys Patterson Smith,*
my parents


To
*Heather Deborah and Jody Brant II,*
our children


and


To
*Maria*
the one and the many

# FOREWORD

Her head is tilted to the right. Her greenish eyes are cast downward in an expression of gentle concern. The mantle that covers her head and shoulders is of a deep turquoise, studded with gold stars and bordered in gold. Her hair is black, her complexion olive. She stands alone, her hands clasped in prayer, an angel at her feet.

So Jody Brant Smith writes at the outset of this remarkable book. Or better yet, so he speaks. It is difficult having read *The Image of Guadalupe* word by word, image by image, fact by fact, not to imagine Dr. Smith articulating this description, and others of equal narrative richness, before an audience of rapt listeners, perhaps a courtroom audience immersed in the building of a phenomenal case.

One imagines growing accustomed to the rhythmic hardwood echo of Dr. Smith's footsteps as he tells the jury exactly what he intends to accomplish, speaking with the ease and confidence of one for whom the hardest part—research of the leave-no-stone-unturned variety—is behind him, at long last.

The great power of Smith's argument in *The Image of Guadalupe* is that it doesn't sound or feel like an "argument" at all, surely not the forced air rhetoric we inevitably associate with C-SPAN congressional coverage, nor the pontification of the self-satisfied professor who no longer notices how few students remain awake. Rather than relentlessly pushing any single theory, Dr. Smith wears different hats—historian, philosopher, detective, mythologist, scholar of religious concerns—at different times, as he traverses the complex spiritual landscape known to millions of pilgrims as, simply, "Our Lady." The reader becomes an active participant as the author grapples with the full story of a life-sized image of the Virgin Mary that appeared on a cactus cloth "tilma" or cape of Juan Diego, an Aztec peasant, in 1531.

By differentiating the picture from the question of origin, and these from the issue of historical context, Dr. Smith brings a rare sophistication to his empirical approach—rare because all too often studies of supposed anomalous events have a way of degenerating into hostile contests between two polarized camps of true believers who show little appreciation for the value of such distinctions. By contrast, Dr. Smith intentionally seeks to balance the competing claims of different kinds of evidence, availing himself of authorities from various relevant fields—entomology, biophysics, opthamology, computer engineering—all contributing to the book's comprehensive vantage point.

Thus the author earns his authority to speculate that a new science—which he terms *archaeoastronomy*—might shed crucial light on the possibly paranormal origins of the mysterious Guadalupe Madonna.

But let us not go too far too fast. We're wise to follow Dr. Smith's example and start simply—with the extraordinary Woman herself. How marvelous to linger with the alluring portrait presented in the author's opening remarks:

> Her head is tilted to the right. Her greenish eyes are cast downward in an expression of gentle concern. The mantle that covers her head and shoulders is of a deep turquoise, studded with gold stars and bordered in gold. Her hair is black, her complexion olive. She stands alone, her hands clasped in prayer, an angel at her feet.

*Keith Thompson*

# PREFACE

The original publication of this book occurs nearer four than three years after its beginning, due to the richness of the materials available. In early stages, friends such as Wayne Walden, Mae Chenier, Barbara Bullock, and, in Canada, author Merle Shain were of great help as well as, later on, Mrs. Pearl Zaki, Ramiz and Lulu Gilada, and Mr. Harry John. The Museum of the American Indian in New York City allowed me to study what appears to be the oldest surviving pre-Columbian book, the Codex Saville. When photographs were not of good enough quality for reproduction, the museum staff generously and without fee furnished me an excellent print of this most interesting artifact.

Naturally a great many people in Mexico have helped me in varied but important ways. Most of all, I particularly want to thank the always helpful Cardinal Corripio Ahumada, the archbishop of Mexico City, as well as Abbot Schulenburg of the Basilica of Our Lady of Guadalupe and his staff, especially Mrs. Lopez for translation assistance, and Moñsenores Cervantes, Salazar, and Montemayor of the Center for Guadalupan Studies in Mexico City, each of whom has played a crucial role in a still-unfolding adventure. Others whose help was much appreciated are Mrs. Madrazo and her sister, and Mr. Ochoa her cousin. Next, I wish to thank three gentlemen who have shared their basic research with me: ophthalmologist Dr. Torija, criminologist Dr. Palacios, and systems engineer Dr. Aste. All are excellent people, but I must reiterate my affection for the last two, since they additionally made available their automobiles and their homes.

Filial affection is the correct description of my feelings for my dear friend Dr. Philip Serna Callahan, professor of entomology at the University of Florida and a biophysicist with the United States Department of Agriculture. Like my initial contact with my editor, my relationship with Dr. Callahan was a rapport built on mutual faith. My appreciation is also extended to several other friends: my college colleagues Dr. Consuelo Teichert, who not only translated old sources but also generously contributed in terms of letters and phone calls, and Dr. James A. Guest, multiple linguist, who translated the *Primitive Relation*, and my close friends Professor John Hofen and Mrs. Philomena Marshall, both for their help with German sources. Mrs. Moger, who typed the Nican, I also thank.

The Guadalupe vista is panoramic. I have tried to indicate in this work something of the surface and of the perimeters, but I have purposely avoided, in keeping with the nature of an introductory work, exploring the depths. I believe the burgeoning field of archaeoastronomy is highly relevant to the subject of Guadalupan origins, especially the immediately Mexican origins. In chapter 2, I have given some references that were found in four Aztec *anales* regarding the Image of Guadalupe. I have by choice excluded an extremely interesting reference date, 1556, and coordinated the Tepeyac apparition with that of another, the 1531 appearance of the stupendous Halley's Comet. The difference in the dating can be explained by the difficulty in reconciling the Aztec and European calendars and by the fact that in 1531 the Julian calendar, not our modern Gregorian, was in effect. To the best of my knowledge, these points have not been made before.

*Jody Brant Smith*
October 1982

# Preface

# to the Revised Edition

In the original preface I suggested that a new science—archaeoastronomy—is relevant to the matter of Guadalupan origins. I want now to clarify briefly that supposition, while also adding a few words of clarification regarding the difficult matter of dating the apparition. In chapter 2 I list several anonymous sixteenth-century *anales* (yearbooks), each of which makes note of the Mexican apparition of Our Lady of Guadalupe.

Here is another particularly fascinating account: *1556,12 tecpatl: Descendió la Señora a Tepeyacac; en el mismo tiempo, humeó la estrella.* [*1556,12 tecpatl*: "Our Lady descended to Tepeyacac; at the same time, there came a smoking star."][1] This source confirms the coincidental apparitions of Our Lady of Guadalupe and the periodic Halley's comet, in the year 1531 (see original preface, above). "Smoking star" is a rendering of the Spanish translation of the Nahuatl word *citlalimpopocal*, which means "comet."[2]

Several additional matters are noteworthy. A heretofore ignored section of the Codex Telleriano-Remensis (see photograph 24) shows a dramatically accurate illustration of a solar eclipse, with a comet in the background. The arabic numerals are clearly visible as 1531.[3] This provides independent confirmation for the date of Halley's comet. Taken with the yearbook entry citing both comet and Guadalupan apparition in the same period, we may infer that both events occurred in the same year and that both were outstanding in the mind of the anonymous Aztec

---

[1] G. Velazquez, cited in J. B. Ugarte, *Cuestiónes Históricas Guadalupanas* (Mexico City, 1946) 27.

[2] Anthony F. Aveni, *Skywatchers of Ancient Mexico* (Austin: University of Texas Press, 1980) 27-28.

[3] Ibid., 9, figure b.

chronicler. The reader is asked to turn to chapter 2 for an explanation of the anomalous dating of the years. The Indians did not immediately comprehend the European system of notation. This is immediately obvious in photograph 24, where we see a correction from "1548" to "1531." One encounters similar "drawn through" corrections on many Indian documents of the same early colonial period.

Other additions include sections on the radically alternative *Apologia del Dr. Mier*, which proposes an entirely different origin for the Guadalupe Image, and a critique by seventeenth-century Mexican scholar Bercero Tanco, together with related materials. These sections are added to the appendixes.

Within the book proper are expanded sections on the controversial "faces," and so forth, in the eye(s) of the Madonna and more "impossible coincidences" in chapter 3. The author considers the phenomenon of coincidence or synchronicity, to constitute the irreducible miracle of the Guadalupan experience.

Finally, a point of clarification. The original preface (above) incorrectly implies that the Julian calendar in effect in 1531 accounts for the dating inconsistencies. The significance of the Julian entry is that (a) as with the Halley's comet entry, the correspondence seems never to have been recognized before, and (b) the traditional date of the Guadalupan painting, December 12, cannot be correct. December 12, 1531 (Julian), corresponds by intercalation to December 22, 1531 (Gregorian), because, even by the sixteenth century, ten days of error had crept into the Julian system. One of several implications of this is again associated with archaeoastronomy: December 22 is at the time of the winter solstice. And of course we may observe that in the painting on the tilma, Our Lady occludes the sun.

*Jody Brant Smith*
July 1984

# PREFACE
# TO THE 1994 Edition

For the 1994 revised edition of *The Image of Guadalupe*, eleven years after the original publication date, I want to use the preface both to elaborate briefly and to emphasize some points made earlier.

Any introduction to the mysterious Guadalupe Madonna—*La Imagen*—requires note of three interrelated but distinct categories: the picture itself is one; another is the context in which tradition places the origin of the picture, that is, the apparition; and a third is the historical context as it may have relevance on either or both of the first two topics.

Without doubt, the picture of Mary itself is of immediate interest: it appears to have evaded deterioration, although placed on a poor-quality, unsized piece of cloth for a time now approaching five hundred years. While there are several possible explanations for this longevity, that of supernatural origin, in accordance with a not unimpressive gathering of early sources, cannot itself be excluded. The scientific testing discussed in the latter sections of the book, while not exhaustive, have certainly produced nothing inconsistent with the explanation offered by both early Spanish and early Indian depositions.

Employing without prejudice to either extreme a theory of knowledge that can accommodate the supernatural (or at least, the paranormal), another major section of the book deals with the historical "hard data," the primary sources referred to above. This involves examining what sixteenth-century contemporary witnesses had to say about the putatively miraculous creation of the picture. That something remarkable did happen in 1531, something that indelibly impressed the indigenous Mexican culture, cannot now be reasonably discounted. Even today, Aztec documents laconically chronicling the Marian apparition are being freshly discovered. The putative connection with *La Imagen* may remain ambiguous to

some, but the event is, as converted skeptic Bartolache affirmed in the
1790 title *Satisfactory Manifestation*, genuine.

The most unexpected and obvious impressive correlation deserving
further discussion is the visit of Halley's Comet in 1531—some two
orbits before its recorded discovery—which is chronicled in the Nahuatl
Aztec language as "the smoking star," along with the apparition of "Our
beloved Lady of Guadalupe."

The third and final category will seem to many—including myself—to be exciting all by itself: Marian portraiture is the most nearly
universal subject in all Western painting. Virtually every artist of the last
two thousand years has produced at least one Madonna. Now the central
question becomes: Was one ever produced *from life*? If so, does that
protocopy still survive? Or alternatively, was one done only to be lost,
while the accurate memory still carefully passed from generation to
generation of artists? And might the Guadalupe Madonna—the name is
Spanish/Arabic, not Nahuatl—be of some significance here?

One further point in this presixteenth-century context: What about the
ancient legend of *acheiropoietos* images of both Mary and of Christ? The
Greek word *acheiropoietos* (ἀχειροποίητος) literally means "not-made-
with-human-hands" and appears at least as early as the first century, for
example, in the New Testament, both in the gospel (Mark 14:58) and in
the epistles (2 Cor 5:1; Col 2:11). There is a fascinating serial, discussed
in the pages that follow, beginning with a mid-fifth-century discovery of
a "miraculously wrought portrait of the Virgin Mary." What is the
relation of the Image of Guadalupe to that ancient image, in one way if
not another?

The Image of Guadalupe was said by Pius XII to have been "painted
with brushes not of this earth." Pope John Paul II made his very first
official visit outside the Vatican (where there is a beautiful Guadalupean
monument) to none other than the Basilica of Guadalupe in Mexico City
in early 1979. Although, as with all albeit recognized apparitional events,
the church requires no member's assent or belief, this beautiful picture
is said to draw crowds that number yearly second only to the Vatican
itself.

Hagiography may provide suitable antecedents for an unbeliever's
"rationalization." Attention to the animistic and/or pantheistic worldview
of the Indians or archeoastronomical correlations with comet or eclipse
may encourage some to call for "Occam's razor." Nevertheless, it is not

imprudent to recall how repeatedly "Occam" has been misused, as for example the clearly more economical assessment of naive empiricism that the sun and the moon revolve around a stationary earth because they so obviously *seem* to do so.

To me at least, while urging against a priori rejection of the possibility of the miracle of Guadalupe itself, multiple ramifications persist that may subsist in the phenomenon. The Jungian or Koestler-like synchronicities will always be matters of individual astonishment for the author (see chapter 3 for a sample). The 1981 discovery, which I am just now publishing, of the frankly puzzling coincidence between the number of stars in the Guadalupe Madonna's blue mantle—forty-six—and the number of chromosomes in the human being is another.

So it would be a mistake to presume—should anyone have "an ax to grind"—that this book is parochial; the wide spectrum of interested parties includes not only "impossible coincidences" but also other areas beyond the scope of this short introduction. To name but a few, these might include "Ley Lines" and/or geomagnetic nodal points and related phenomena on the one hand and the interest and affirmation of scholars of other, non-Western traditions on the other. (For example, Evans-Wenz, author of the famous translation of *The Tibetan Book of the Dead*, is said to be positively impressed by the Guadalupe documents.)

In conclusion, I want to thank Michael Pallazolo, Laura Patterson, and Katherine Kavich of Cosgrove-Menerer Productions for their fine Guadalupe presentation for the television series "Unsolved Mysteries." A special thanks goes to Ms. Germana Wakefield and to Dr. George A. Jones, both of whose help and encouragements were inadvertently omitted recognition in the earlier preface.

*Jody Brant Smith*
August 1994
Pensacola

# Our Lady
# of Guadalupe

All over the world myths have risen to conscious popularity because we can no longer understand the dreamscape of our everyday waking life. The myth is something that never was but is always happening. It serves as a kind of DNA of the human psyche, carrying within it the coded genetics for any number of possible evolutionary and cultural paths we might yet follow.
　　—Jean Houston, *Travelogues at the Edge of the West*

Her head is tilted to the right. Her greenish eyes are cast downward in an expression of gentle concern. The mantle that covers her head and shoulders is of a deep turquoise, studded with gold stars and bordered in gold. Her hair is black, her complexion olive. She stands alone, her hands clasped in prayer, an angel at her feet.

She is Our Lady of Guadalupe, a life-sized image of the Virgin Mary that appeared miraculously on the cactus cloth *tilma*, or cape, of Juan Diego, an Aztec peasant, in 1531, a mere dozen years after Hernán Cortés conquered Mexico for the king of Spain. For four hundred and fifty years the colors of the portrait have remained as bright as if they were painted yesterday. The coarse-woven cactus cloth, which seldom lasts even twenty years, shows no signs of decay.

Today Juan Diego's tilma is preserved behind bulletproof glass in a magnificent new basilica in Mexico City, built especially to house the picture. It is placed where pilgrims can view it from as near as twenty-five feet. Yearly, an estimated ten million bow down before the mysterious Virgin, making the Mexico City church the most popular shrine in the Roman Catholic world next to the Vatican.

It is not only Catholics who regard Our Lady of Guadalupe with awe and wonder. The highly respected philosopher F. S. C. Northrop, in *The*

*Meeting of East and West,* wrote that the Image "conveys in some direct and effective way a basic and intuitively felt element in the nature of things and in the heart of human experience." After describing the unending stream of worshipers dropping to the floor at their first sight of the Image, then moving forward on their knees, he said: "Nothing to be seen in Canada or Europe equals it in the volume or the vitality of its moving quality or in the depth of its spirit of religious devotion."

Pope John Paul II, in his address in the Basilica of Our Lady of Guadalupe on January 27, 1979, acknowledged the unwavering appeal of this unique portrait. Addressing the Virgin directly, he said:

> When the first missionaries who reached America . . . taught the rudiments of Christian faith, they also taught love for you, the Mother of Jesus and of all people. And ever since the time that the Indian Juan Diego spoke of the sweet Lady of Tepeyac, you, Mother of Guadalupe, have entered decisively into the Christian life of the people of Mexico.

In the 1950s an archaeologist, Father Ostrapovim of the Russian Orthodox Church, was shown a copy of the painting but was not told of the Image's history or origin. He deduced that the painting was "presumably Eastern Asiatic, definitely of the Byzantine type." The early Byzantine portraits of the Virgin are believed to approximate her actual appearance most closely. How had such an authentic portrait happened to turn up in Mexico, half a world away?

According to sixteenth-century documents in Nahuatl, the native Aztec language, Juan Diego, his wife, and his uncle had been among the first Aztecs to be converted by the Christian missionaries who had accompanied the Spanish soldiers to Mexico. Juan was fifty years old at the time of his conversion, an advanced age in an era when few lived past forty. A religion that promised redemption and eternal life was far different from the harsh beliefs of the Aztecs, whose gods demanded human sacrifice. When Juan was thirteen he may have witnessed the bloody ceremony dedicating a new temple in nearby Tenochtitlan (Mexico City) in which some eighty thousand captives were put to death.

Juan's wife died two years after her conversion. They had no children and Juan was left alone to take care of his aged uncle, who had been like a father to him. They lived in a hut with a thatched roof and dirt floor.

Members of the Aztec servant class, they were among the poorest inhabitants of their small village five miles from Mexico City.

On Saturday, December 9, 1531, Juan Diego left his village before daybreak so he would be in time to hear Mass celebrated at the church of Santiago in the nearby village of Tlatilolco. On his way, as he passed around the base of a hill called Tepeyac, near which there had once been a shrine to the Aztec mother goddess, he heard a burst of birdsong. At this bleak time of the year, few birds remained, and Juan looked up to see where the song was coming from. On the summit of the hill he saw a bright light.

Suddenly the melodious birdsong ceased as abruptly as it had begun. From the barren rocks at the top of the hill a voice called him: "Juan! Juanito!"

He climbed the hill quickly and saw on the summit a young woman who seemed to be no more than fourteen years old, standing in a golden mist. She beckoned to him and he knelt before her radiant presence. "Juanito," she said, "the most humble of my sons, where are you going?"

"My Lady," he replied, "I am on my way to church to hear Mass."

"Know then," she continued, "that I am the ever-virgin Holy Mary, Mother of the True God. I wish that a temple be erected here without delay. Go to the bishop's palace in Mexico City and tell him what I desire."

Assuring her that he would carry out her mission, Juan Diego descended the hill and continued along the road leading to Mexico City.

At the bishop's palace the servants kept him waiting for hours, but at last he was ushered in to the bishop's presence.

Juan de Zumárraga, a Franciscan who was to be formally elected bishop two years later, had come to Mexico in 1528 at the command of the king of Spain, Charles V. He was a powerful but kindly man, who used his influence to lessen the cruelty with which the Spanish soldiers treated the Indians.

He listened sympathetically as his interpreter translated Juan Diego's words from Nahuatl into Spanish, and in reply told Juan Diego to visit him again at some unspecified future date. It was clear to the Indian that he had not been believed.

On his way home, with the sun setting in the west, he climbed once more to the top of Tepeyac and again saw the Virgin Mary. He told her that he had delivered her message but that the bishop did not seem to believe him. He begged her to entrust her mission to someone more important who would be more likely to be believed. But the Virgin insisted that the humble Indian was her chosen messenger. "I command that you go again tomorrow," she said, "and see the bishop. Go in my name, and make known my wish that he has to build a temple here. And again tell him that the ever-virgin Holy Mary, Mother of God, sent you."

Juan Diego promised to follow her instructions and return the following afternoon, at sunset, to give her the bishop's reply.

The next day, a Sunday, Juan again left his home before dawn to go to the village church. It was nearly ten o'clock before the morning services were completed and Juan was able to leave for the bishop's palace. On arriving there, the humble Indian found it even more difficult to be admitted to the bishop's presence. Bishop Zumárraga had not expected the Aztec to return so soon.

This time the bishop questioned Juan Diego more closely. The vivid detail with which the Aztec described his two meetings with the supposedly Heavenly Lady led him to believe that Juan was telling the truth. But perhaps the Indian was deluded. Zumárraga would have been happy to build another church to the honor and glory of the Blessed Virgin, yet if Juan Diego's story was a hoax, the church could lose a great deal of the ground it had gained with the Indians.

What was needed was proof. He urged Juan to bring a sign from the Heavenly Lady that she had indeed spoken to him, and the bishop would then eagerly comply with her request. Juan agreed without hesitation. As soon as the Indian left, Zumárraga ordered two servants to follow him and report back, but as they neared Tepeyac hill, they lost sight of him. They thought that he had deliberately eluded them, and they reported back to the bishop that he was not to be trusted.

Juan Diego had no idea he was being followed. On the top of the hill he gave the Blessed Virgin the bishop's message, and she told him to return the next day when she would give him the sign the bishop had requested.

That was not to be, because when Juan reached home he found that his uncle was desperately ill. The next day a Monday, his condition had worsened. The uncle asked Juan to go at daybreak to the village church and bring back a priest to hear his confession, for he was certain he was dying.

Before dawn on Tuesday, Juan Diego went to summon a priest. As he neared Tepeyac hill, he decided to skirt the hill to avoid being detained by any further meeting with the Virgin Mary. But as he rounded the hill, he saw her descending to the plain. He was frightened that he had disappointed her and worried about his uncle, but she reassured him that his uncle would recover and that she was still anxious to provide the bishop with the sign he had requested. She instructed Juan to climb the hill to the same place where he had first seen her and spoken with her, and there to pick some roses and bring them back to her.

Juan climbed the hill with misgivings. It was early December and the barren hilltop was touched with frost. If there had ever been any roses blooming there, they would not be there now.

But when he reached the hilltop he found several varieties of Roses of Castile in full bloom, the petals touched with morning dew. He gathered the roses and put them in his tilma, his loose cape made of two pieces of cactus cloth stitched together. He brought the blossoms to the Virgin and she arranged them in the tilma, saying:

> This is the proof and the sign you will take to the bishop. You will tell him in my name that he will see in them my wish and that he will have to comply with it. Rigorously I command you that only before the presence of the bishop will you unfold your mantle and disclose what you are carrying. You will tell him that I ordered you to climb the hilltop, to go and cut the flowers; and all that you saw and admired, so you can induce the prelate to give his Support that a temple be built and erected as I have asked.

When he reached the bishop's palace he again had difficulty gaining admittance, until the bishop's servants saw the out-of-season roses peeking from the folds of the tilma. They informed the bishop of the gift the Indian was bringing him, and Zumárraga, guessing that what Juan Diego carried was the proof he had requested, ordered that he be admitted immediately.

When Juan Diego entered, he knelt before the bishop and described his last encounter with "the Lady from Heaven." He then stood and untied his tilma from around his neck so that the roses fell on the floor in a heap. Suddenly, on the cactus cloth of the tilma, there appeared a brightly colored image on the Virgin Mary. The bishop fell to his knees, as did all those present.

The next day Juan Diego took the bishop to the hill where the apparition had appeared and where the Holy Mother had asked that a church be built in her name. The church at Tepeyac was designated the Shrine of Our Lady of Guadalupe, in honor of a village in Spain where a small statue of the Virgin had been discovered two hundred years earlier.

News of the miraculous appearance of the Virgin's image on a peasant's cloak spread quickly throughout New Spain. Indians by the thousands, learning that the mother of the Christian God had appeared before one of their own and spoken to him in his native tongue, came from hundreds of miles away to see the image hung above the altar of the new church.

The miraculous picture played a major role in advancing the Church's mission in Mexico. In just seven years, from 1532 to 1538, eight million Indians were converted to Christianity. In one day alone, one thousand couples were married in the sacrament of matrimony. Throughout the four hundred and fifty years since its first appearance, adoration of the Image of Guadalupe has remained the most striking aspect of Roman Catholic worship in Mexico. Northrop writes of seeing the Image "on the windshield of taxicab after taxicab in Mexico City and in the front interior of almost every bus."

Less pervasive but perhaps more remarkable is the presence of the Image of Guadalupe in churches around the world. Copies of the original or paintings depicting the miracle on the hilltop hang on the walls of churches in Madrid, Rome, Jerusalem, Paris, and even in Taiwan. In the western hemisphere Our Lady of Guadalupe has been crowned in ceremonies in New York City, Newark, New Jersey, and in Cuba, Nicaragua, Uruguay, and Argentina. In every country in Latin America copies of the miraculous picture hang in the churches.

For those millions who come to kneel and pray before the original Image, protected behind bulletproof glass in the new cathedral in Mexico City, Our Lady of Guadalupe with her expression of motherly concern, her olive skin, and dark hair, arouses unique feelings of trust and closeness. Many ask for her intercession in heaven to solve their problems on earth, and many return to give thanks when their petitions appear to have been successful.

Tales of the miracles she has wrought abound. In the early seventeenth century, when floods almost destroyed Mexico City, the Image escaped unharmed. In 1921, during the Mexican revolution, a bomb was planted in some flowers placed before the altar of the basilica. The Image hung close behind the altar. Although the bomb exploded, damaging the altar, no one was hurt. Even the glass in front of the picture was unbroken.

Forty years after the Virgin of Guadalupe appeared before Juan Diego she is believed by many to have been responsible for a turning point in Western history. Copies of the Image circulated throughout Europe. Possibly the first and most important of those copies was given to Giovanni Andrea Doria (grandnephew of the more famous Admiral Andrea Doria) by the king of Spain. The younger Doria carried the framed picture with him when he took command of a squadron of ships sailing from Genoa to the Gulf of Lepanto (now Corinth) where some three hundred Turkish ships were drawn up, blocking entrance to the gulf. The Christian force, which also numbered some three hundred ships, came around the headland north of the gulf and attempted to meet the Turks head-on, but they were outmaneuvered by Turkish forces on the right. Doria's squadron was completely cut off from the rest of the Christian flotilla. At this crucial hour Doria is said to have gone to his cabin, knelt, and prayed to the Image of Guadalupe to save him from certain defeat.

At nightfall, the tide of the sea battle miraculously began to turn. When one Turkish squadron was captured, the others panicked. Most of the Turkish fleet was destroyed, and fifteen thousand Christians, enslaved in the Turks' galleys, were freed. The Battle of Lepanto, the last great naval battle fought under oars, marked the end of the Ottoman Empire's expansion into the western Mediterranean.

The victory at Lepanto and the thousands of smaller, less historic miracles imputed to the Virgin of Guadalupe can be explained in large part as the result of natural causes. The one miracle that allows of no such natural explanation is that of the Image itself. Why does it show no sign of cracks or other deterioration after so many years? Why do the colors remain so bright? Why didn't the crude fabric on which the Image is imprinted disintegrate long centuries ago?

The search for answers to these and related questions is the subject of this book.

# Not Made
# with Human Hands

A stone was cut out by no human hands and struck the statue, struck its
feet of iron and earthenware and shattered them.   —Daniel 2:34

Although I am not a Roman Catholic, on the morning of December 13,
1978, I stood with the throngs outside the Basilica of Guadalupe in
Mexico City, waiting to view the Virgin of Guadalupe at firsthand. What
brought me there had nothing to do with the casual curiosity of the
people disembarking the tourist buses, nor the already confessed faith of
the conventional believer offering overt worship. It could indeed be
argued that my propellant for that initial visit was and would remain
unmitigated the center of my particular faith as regards the Guadalupe
mystery.

I knew of the story of Mary's appearance to the Aztec, but not on a
conscious level, for my first and only encounter with the claim was a
short account as a part of an anthology read more than ten years earlier.
The reason, much more recent, for my being on my first visit to Mexico
was the much-publicized public exposition of the artifact often called the
Holy Shroud that same autumn, in Turin, Italy.

I had been seized by a desire to see the rarely unveiled Shroud of
Turnin when I earlier that year had read an article about it in *Guideposts*
magazine. I had before only the vaguest awareness of it, even less than
the Image of Guadalupe, because I had never read anything about it
before.

It was only after I had with considerable effort made plans to go to
Turin, securing time away from my professional responsibilities, that
quite suddenly, with no conscious predeliberation, there formed a single
question in my mind: Is there anywhere anything else in any way said to
be like the Shroud of Turin? Once this singular query had been framed,

I found myself making an immediate reply, and in the affirmative: "Yes!" I exclaimed to myself. "The picture of the Virgin Mary called Guadalupe." And Mexico was a much more feasible destination than Italy.

I set my sights on Mexico. I decided (giving myself squirming room by way of procrastination) to wait until the end of the semester to make the trip. In part, I thought, this strange compulsion would abate in the intervening weeks, and I could return from my acharacteristic mysticism to a stolid if somewhat smirky type of skeptical worldview.

Developments toward going increased rather than decreased. These were not only in terms of quantity but *quality* as well. There were no more striking questions of a nature as to almost seem auditions. But this gambit was replaced by something far more vivid and unforgettable: chance alignments or parallels, without exception of a nature metaphysical, religious, and mystical came into my experience. I relate some of these elsewhere in this book. But here I seek to state and explain a bit of how and why a conventionally religious, protestant intellectual, trained in the predominantly agnostic (at best) philosophy of the twentieth century West, could find himself reflecting on such matters while waiting for the doors to open on a miracle.

Today copies of the famous Virgin of Guadalupe painting are much more common in North America than fifteen or twenty years ago. This is because of the work of many fine persons who are unlikely to have received the empirical credit they deserve. When, having thought affirmatively about the Guadalupe image as tentatively analogous to the Turin Shroud, I searched for a photograph, one was not immediately forthcoming. Now, while waiting a first chance to inspect such a copy, I was busy learning all I could about putative apparitions of the Madonna, since I remembered that the "painting" on the peasant's *tilma* was popularly credited to just that supernatural category.

While a high school student living in Miami, I experienced a deep sense of the mystical. When, two years later, I began college at Mercer University, even a seriously religious commitment on the part of the institution could not deflect a feeling that, layer by layer, the intellectual protocol of systematizing higher education was clouding over my sense of the numinous. I already held a considerable interest in such phenomena as telepathy and such, and upon being introduced to a vastly greater context with the surroudning context of college life, I deepened my studies in parapsychology generally. This was almost exclusively extracurricular research, however, for the 1960s were notoriously less tolerant of unconventional explorations (if they were discussed in class) than is popularly remembered in the 1990s. There remains in my files an unpublished thesis on P. P. Quimby completed in the mid-1960s but deemed outside the scope of legitimate scholarship: it was rejected, as was my first master's degree candidacy, in 1967. Nevertheless, this was far from an all-bad period. I met personally such famous researchers as J. B. Rhine, J. G. Pratt, and Henry Puharich, who introduced me to Ambrose and Olga Worrall. Correspondences were established. I learned much from such pioneers as Bernard Grad and Sister Justa Smith, and discussed psychology and philosophy (via-à-vis the P. P. Quimby thesis as entre) with Gardner Murphy, C. D. Broad, and H. H. Price. It was also during these closing years of the 1960s that I read a little book called *A Lady Clothed with the Sun*, my first and (prior to the fall of 1978 and the Turin Exposition) only notice of the Guadalupe story.

I clearly recall my disappointment when after much fruitless preliminary search I first cast eyes on a Guadalupe Image photograph. I was disappointed. This seemed much too detailed to qualify within the scope of the supernatural: unlike the ephemeral Holy Shroud, one could easily discern all kinds of surface features in this painting. That, at least, was my immediate reaction. While this search was proceeding, however, and while I was concurrently reading about Marian apparitions, the major "coincidence" was brewing outside the boundaries of conscious awareness. I had been introduced at this time to the, apparently stupendous *Zeitoun* event, which is said to have transpired between 1968 and 1970 near Cairo, Egypt. This phenomenon is at once so potentially awesome but also so complicated to unravel as to be quite outside the scope of the present book, let alone the present chapter. But I introduce it here anyway, for it supplies the needed context for the coincidence: while on a virtually day-to-day basis corresponding with persons cognizant of the *Zeitoun* story, I one day obtained for my teaching a thirty-five-millimeter, carousel-type slide projector. This particular item was one of a great number of identical pieces of equipment, selected by a process we are pleased to refer to as "at random." While looking at the wheel-tray, I noticed one slide had been left in it, unclaimed. The subject of the slide was van Gogh's *Starry Night*. For me—if perhaps only for me—this was totally unexplainable in mundane terms. Not only was the projector merely one of many used by the faculty, but furthermore, I felt, this was unconventional by virtue of the location of the art department—across the street, across the campus—from my building and (philosophy) department. To complete this brief odyssey, I will simply say something explicitly that you the reader may have inferred: the Zeitoun, Egypt, event or apparition or whatever it may have been was signally characterized by a sky of whirling stars.

So, although I was deeply moved, I was not particularly surprised when on visiting the Zeitoun church over which the Virgin is believed to have appeared, I saw among a group of attendant icons a large color reproduction of Our Lady of Guadalupe-Tepeyac. No one there was aware either of the icon's name or its rich history in New Spain.

The church at Zeitoun was Coptic Christian, and its own particular name was Saint Mary's. Coptic tradition, which traces its origins back to the evangelizing of Mark, claims that this very church is built on the crossroad traversed by the Holy Family as they came to their sojourn in Egypt after the Nativity.

There is another long tradition in the church of belief in creation without human intervention, beginning with the virgin birth of Jesus. Indeed, as was pointed out by Irenaeus, an early Christian writer of the second century A.D., the Virgin Mary can be considered the mother of all things created supernaturally.

The first work of art that was said to be "not made with hands" (the Greek word is *acheiropoietos*[1]) was a portrait of Christ, discovered around the year A.D. 550 in Edessa (now Urfa, Turkey). It came to be known as the "Image of Edessa" or the Mandylion. In the tenth century it was taken to Constantinople. Three hundred years later, during the Fourth Crusade, it disappeared.

---

[1]The Greek ἀχειροποίητος appears in the N.T. at Mark 14:58; 2 Cor 5:1; and Col 2:11.

A clear distinction needs to be made between an image said to be supernatural by virtue of its *acheiropoietos* nature on the one hand with, by comparison, one that is accorded supernatural status by virtue of some other quality or ostensible power of action, e.g., miraculous by virtue of its power to heal. This distinction is made when one recalls the adjective Anna Brownell Jameson used in referring to a certain type of Marian iconography as *miraculissima*. This would be a more inclusive concept that could include other species of demonstration, additional to the *acheiropoietos*.

Artists' copies of the Mandylion made before its disappearance show a blurred image of Christ's head ranging from sepia to rust-brown. The color, wrote Ian Wilson in *The Shroud of Turin*, is "virtually identical to the coloring of the image on the Shroud." Some people believe the Mandylion and the Shroud of Turin are one and the same, a view that Wilson strongly supports by virtue of his own theory that the Shroud was *folded*, then tucked away in a frame with only the upper part visible.

The two images of Christ belong unquestionably with that group of objects believed to have been created supernaturally. They bring to mind the fabled portrait of Christ named "Veronica's Veil." According to medieval legend, a woman named Veronica gave Christ a cloth to wipe the sweat from his face while he was on the road to Calvary. The imprint of his face was left on the cloth.

Buddhist tradition describes a similar imprint made by the Buddha on a cloth. Such a tradition, of ascribing a supernatural source for objects of veneration, including portraits of religious figures, is an ancient one.

Whatever the power of such objects to answer humanity's spiritual needs, in recent years there has been a widespread effort to establish their authenticity by the use of new scientific techniques of dating and chemical analysis. At the time I first viewed the Image of Guadalupe, the Shroud of Turin was front-page news throughout the Western world as a result of research by teams of European and American scientists. One of the scientists appeared to have established beyond question that the Shroud originated in the eastern Mediterranean, in Turkey or Palestine, in the first century A.D., although the Shroud's documented history went no further back than the fourteenth century in France and Italy.

I could not help but wonder what sort of light scientific research would shed on the Image of Guadalupe. The Guadalupan Image and the Turin Shroud are, of course, very different. The image said to be Christ on the Shroud is faint and shadowy and in one color only, while the Image of Guadalupe is vivid and lifelike, with several bright colors. Clearly a different technique produced each image, but the scientific methods being used to authenticate the Shroud might apply equally well to Guadalupe.

Efforts to confirm or repudiate the supernatural origin of the Image of Guadalupe are not new. As long ago as 1556, twenty-five years after Juan Diego untied his tilma in Bishop Zumárraga's palace, a formal investigation was launched by the church authorities. The investigation appears to have been an outgrowth of rivalries between the Dominican and Franciscan brotherhoods and their divergent understanding of what constituted a proper form of worship for the millions of new Indian converts.

In a sworn statement representing the views of Francisco de Busta-mante, a provincial priest, a representative of the Franciscan viewpoint named Juan de Massaquer reported that

> the aforementioned Fray Francisco de Bustamante said . . . that seeing the multitude of people that go there because of the fame of that image, painted yesteryear by an Indian, performing miracles was again undoing what previously had been done.[2]

The shrine that held the Image of Guadalupe had been erected on a hill directly in front of the spot where there had been an important temple dedicated to the Aztec virgin goddess Tonantzin, "Little Mother" of the Earth and Corn. Tonantzin had been one of the most popular figures in the Aztec pantheon of gods and goddesses. So Bustamante's concern that the Aztecs, by their adoration of the Image of Guadalupe, might be reverting to their pagan beliefs, was soundly based. Another priest, Alonso de Santiago, testified that

> and all the other religious had tried with great insistence that the natives of this land did not have their devotions and prayers represented in painting or sculpture, in order to prevent them from exercising their old rites and ceremonies to their idols, and that this new devotion to Our Lady of Guadalupe seems to be an opportunity to return to the practices which they previously had.[3]

De Maseques went on to supply the name of the reputed artist, saying that the Image "was a painting that the Indian painter Marcos had done."

---

[2]*Informacion de 1556*, in F. de Jesus Chavet, *El ulto Guadalupano del Tepe-yac*, 6th ed. (Mexico City: Centro de Estudios Bernardo de Sahagun, 1978) original folio.

[3]Ibid., 220 (folio 6a). This theme was soon to be reiterated by Sahagun himself in the *Historia General* of 1569.

There was an artist of that name active in Mexico around that time. But why didn't Marcos Cipac, as he was known, simply attend the formal investigation and put the matter to rest? It is unlikely he could have been bribed to stay away, for any deceit involving the Holy Mother of God would have been dangerous for a Spaniard, let alone an Aztec. If Marcos was already dead, a fellow artist could easily have come in his place to present testimony, just as Bustamante's testimony was presented by De Maseques.

In the investigation of 1556, no mention was made of Juan Diego and the legend of the Virgin's three appearances before him. Was the omission deliberate, or in the intervening twenty-five years had the effort to exterminate Indian idol worship also extended to the legend of a miracle involving an Indian peasant?

The other side of the controversy was represented by a member of the Dominican order, Alonso de Montufar, who replaced Zumárraga as archbishop of New Spain. On September 8, 1556, some weeks before the formal investigation and on a date widely regarded as the birthday of the Virgin Mary, Archbishop Montufar preached a sermon in the small church on Tepeyac hill. Standing in front of the Image of Guadalupe, he asked that the Image be accorded special status as an object of worship. He meant by this to exempt the Image from the ban on icons then being enforced throughout New Spain in an effort to stamp out whatever remained of Aztec idol worship.

A world-renowned scholar of the Nahuatl language, Ángel Garibay-Kintana, discovered Montufar's sermon in the mid-1940s and published it in 1955 and 1961 in Spanish and English translations. In his comments on the sermon, Garibay called attention to an unusual alteration in the order of service of the Mass celebrated at Tepeyac:

> In the sermons of that epoch it was customary to take as the text for the day of the Nativity of the Virgin the Gospel . . . which was and still is the genealogy of Christ given in the first chapter of Saint Matthew. The text is "Mary, of whom Jesus was born, who is called the Christ." We have hundreds of sermons in our libraries which take these words as the basis for the explanation of the dignity and offices of Mary in the work of redemption. On this occasion Archbishop Montufar broke that tradi-

tion, quoting Saint Luke in the text "Blessed are the eyes which see what you see."[4]

The substitution of this passage from Luke for the traditional first chapter of Matthew, said Garibay, "gives us a clue by which we can penetrate the mind of the prelate. Our Lord congratulates the Apostles because they are seeing something new . . . new and extraordinary, not to be confounded with any other deeds . . . the presence on earth of the Son of God made flesh." Thus, Garibay concluded,

> If Archbishop Montufar, who was a good theologian and a strict Thomist, took the liberty on this occasion of applying these words to Mary and to her Image, it was because he saw something in it that could not be found in any other images. He believed, twenty-five years after the event, that he had found something truly extraordinary.[5]

In the years following his sermon at Tepeyac, Archbishop Montufar authorized construction of a considerably larger building there, replacing the tiny chapel in which he conducted the Mass in 1556. The new church was completed some time before the end of 1567, according to the best-known historian of that period, Bernardino de Sahagún.

Sahagún, a Franciscan, was not enthusiastic about the Virgin of Guadalupe. He complained, with some reason, that certain priests insulted the Virgin Mary by referring to her by the Nahuatl name of Tonantzin, the name of the Aztec goddess, which means, literally, "Our Mother." Sahagún thought the term for the Virgin Mary in Nahuatl should be "Dios-nantzin," meaning "God's Mother."

In 1570 Montufar sent to King Philip II of Spain an oil-painted copy of the Image of Guadalupe. Archbishop Montufar had commissioned the painting from an artist who remains unidentified. The fact that Montufar sent a copy of the Image to the king of the world's then most powerful

---

[4]Angel Maria Garibay-Kintana, *La maternidad Espiritual de Maria* (Mexico City, 1961). See Helen Behren, *The Virgin and the Serpent* (Mexico City, 1966) 183-84, for English translation.

[5]Garibay-Kintana, *La maternidad*; see also *Informacion de 1556*, folio 15b, pp. 240-42.

nation proves that, despite the controversy surrounding the Image, it was indeed being venerated by many.

This copy of the Image is believed to be the one Giovanni Doria carried aboard ship during the Battle of Lepanto in 1571. The portrait was given by a cardinal of the Doria family to the Church of Santo Stefano in Aveto, Italy, where it remains to this day.

In recent years, scholars have brought to light several sixteenth-century documents that confirm that there was widespread belief in the miraculous appearance of the Virgin on Tepeyac hill. Most of these documents now reside in the Mexican National Library, in the library of the Basilica of Guadalupe, or in museums and libraries in New York and Paris. Among the most impressive of these documents are four Aztec *anales*, or yearbooks, discovered by two twentieth-century Mexican priests, Mariano Cuevas and Bravo Ugarte. Here are some representative selections referring to the Image of Guadalupe:

(A) 5 tecpatl, 1510: in this new year came a president to Mexico; in the same year arrived first a prelate of the status of bishop, his name Fray Francisco de Zumárraga, of the religious order of Saint Francis; and then appeared Our Beloved Mother of Guadalupe.

(B) 1510 tecpatl: then came a new president to Mexico, governor; in the same year, Our Beloved Mother of Guadalupe appeared and manifested Herself to the poor Indian named Juan Diego.

(C) In the year of 1555, Saint Mary made herself to appear on Tepeyac.

(D) 13 acatl, 1531: the Castilians founded Cuitlaxcoapan, Puebla de Los Angeles, and to Juan Diego was manifested the Beloved Lady of Guadalupe in Mexico, that is named Tepeyac.[6]

The inconsistency in dating is explained by the difficulty in reconciling the Aztec and European calendars. Furthermore, the Nahuatl language used picture symbols rather than letters or numbers. According to historian Bravo Ugarte, the authors of these sixteenth-century annals no more knew "how to write arabic numerals, much less how to form a

---

[6]Bravo Ugarte, *Cuestiones Guadalupanes* (Mexico City, 1946) part 1.

running enumeration" than how to write a letter, word, or sentence in Latin-lettered Spanish.

If we check the annals against actual events, we find that "1510" is probably 1528, for that is the year Zumárraga is known to have arrived in Mexico. The second *anale* may refer to 1530, the year the Spanish king, in response to a plea from Zumárraga to put an end to the cruel treatment of the Indians, sent a second administrator to New Spain. This would correspond with the mention of a "new president . . . governor." But none of these annals were written at the time of the Virgin's apparition. They were recorded a decade or more after the event, so the actual dates could well have become blurred by the passage of time.

Further corroboration exists, however. One is a description of the Guadalupan shrine written by Suárez de Peralta, a Spaniard who retired to Spain in 1570. The shrine, he reported, "contained . . . a very holy Image which . . . has performed many miracles. She appeared among the rocks, and the whole country is devoted to her."

Another historical confirmation of the Image's supernatural origin was made by Francisco Verdugo Quetzalmalitzín. In 1558, only two years after Montufar's sermon in the Tepeyac chapel, Quetzalmalitzín made a collection of documents in Nahuatl that directly referred to the Guadalupan apparition and Image. That collection of documents has now disappeared, but reference to it is made in the famous Boturini Collection.

The most complete and most famous source for the legend of Juan Diego is a document written in the sixteenth century by an Indian named Antonio Valeriano. For many years the document was considered a seventeenth-century fraud, because of doubts concerning the actual existence of the reputed author. Until independent proof of Valeriano's existence could be located, the question remained insoluble. But a few years ago, from an archive in the library of the Mexican National University, there came to light a sixteenth-century civil paper with the clear signature, *Antonio Valeriano*, at its bottom.

Valeriano's story of Juan Diego is called the *nican mopohua* (literally, "an account") and is given in full in an appendix below. A dating device in common use among historians places the document as being written sometime between 1540 and 1580. That dating device is handwriting style, which changes from period to period in all cultures, often in only a few decades. By comparing the penmanship of documents we know were written in the sixteenth century with the penmanship of the *nican mopohua* we can confirm that it, too, is a product of the sixteenth century.[7]

Another sixteenth-century document, known merely as the "Primitive Relation," was unearthed from the Mexican National Library Archives. Nahuatl scholar Garibay appears to have found it first. With the incorporation of the Center for Guadalupan Studies in Mexico City in 1976, this important work received further scholarly attention by Father Mario Rojas, leading to wider recognition of its value in authenticating the origin of the Image of Guadalupe.

The "Primitive Relation" (which also appears in full in an appendix, below) was probably written about 1573 by the historian Juan de Tovar, who transcribed the story from a still-earlier source. Garibay believes the earlier source was Juan González, who is thought to have been Bishop Zumárraga's translator in 1531.

The thirty-nine short paragraphs of the "Primitive Relation" begin with these words: "This is the great marvel that Our Lord made through the medium of the always virgin Saint Mary." One of the two quotations in the text ascribed to the apparition of the Virgin Mary gives a direct clue to the time when the "Primitive Relation" was written. The apparition, addressing Juan Diego, said:

> My little son, walk to the center of the great City of Mexico. Say there to the Spiritual Governor, the Archbishop, that here in Tepeyac they make me a habitation, they raise me a little house, so that the faithful Christians can come pray to me. There in it I will convert, when they make me their counsel.[8]

The "Primitive Relation" had to be written down sometime *after* Zumárraga became the archbishop of Mexico.

---

[7]There are many studies of the Valeriano/*Nican Mopohua* authorship thesis. See, e.g., Enrique R. Salazare, P. Maurilo N. Montemayor, and P. Luis Medina-Ascensio, eds., *Documentarios Guadalupano 1531–1768* (Mexico City: Editorial Tradicion, S.A., 1980). See also by the same editors, *Primer Encuentro nacional Guadalupano* [1976] (Mexico City: Centro de Estudios Guadalupanos, A.C., 1978).

[8]See the appendix, below.

A final sixteenth-century source documenting the miracle on Tepeyac was found in Peru in 1924 by M. H. Saville, an anthropologist. It is a pictorial calendar, known as the Codex Saville or the Codex Tetlapalco (from the city where it was found). It now rests in the collection of the Museum of the American Indian in New York City.

The Saville-Tetlapalco codex records Aztec history from the year 1430 to 1557. Discs on the calendar were used by the Indians to signify the passage of time. Mariano Cuevas, who translated the calendar in 1929 (his translation appears below in an appendix), gave the following comments regarding a female figure located next to the top of the disc representing the year 1531:

> A virgin with her hands folded near her heart; her head bent toward her right shoulder, dressed in a salmon colored tunic and a greenish blue mantilla—see the unique design as in the original—is the Virgin of Guadalupe as venerated at Tepeyac, four miles north of Mexico City, and some six miles south of San Marcos. By painting it a little lower than the year 1532, it is well indicated that her year was 1531.

Given this historical evidence, much of which has come to light only recently, there seems little reason to doubt that the Image of Guadalupe was a recognized object of worship in the sixteenth century, beginning in 1531 or soon thereafter.

A second official inquiry concerning the Image of Guadalupe was conducted by church authorities in 1666, one hundred and ten years after the first. The first part of the inquiry consisted of testimony from the oldest Aztecs still alive regarding their memories of the origin of the painting. They were unanimous in maintaining that there had been an apparition of the Virgin Mary and that she had made the miraculous Image. After the Indians, several artists and scientists were called on to testify. Was there indeed anything unusual, perhaps supernatural, about this painting of Mary?

Nineteen years earlier, in 1647, the Image had been covered with glass for the first time. In 1666, the Image was taken down from the wall, the glass cover removed, and master artist Juan Salguero and his associates were given one hour to examine the painting.

The artists were truly surprised by what they saw. As has been rumored, the picture was painted not on canvas but on an Aztec tilma.

The artists could clearly see the vertical seam running down the middle of the cloak, which must have made the painting of a portrait even more difficult. The tilma was made of roughly woven cactus cloth, which was known to disintegrate in only a few years. Salguero concluded that this tendency to rapid disintegration should, if anything, have been hastened by the fact that the paints had been applied directly to the bare, burlap-like surface. The cactus cloth had not been sized or prepared in any way. It had not been stretched to provide a firm surface for the painter, as a canvas would have been, and there was no filler in the holes between the loosely woven fibers.

The artists on that March day in 1666 were genuinely perplexed. Not only should paint applied to an unprepared surface—any surface, and certainly the inferior cactus cloth—have rotted it within a few years, but how had the artist contrived to paint such a beautiful human face on so rough a cloth?

And what about the colors of the painting? Could they be paints, and if so, what kind of pigments could be made to adhere to such a rough, unprepared surface? Futhermore, these paints—if they were paints—were not faded or browned, nor were they cracked or peeling. In a painting more than a hundred years old, this was unimaginable.

Of the many mysterious aspects of the Image, none was more of a puzzle than the apparent mixing of media. Salguero and his fellow artists saw at once that the coloring material used for the portrait was of several different types. How had watercolor and oil, if that was what they were, been blended so perfectly with each other?

The artists concluded, in sworn testimony, that

it is impossible for any human craftsman to paint or create a work so fine, clean, and well formed on a fabric so coarse as this tilma or ayate, on which this divine and sovereign painting of the Most Holy Virgin, Our Lady of Guadalupe, is painted.[9]

Fifteen days later, on March 28, 1666, the Physicians Royal came to examine the painting. They were particularly struck by its remarkable

---

[9]Francisco de Florencia, S.J., *La Estrella del Norte* (Madrid, 1685); for an English translation, see Donald Demarest and Coley B. Taylor, *The Dark Virgin* (New York: Devin-Adair Co., 1956).

state of preservation, not only regarding the colors and the untreated cactus cloth, but also because of another factor that should have destroyed the painting long ago—namely, the high concentration of niter (potassium nitrate) or saltpeter in the humid air. The church was situated next to a large lake, Texcoco, which was known to contain the caustic chemical. Moisture-laden breezes off the water carried the chemical into the atmosphere, and caused serious damage to another noted painting that hung in a nearby church.

The Physicians Royal also made a close examination of the back of the tilma—and they were probably the last to do so. A few years later the tilma was permanently backed by a half-inch-thick plate of silver.

The back of the tilma provided another source of astonishment, for on the back was an oval patch of green, a color that was nowhere to be seen on the front. One observer described the color as comparable to "the leaves of lilies."

The scientific observers noted still another odd property of the picture. When the tilma was held up to the light, it was so thin they could see right through it. As explained in their testimony, this did not mean that they saw through the holes in the loosely woven cloth but rather literally *through* the translucent fibers themselves. What was even stranger, the green on the back could not be seen from the front, although the Image of the Virgin was clearly perceivable from the back. The Physicians Royal commented:

> The understanding wavers, discourse confuses, and the prodigy refers itself to the realm of mystery, as Aristotle, prince of philosophers, asserts an incontrovertible principle: *Idem in quantum idem, semper est natum facere idem* [The same element in the same matter, with the same arrangement, can only produce the same effect]. Why does the color green, which this tilma bears on its opposite side, not pass through at all to the front side? God alone, Who made it, knows why.[10]

The inquiry of 1666 came in response to the widening fame of the Image of Guadalupe. Acceptance of the image as divinely created—"not with human hands"—had been growing, primarily because of the writings of two priests. The first, Miguel Sánchez, in 1648 published in Mexico

---

[10]See Florencia, *La Estrella*; and Demarest and Taylor, *The Dark Virgin*.

City in Spanish a work entitled *Image of the Virgin Mary, Guadalupan Mother of God*, a theological discussion of the story of the apparition and the creation of the Image. The following year a fellow priest named Lasso de Vega published the *Huey Tlanahuicoltica*, which told the story of the miraculous painting in the Indians' native language, Nahuatl. Vega, in his preface, insisted he was presenting an edited version of the story as it appeared in earlier Nahuatl sources (probably the *nican mopohua* mentioned above and given in full in an appendix, below), but many skeptics accused him of merely producing a popular, Indian-language edition of Sánchez's book. Slowly, that view has changed; now it is generally believed that Vega did in fact use earlier Nahuatl sources for his book. Yet largely because of the apparent hundred-year delay in publishing the story of the Image of Guadalupe, many thoughtful people throughout the years had come to regard the entire narrative as no more than an interesting fable.

The next extensive study of the Image was made about a hundred years later by the famous painter Miguel Cabrera. Cabrera's book *American Marvel* appeared in 1756. In it he confirmed everything that had been observed by the 1666 specialists and was equally at a loss to explain how the painting had been produced or why it had endured so remarkably well. He noted that the cactus cloth was indeed without sizing and that the only other material present in the picture was a more-expensive cotton thread used to sew the two vertical panels of the tilma together. In chapter six of his book he marveled at the gold powder used for the stars on the Virgin's blue mantle, since gold powder was unknown to artists of the sixteenth century.

In chapters five and seven Cabrera undertook to answer objections raised against the reputed miracle. Is the figure of the Virgin off-center? Of course it is, in order to avoid the vertical seam of the tilma. Isn't one leg shorter than the other? Yes, to capture ingeniously the Virgin's slight bow. Aren't the hands too small in relation to the overall figure? At first it may seem so, Cabrera admitted, but a closer study shows that such delicate hands are typical of ladies of all countries.

Cabrera also noted the surprising mixture of media in the Image. He wrote:

I find it . . . extraordinary that in a painting there are together on one surface four distinct species of media such as we find miraculously

united in . . . Our Lady of Guadalupe. The science of aesthetics can
deal with each of these media separately, but there is no authority . . .
who can adequately treat of their coming together on one canvas. . . .
Specifically the head and hands were executed in oils; the tunic . . .
angel and clouds . . . in tempera; mantle . . . watercolor; the background, a fresco.[11]

"Oil painting," Cabrera explained, "is executed in oleaginous pigments which when dry blend and attain harmony only when the surface
of the canvas is properly prepared by sizing—then this is the most marvelous medium available." He further noted that

the second, tempera, employs pigments of all colors with gum, glue, or
similar bases. The third, watercolor, is executed on fine, white material
and necessitates soaking the obverse side of the surface so that the
whole is permeated with color. Fresco painting, the fourth medium, is
a plastering and coloring of the surface in the same action. . . . it
requires a firm, solid surface such as a board, stone, etc. . . . each in
itself demands . . . a peculiar mastery of technique.

Concluding, Cabrera said that

the most talented and careful painter, if he set himself to copy this
sacred Image on a canvas of this poor quality, without using sizing, and
attempting to imitate the four media employed, would at last, after great
and wearisome travail, admit that he had not succeeded.[12]

The portrait artist spoke from not only his own experience, but also
that of his contemporaries:

this can be clearly verified in the numerous copies that have been made
with the benefit of varnish, on the most carefully prepared canvases, and
using only one medium, oil, which offers the greatest facility. . . . there

---

[11]Miguel Cabrera, *Maravilla Americana* (Mexico City, 1756) 1822 (chap. 7);
for an English translation, see Demarest and Taylor, *The Dark Virgin*, 153-55.
[12]Cabrera, *Maravilla Americana*.

has not been one which is a perfect reproduction—as the best, placed beside the original, clearly shows.[13]

About three decades later, José Ignacio Bartolache, a skeptical physician who was also a priest, insisted on retesting Cabrera's thesis. In 1789 he ordered eleven copies made of the painting. Procuring the services of the best artists in Mexico, he insisted that each utilize only color agents that were known to have been in use in the sixteenth century. For example, for the color red he would allow only mercuric oxide or sulfate, or a vegetable dye from the "dragon's-blood flower," or plants such as the "ruby," the "brazil-red," and the sunflower.

What were the results of Bartolache's experiment? All reproductions had been painted on cactus cloth of the same inferior type as the original, and in 1796, seven years later, all eleven copies still survived. Clearly something in the pigments used had preserved the crude material. Perhaps the mystery of the cactus cloth was solved. How, then, to explain that, while all the copies had survived, *one* was in such a poor state, peeling and coated with fungus, that it had to be removed from public display. This copy was the one which had been placed beside the original at the Basilica of Guadalupe on Tepeyac hill.[14]

Studies made of the Image since Bartolache's time have yielded similar results. Since 1950, two artists, Francisco Campa Rivera and Francisco de Guadalupe Mojica, have examined the painting. Their reports are in the Basilica Archives. Rivera, who came to Mexico from Barcelona in 1941, studied the painting in 1954 and 1963. Like Bartolache, Rivera was interested in the media used. In addition to attempting to make copies using oils, watercolors, and tempera, he experimented with pastels and inks of various colors. Whatever media he used, the color could be seen on both sides of the cactus cloth, which was not true of the green patch on the back of the original. Nor was Rivera able satisfactorily to determine the composition of the original colors of the painting. Rivera and Mojica agreed that by the mid-1950s a number of additions had been made to the painting. When these had occurred they were unable to tell, but none of the alterations affected significant areas of the Image. They

---

[13]Ibid.

[14]José Ignacio Bartolache, *Manifiesto satisfactorio* (Mexico City, 1790).

involved such things as the black outlining of dress and mantle and restoration of the gold rays around the entire figure. The angel and crescent at the Virgin's feet were also believed to be additions, as were the fleur-de-lis on the robe, since if they had been part of the initial painting they would have been shown as falling between its folds.

Mojica is convinced of the miraculous nature of the Image. Because of many details—for example, the robe is of the same style as that worn by ancient Palestinian women—he believes the Image is an authentic portrait of the Virgin. Rivera, an established critic as well as an artist, began with the hypothesis that the work was anonymously done by a European master or by a Mexican strongly under European influence. He reluctantly abandoned this view when he was unable to identify any Spanish, Flemish, or Italian characteristics in the painting.

In addition to Marcos Cipac, mentioned in the inquiry of 1556, two other Aztec artists known to have been active in the sixteenth century have been suggested as possible creators of the painting. Their names are Pedro Chachalaca and Francisco Xinmamal. But they are, ultimately, only names. We have less reason to associate them with the picture than we do Cipac.

Over the centuries, there have been repeated attempts to disprove the miraculous origin of the Image of Guadalupe. Most frequently cited are the comments of a famous nineteenth-century Mexican historian Joaquín García Icazbalceta, in his introduction to a new edition of the official inquiry of 1556. Icazbalceta chose to interpret several key phrases of that inquiry as indicating that veneration of the Image of Guadalupe merely echoed the earlier worship of Aztec idols, and that the legend of Juan Diego and the apparition of the Virgin was a later invention. The fact that the 1556 inquiry made no mention of the apparition is taken as proof by Icazbalceta and by twentieth-century historian Jacques LaFaye that the story had not existed at that time.

Both Icazbalceta and LaFaye cite a section of the inquiry in which the delegate, speaking for the Franciscan Bustamante, mentions the shrine at Loreto, Italy. Loreto has been revered as a shrine to the Virgin Mary since medieval times. LaFaye believes Bustamante would, on mentioning Loreto, have gone on to compare it with Tepeyac, if the apparition story had been known in his day.

This seems reasonable enough, but I believe the inquiry's silence about the apparition at Tepeyac was deliberate. It is equally feasible that

Bustamante's reference to Loreto was an indirect way of disputing the Tepeyac story. The point is not that the Loreto shrine was the site of an apparition, but that it was the site of an *approved* apparition. In fact, in the 1556 inquiry, Bustamante explicitly notes the sound basis on which the church awarded official sanction to the Loreto shrine. The Franciscan's dilemma was not, I think, whether to endorse one and not the other; it was to decide whether or not to endorse both.

Like many people, I share some of Bustamante's skepticism. But, while he had no qualms about accepting the existence of an apparition at Loreto and the reality of the supernatural in general, today most of us shy away from the very idea of the supernatural, though we have a most imperfect notion of what it might be.

Recent scholarship, especially among the Indian Archives of the Mexican National Library, has brought to light several documents that, if they do not prove the actuality of the Virgin's supernatural appearance before an Indian peasant in 1531, do at least confirm the fact that belief in that appearance was widespread in those times. No historiographical technique can irrefutably prove that any past event did in fact occur, be it mundane or supernatural. The historian can locate, collate, and compare materials relevant to a disputed question, but there his own special expertise ends.

In the case of the Virgin of Guadalupe, however, the Image itself survives. As my research drew me deeper and deeper into the subject, I became convinced that firsthand scientific examination of the painting would enable us to transcend the limitations of the historian.

# Impossible
# Coincidences

Synchronicity . . . means the simultaneous occurrence of a certain psychic state with one or more external events which appear as meaningful parallels to the momentary subjective state—and, in certain cases, vice versa.
    —Carl Gustav Jung, *The Structure and Dynamics of the Psyche*

The scientific findings regarding the shroud of Turin were very much on my mind as I flew back from Mexico City to my home in Pensacola, Florida. I considered it a good omen when I picked up the candy on my dinner tray and found printed on the wrapper "Made in Turin, Italy."

There are other instances of coincidence that seem to me of another order, something other than "coincidence" as a synonym for "chance" (a concept that in itself appears to me far from unambiguous). Although outwardly an ongoing narrative relating the chronological developments that led to the remarkable photographic study of the Image of Guadalupe, I try to illustrate a few instances of a phenomenon that to me is and always has been at the very heart of the Guadalupe mystery. I am referring to what Carl Jung in a book with Wolfgang Pauli referred to as *synchronicity*. There were many such synchronicities other than those specified in the present work, which is primarily historical and not epistemological.[1]

What I am relating here was learned well after the original adventure was over, but it seems no less worthwhile on that account. As mentioned above, my home—since 1970, in fact—has been Pensacola, Florida. I

---

[1] I supply a partial list of recommended studies for the interested reader in the bibliography.

learned that the supposedly Indian name *Pensacola* was not applied by the Spanish who first settled the area, but waited for about another one hundred years before taking that title. The expedition of 1559 that established our town immediately named it *Saint Mary*. Admittedly this "chance" is something I find fascinating in itself. Many prominent citizens of Pensacola today still are unaware of the original name of our "First Place City," and express incredulity on being informed about it.

I have no ready way to demonstrate it (given the near-universal admiration that sixteenth-century Spanish culture imbued on the Madonna), but I think it is of further interest to recall that Tristan de Luna set sail from Mexico, not Spain, and I think actually originated in none other locale than the City of Mexico itself. Only three years before, in the same city, appeared the much-discussed *Información* of Archbishop Montufar, regarding of course the Guadalupe-Tepeyac Image of Saint Mary.

Another interesting coincidence concerning my home area of Northwest Florida is that the town just east of Pensacola is named Mary Esther. The pregnancy of this particular discovery will become obvious later in this book. Here I only want to say that my efforts to uncover the origin of the name were only partially successful: it was not old, but apparently of the early nineteenth century, and appears to have been applied by a Presbyterian minister after the names of his two daughters.

One more account that seems to me worth noting is another subsequent discovery I made, namely, that the body of water immediately west of Pensacola is named *Perdido* (Lost) Bay, and was named by the famous Mexican scholar of the late seventeenth century, Sigüenza y Gongora. Sigüenza y Gongora, as I discuss later, was a most significant link in the transmission of Aztec documents concerning the origins of the perhaps miraculous Guadalupe-Tepeyac Image. Sigüenza y Gongora had visited the area as a part of an expedition (unrelated to the Guadalupe mystery, so far as I know) continuing exploration of Northwest Florida in the late 1600s.

Following my first visit to the Basilica of Guadalupe in December 1978, I had arranged a meeting with an associate of the abbot of the basilica, Monseñor Enrique Salazar, to discuss the possibility of subjecting the Image of Guadalupe to scientific study. Immediately after my meeting with the monseñor, I met with Abbot D. Guillermo Schulenburg-Prado. He and Monseñor Salazar assured me that, if I could satisfy certain requirements, they would permit the famous tilma to be removed from its frame and protective glass. This could not be arranged, however, until sometime after Pope John Paul II's visit to Mexico, scheduled for January 1979.

What I had specifically proposed was that the Image be subjected to such techniques as computer enhancement and infrared photography. The last time the glass cover had been removed for photographic purposes was in 1961 when official color prints of the portrait had been made. Only one infrared photograph of the Image had ever been taken—by Jesús Catano in March 1946—and great strides had been made since then in the techniques of infrared photography.

Feeling that I had at least tentative approval from the church authorities, on returning to Florida I proceeded at once to try to locate a scientist with special expertise in the field of infrared studies, someone who could direct the research and interpret the results. Among many possibilities, Dr. Philip Serna Callahan had come to my attention through his study of "unidentified flying objects" (UFOs). An entomologist with a background in biophysics, Dr. Callahan had advanced the hypothesis that, in certain cases at least, UFOs were atmospherically lighted insect swarms.

After spending several weeks trying to locate him, I discovered Dr. Callahan also lived in Florida. I was naturally anxious as I dialed the phone to speak with him. I had no reason to suppose Callahan would have any interest at all in a religious object hanging on the wall of a church in Mexico City. But, after I briefly described the Image and the research that I proposed be done, he told me of his interest in the Shroud of Turin. He was soon to present his conclusions in a lecture. From his enthusiasm I knew at once that he was the man I was looking for, and I arranged to visit him within the week.

Although he had given me directions for finding his house, dusk had already fallen on that day in late January 1979, and I had not yet arrived at his door. My thoughts drifted back to another appointment, a few weeks earlier: I had huddled off the bridge spanning the Pascagoula River, stung by the freezing wind, the hurtling rush-hour traffic darting by in the darkness. I was looking for an interesting individual, but I had no thoughts of Guadalupe.

Callahan had an unlisted telephone, so I went to the local police station for help. All seemed lost until the county constable, overhearing my dilemma, stepped out of an adjoining office and personally located my "appointment." As I left the police station I saw a church across the street. Looking up, I noted a shimmering golden disc, inside of which was painted a large picture of the Virgin Mary. The name of the church was Our Lady of the Victories. Several months later I uncovered the fact, already related here, that a copy of the Image of Guadalupe had been on board Doria's flagship during the Battle of Lepanto. The Christians' triumph in that battle led Saint Pius V to name the day, October 7, 1571, "Our Lady of Victory."

It had been a good omen, for my interview with Dr. Callahan that evening was everything I hoped it would be. Together we mapped out detailed plans for research with two specific objectives in mind—one, to discover whether there was an undersketch beneath the picture, because if such an undersketch existed, the Image was undoubtedly the work of a human artist; and two, to determine the composition of the coloring materials used.

Infrared photography of old or possibly valuable paintings is today a routine procedure, of immeasurable help in determining the age and authenticity of the paintings. Pigments used by artists in bygone times were quite different from those in use since the late nineteenth century when aniline dyes were first introduced. Aniline-derived pigments, when seen under infrared light, are radically dissimilar from earlier pigments, although the difference cannot be seen in ordinary light.

Throughout my dealings with the church authorities, which occupied my time in the early months of 1979, I kept in constant touch with Dr. Callahan, informing him of the progress of our negotiations. I did not learn until months afterward that at one point, on reviewing his numerous previous commitments and after a bout with flu, he had seriously considered bowing out of the Guadalupe investigation. What changed his mind was something he unearthed one evening while reading his family history. Callahan's Irish father had married a Mexican woman whose ancestry traced back several hundred years. On that night Dr. Callahan first learned that a direct ancestor had been one of the group of missionaries and explorers who set out from Mexico for the sparsely settled northern territory known as California. At the end of their journey they founded the city of San Francisco, but before leaving the vicinity of Mexico City they had stopped at a shrine to pray for the success of their mission. The shrine was the Church of Our Lady of Guadalupe on Mount Tepeyac.

How should I account for the fact that out of the scores of scientists I considered, the one I chose to help me in the Guadalupe project was one whose ancestor had clearly accepted the supernatural origin of the Image?

Now, however, it was our task to assure the church authorities that no harm would come to the painting as a result of our proposed examination. I wrote to Mrs. Georgina López-Freixes, assistant to the abbot of the basilica, in January, requesting copies of certain documents in the basilica library, which she sent to me without delay. Throughout, Dr. Callahan and I were helped immeasurably by Mrs. López; by the abbot, Dr. Schulenburg Prado; and by the basilica's executive, Monseñor Enrique Salazar. Later, Father Luis Medina, president of the Center for Guadalupan Studies, and Father Maurilio Montemayor, secretary of the Center, gave us invaluable assistance.

Pope John Paul II visited the basilica on January 27, 1979. Shortly after his unprecedented visit, the abbot and Monseñor Salazar informed us that sometime in the spring the sacred Image would be taken down and the glass and frames removed so we could take the proposed series of infrared photographs and black and white close-ups suitable for computer enhancement.

Our excitement grew. What would we find? Would the infrared photos reveal an undersketch? Would close-up examination show signs of retouching, which would explain the unfading brightness of the colors? Had some kind of sizing or preservative been applied to the coarse cactus cloth?

These questions coursed through my mind day after day as I waited for a definite date to be set.

# IN SEARCH OF MARY

Lycomedes! What meanest thou by this matter of a portrait? Can it be one of thy gods which is painted here? I see that thou art still living in a heathen fashion.  —*Apocryphal Acts of John*

Once the details of our project received official approval, I could return to my historical research. I was particularly eager to discover whether any portrait of Mary from life had ever been made. If such a portrait existed, and if the Image of Guadalupe had truly been created without human intervention, the two pictures should resemble each other.

Many people believe erroneously that widespread veneration of the Virgin Mary did not begin until the late Middle Ages, when magnificent churches such as the cathedral at Chartres were built in her name. But the date accepted by most scholars as the beginning of veneration for the Holy Mother is the Council of Ephesus, A.D. 431. Even earlier, in A.D. 245, the Christian philosopher Origen termed Mary *theotokos*, meaning "Mother of God." And beneath the new Roman Catholic Church of the Annunciation in Nazareth archaeologists have found "Hail Mary" scrawled in Greek on walls dating back to the second century. Although this evidence of extremely early veneration of Mary is tantalizing, it provides no assurance that an authentic portrait of the Virgin ever existed. There is generally believed to have been a prohibition among the earliest Christians against all religious art, whether painting or sculpture. In the early fifth century Saint Augustine of Hippo wrote: "As to Christ and His Mother, we know nothing of their true appearance."

Despite what is generally believed and Augustine's blunt statement, however, evidence to the contrary exists. Around A.D. 326, Eusebius of Caesarea, a confidant of Constantine the Great, wrote the first comprehensive ecclesiastical history. In that history Eusebius says that not only had there apparently been religious art in the first century of Christianity, but that statues and paintings of Christ himself had been made:

They say that the statue is a portrait of Jesus . . . nor is it strange that
those of the Gentiles who, of old, were benefited by our Savior, should
have done such things, since we have already learned that the likenesses
of Peter and Paul, and of Christ Himself, *are preserved in paintings*, the
ancients being accustomed, as it is likely according to the habits of the
Gentiles, to pay this kind of honor indiscriminately to those regarded by
them as deliverers.[1]

Eusebius himself may have seen the statue he refers to, for others
report seeing such a statue in a church at Paneas, in Palestine. The statue
was said to show Christ healing the woman with the blood disease, as
recounted in the Gospels.

Eusebius further mentions paintings of Peter and Paul. Coin-sized
medallions of these saints continue to be found in Rome even today.
From the style of composition and manufacture, we know that they date
from before A.D. 325.

André Grabar, a historian of Christian art, believes that both the
medallions and the Paneas statue are of no later than second-century origin. He observed:

However astonishing it may seem, these images appear at a period
which has left us no equivalent images of Christ. Is this a matter of
chance? Or is it, rather, that the monuments that have been preserved
testify to the actuality? In the latter event, the monuments would be of
considerable moment (corresponding to a first blaze of Christian
portraiture, afterwards extinguished). They would be comparable to
written testimony on Christian portraits from the beginning of our era.[2]

Grabar's suggestion that the early Christians made portraits of the
church fathers and of Christ, soon thereafter lost or destroyed in a wave
of iconoclasm like that which swept Christianity in the eighth and ninth
centuries, allows us to entertain the possibility that there may have been,
after all, an authentic portrait of the Virgin Mary.

---

[1]Augustine, *On the Trinity*, in *Patrologia Latina* 5.428.8; italics added.
[2]Eusebius, *Ecclesiastical History* (Cambridge MA: Harvard University Press,
1926) 192.

Generally accepted as the earliest Madonna is a crude painting, found in the Catacombs of Priscilla in Rome. Usually dated from the late second century, it gives us little idea of Mary's appearance.

More interesting is an ancient legend brought to light in 1887 by Christian historian Ferdinand Gregorovius. The legend may well contain more than an atom of truth. As Luke lay dying in the Greek city of Thebes, he revealed that for years he had carried with him a genuine portrait of Christ's mother. He asked a disciple, Ananias, to take care of the portrait after his death. Ananias carried the precious icon to Athens. Although most Athenians in the first century A.D. were still pagans, they readily venerated the beautiful painting, calling it either the "Athenaia" or the "Athea." The portrait remained in Athens until sometime during the reign of Theodosius the Great (A.D. 375–395). A Christian named Basilius Soterichus received a vision of the Virgin, who instructed him to move her portrait to a new home, remote from the troubled cities of the eastern Roman empire. Soterichus, with a band of pilgrims, traveled hundreds of miles over land and sea seeking an appropriate location. At length they arrived at Mount Sumela in what is now northeastern Turkey near the fabled city of Trebizond (Trabzon). There they founded the Panagia monastery. *Panagia* literally means "All Holy," but in the Eastern Orthodox Church it has in fact become synonymous with the Virgin Mary.

The existence of a monastery on Mount Sumela is confirmed by Greek church history. Historical documents also confirm that it was founded in the late fourth century and was dedicated exclusively to Mary. There is growing evidence that the icon itself may have survived for hundreds of years and that it exists even today. An artifact that I am now researching is located in a remote corner of the Greek peninsula and corresponds remarkably to this legend.

What we do know, on the authority of the sixth-century chronicler Theodor Lector, is that a beautiful portrait of Mary was found in the mid-fifth century in Jerusalem. The discoverer of the portrait, according to Theodor, was the empress of the eastern Roman Empire, Athenaïs Eudokia, wife of Theodosius II. A common belief at the time and for centuries thereafter was that the portrait had been painted by Luke, who was also held to be, in the legend related above, the painter of the portrait that supposedly was carried to the Panagia Monastery in northeastern Turkey.

Could that portrait have turned up seventy-five years later in Jerusalem? It is not impossible.

Eudokia was noted for her blond beauty and for her intelligence. She wrote allegorical poetry, some of which survives to this day, and was instrumental in founding Constantinople's first college. Her father was an Athenian rhetorician named Leontius. Once converted, she was an ardent Christian, gaining fame as the era's most noted collector of early Christian relics.

According to historical records, the empress made at least two journeys to Jerusalem. The first, in A.D. 438–439, was a pilgrimage made in gratitude for the successful marriage of her daughter Eudocia to the western Roman emperor two years earlier. It seems unlikely that the empress discovered the portrait during that first visit, for there is no mention of it in the chronicles of the period when Eudokia shared the role of empress with Theodosius's sister Pulcheria. Eudokia fell from imperial favor around the year 440. She left Constantinople for Jerusalem and lived there until her death in 460. One account specifically notes that Eudokia sent the famous portrait to Saint Pulcheria, her sister-in-law, who died in 453.

We may never know if it was Empress Eudokia who found the *Theotokos Hodegetria*, as the icon was named. There is more general agreement that she did indeed find relics of Saint Peter and Saint Stephen during her first visit to Jerusalem in 438, and that she returned triumphantly with them to Constantinople. Perhaps because of the empress's fame as a relic hunter, the discovery of the early portrait of the Virgin was linked with her name.

Pulcheria, while she was coempress, is credited with the building of at least three churches in Constantinople dedicated to the Virgin Mary, one of which was named the *Hodegetria*. The name *Hodegetria* means "Pointer of the Way." Some believe the portrait was located there where it "pointed the way" to a miraculous well. Others who have looked at Byzantine paintings, such as Our Lady of Saint Vladimir in Moscow, that are believed to resemble the *Hodegetria* in general outline point out that in all cases the Virgin points with her finger toward the Child she holds in her arms. In the Guadalupan Image, as noted earlier, the Christ Child is absent.

Bishop Photius, a ninth-century patriarch of Constantinople, has given us a description of the icon found in Jerusalem four centuries earlier:

A Virgin Mother carrying in her pure arms, for the common salvation of our kind, the Common Creator reclining as an infant . . . turning her eyes on her begotten Child in the affection of her heart.[3]

The *Hodegetria* remained for centuries in the magnificent Church of Saint Sophia, built by Justinian and dedicated on Christmas Day 538. According to a letter of Pope Innocent III written on January 13, 1207, as a result of the turmoil caused by the fourth Crusade, the famous portrait was moved to another church, Christ Pantocrator, which was favored by Roman Christians. The move was made just in time, for, as historian Edward Gibbon comments,

> trampled under foot [were] the most venerable objects of Christian worship. In the Cathedral of S. Sophia, the Ample Veil of the sanctuary was ripped asunder for the sale of its gold fringe; and the altar, a monument of art and riches, was broken into pieces and shared among the captors. . . . a prostitute was seated on the throne of the Patriarch and . . . [they] sung [*sic*] and danced in the Church to ridicule the hymns and processions of the Orientals.[4]

A generally accepted view is that the *Hodegetria* was destroyed when the Turks conquered Constantinople in 1453. A contemporary record tells of a Turkish commander who had one of his soldiers flogged because he "ripped into four pieces and dragged through the mud" a painting sacred to the Christians.

Yet there is widespread belief in two alternative histories of the famous painting. Robert de Clari, a French historian who accompanied the fourth Crusade in the sacking of Constantinople in 1204, reported that:

> Murzuphius . . . took with him the icon, an image of Our Lady which the Greeks call by this name [*Hodegetria*] and which the emperors carry

---

[3]André Grabar, *Christian Iconography* (Princeton NJ: Princeton University Press, 1961) 68.

[4]V. Lacurdhas, ed., *Photius Homiliai* (Thessalonika, Greece: Etaireia Makedhonikon Spoudon, 1959) 45 (fourth homily).

with them when they go into battle . . . having as great a faith in this icon that they fully believe that no one who carries it can be defeated.[5]

According to an old Venetian tradition, the famous painting was brought with other booty to Italy and installed in Saint Mark's Basilica in Venice. De Clari lends some support to this view:

> When the Doge of Venice and the Venetians saw that they wanted to make my lord Henry of Flanders emperor, they were against it, nor would they suffer it, unless they should have a certain image of Our Lady which was painted on a panel. This image was rich beyond measure, and was all covered with rich and precious stones. And the Greeks said that *it was the first image of Our Lady ever made or painted*. The Greeks had such faith in this image that they treasured it above everything, and they bore it in procession every Sunday and they worshiped it and gave gifts to it. Now the Venetians were not willing to allow my lord Henry to be emperor, unless they should have this image, so . . . it was given to them and . . . Henry was crowned.[6]

Doge Enrico Dandole, to whom the treasure was reputedly given, died before he could return to Venice, but perhaps a colleague brought it back. In another letter of Pope Innocent III (sometime after 1207), there is some support for this view. The Pope never traveled to Constantinople, yet in his letter he praised the *Hodegetria* for its miraculous beauty, writing that the painting contains something of Mary's own soul. Such a description would seem to indicate that the Pope had actually seen the famous icon.

Another strong tradition holds that the *Hodegetria* still exists. Chroniclers claim that, early in the eighth century, when the iconoclasts were bent on destroying all sacred images, the Hodegetria was taken to a chapel deep in a forest far outside Constantinople for safekeeping. According to official records, it was returned to Saint Sophia in 840. But how can we be certain that the original painting was returned?

---

[5]Edward Gibbon, *Decline and Fall of the Roman Empire* (New York: Modern Library, n.d.) 3:1116.

[6]Robert de Clari, *The Conquest of Constantinople*, trans. Edgar H. McNeal (New York: Columbia University Press, 1936) 89, 12G; italics added.

One story says that in truth the inhabitants of that distant forest kept the icon. A variant of this same story holds that the *Hodegetria* was returned to Saint Sophia, only to leave Constantinople once more, this time never to return, when Princess Anna, sister of Emperor Basil II, was married in 988 to Vladimir, Grand Duke of Russia and a recent convert to the Eastern Orthodox Church. Tradition says that the icon, which survives to this day at Jasna Gora (Hill of Light) near Kraków, Poland, and is known as Our Lady of Częstochowa, is in reality the ancient *Hodegetria*. Just as Mexico's national symbol is Our Lady of Guadalupe, so Poland's is Our Lady of Częstochowa. It is linked to the *Hodegetria* through a Ukrainian named Prince Ladislaus Opolszyk, who is known to have existed historically and to have had ties with both Hungarian and Neapolitan nobility. The story goes that this prince brought the icon from the deep forest—in what would be modern Hungary—to Częstochowa, where a monastery grew up around it. The fleur-de-lis on the robe of Our Lady of Częstochowa is a typical motif of Neapolitan art. As mentioned earlier, fleur-de-lis also decorate the rose-colored robe of the Virgin of Guadalupe.

But an even more remarkable similarity between Our Lady of Częstochowa and the Image of Guadalupe lies in the darkness of the Virgin's complexion. The tradition of a "Black Madonna" has roots in the discovery by an Italian monk in the middle of the fourth century of three statues of Mary. He distributed these statues to churches in Italy and Sardinia. One, at the Santuario d'Oropa, northeast of Turin, exists to this day. About three feet high, it is made of cedarwood. The dark-faced Virgin carries the Christ Child and her arms are lifted in prayer.

Worship of the *Vierge Noire* was especially widespread in medieval France. Shrines to the Black Madonna can still be found at Avioth, Le Puy, Moulins, and Marsat. There are several in Spain also, including a famous one at the Benedictine Abbey of Montserrat.

The dark color of the skin of Our Lady of Częstochowa did not arise from the oxidation of silver in the pigment of the paint but from smoke from charcoal burning. The olive complexion of the Virgin of Guadalupe, on the other hand, seems to have been there from the first and is not the result of aging or smoke. What of the *Hodegetria*, the famous early painting of the Virgin? A seventeenth-century scholar from Chartres, Vincent Sablon, found a description of the portrait in the works of a fourteenth-century Byzantine historian named Necephorus Callistos:

Necephorus, however, says he saw several paintings made from nature by Saint Luke, in which the color of her [Mary's] skin was the color of wheat—which is probably to say that when wheat is ripe it tends to be brownish, or a chestnut color.[7]

Mexicans have for centuries considered this brownish or chestnut color to be the identifying characteristic of the Virgin of Guadalupe. In fact, they have named the image "La Morena," meaning "the dark-complexioned woman".

In this regard they echo an ancient Greek tradition. As noted by Anna Brownell Jameson, a nineteenth-century writer:

To this day, the Neapolitan lemonade-seller will allow no other than a formal Greek Madonna, with olive-green complexion and veiled head, to be set up in his booth.

Jameson continued,

It is the *dark-colored, ancient Greek Madonnas such as this which all along have been credited miraculous* . . . Guido, who himself painted lovely Madonnas, went every Saturday to pray before the little black Madonna de la Guardia and we are assured held this old Eastern relic in devout veneration.[8]

Perhaps the fact the Virgin of Guadalupe was one in a long chain of "dark Virgins" was the reason Archbishop Montufar defended its authenticity in that first inquiry in 1556. The color, whether "black," "chestnut," or "Indian olive," was no accident; for it is by means of this unique characteristic that the ancient Madonnas were "credited miraculous."

Was it because the icon she discovered in Jerusalem was dark of hue that Empress Eudokia, her sister-in-law Pulcheria, and then all Byzantium came to pay it homage? Let us suppose the color of the Virgin's complexion in the *Hodegetria* was, indeed, what produced the ardor in the Byzantines, the connection between them dimmed into forgetfulness with

---

[7]Vincent Sablon.

[8]Anna Brownell Jameson, *Legends of the Madonna—as Represented in the Fine Arts* (London: Hutchinson & Co., 1852) 25.

the passage of a thousand years. Let us also grant that the same skin color appears in both the *Hodegetria* and the Virgin of Guadalupe. What might this have to do with either being an authentic representation of the Virgin Mary? The answer comes by recalling that Mary was a Jew.

Just as I was completing the research for this study, I found myself paging through a compilation of old Hebrew themes, newly translated from ancient sources. My eyes focused on a tiny footnote translating, from the Aramaic Babylonian Talmud (ca. A.D. 475), a statement about the supreme beauty of Queen Esther of Old Testament fame. I was astonished:

Her skin was greenish, like the skin of a myrtle.[9]

Precisely the same can be said about the skin tone of the Virgin of Guadalupe.

---

[9]Robert Graves and Raphael Patai, *Hebrew Myths: The Book of Genesis* (Garden City NY: Doubleday, 1964) 64.

1. The mysterious Image of Guadelupe as it appears imprinted on Juan Diego's tilma that was made from agave cactus. Since 1531, when the Madonna appeared to Diego, this Image has been the source of faith, strength, and awe for millions of people. (*Photo courtesy of Luis Carlos Pèrez Cavilan*)

The Image of Guadalupe is just one in a long line of portraits of the Mother of Jesus which have survived the ages.

2. Another is Our Lady of Częstochowa, preserved in the Shrine of Jasna Gora in Poland. Legend says that this is an actual portrait of the Virgin Mary done by Saint Luke on a cypress tabletop from the House in Nazareth. Science says that this beautiful painting can factually be dated as far back as the fifth or sixth century. (*Photo courtesy the National Shrine of Our Lady of Częstochowa, Doylestown, Pennsylvania*)

3. This beautiful icon is a characteristic *hodegetria* — "pointer of the way" (the Madonna's fingers point to the Infant Christ). It can be found in a church in Thessalonika which bears the intriguing title Sanctuary of the Archeiropoietos ("not made with human hands"). (*Author's photo*)

5. (Above right) Beneath the floor of the church in Salonica lies exposed the mosaic floor of a n earlier Greek structure, possibly a pagan temple. (*Photo by author*)

4. (Above left) The legendary black-faced icon *Acheiropoietos* ("not made with hands") located in Salonica (Thessaloniki), Greece. (*Photo by author*)

6. (Right) Sixteenth-century painting of the Madonna. (*Photo by author*)

7. Our Lady of Guadalupe of Estremadura. This Spanish Black Madonna plays a pivotal role in the evolution of the Guadalupan mystery. (*Turn-of-the-century depiction of ancient statue*)

8. "Virgin Nikopeja" (i.e., "bringer of victory"), purportedly the same sent by the Empress Eudockia to her sister-in-law Pulcheria, now in the Cathedral of St. Mark at Venice. (From Anna Brownell Jameson, *Legends of the Madonna—As Represented in the Fine Arts*. London: Hutchinson & Co., 1850)

9. There are literally hundreds of representations of the Image of Guadalupe in Mexico City alone, this one near the Church of San Diego, adjacent to a museum commemorating the Mexican-American War and the Treaty of Guadalupe-Hidalgo. The Latin words mean "To no other nation has been accorded such honor," and were said by Pope Benedict XIV in 1754 as he proclaimed the Virgin of Guadalupe patron of Mexico. The two cultures of Mexico are never far apart, as can be noted by the Aztec goddess atop the gate. *(Author's photo)*

A group of pilgrims crowds a Mexico City street near the basilica. About five thousand pilgrims pray before the Image every weekday and almost a hundred thousand on Sundays. The basilica draws more than a million people on the feast of Our Lady of Guadalupe, December 12. *(Photo courtesy of National Catholic News Service)*

10, 11. The Tribute Roll, section 1, on amate paper, executed about 1530, showing (closeup) the Virgin of Guadalupe located at the year 1531 by the Aztec Disc. This is the Codex Saville, a pre-Columbian Mexican pictorial calendar, painted on paper made of treated fibers from the agave (maguey) cactus, the same plant that supplied the cloth for Juan Diego's tilma. See the Codex Saville in the Appendices. (*Photo courtesy the Museum of the American Indian, Heye Foundation*).

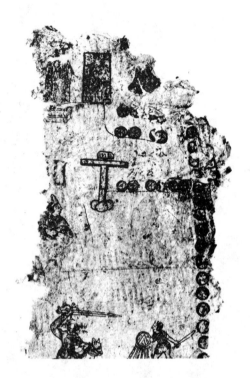

**12, 13. Codex**
*Tellerian-Remensis.*
*(From Anthony*
*F. Aveni,* Skywatchers
of Ancient Mexico)

**14.** The Church of the Well (east of Tepeyac Hill), so named for the still-usable water source it surrounds. This church was built about two hundred years ago, but the site has long been associated with both the first Guadalupan Hermitage and the grave of Juan Diego. (*Author's photo*)

**15.** The old Basilica of Our Lady of Guadalupe, Mexico City. After more than four hundred years the building had settled so much that it was declared unsafe and a new basilica was planned. (*Photo courtesy Religious News Service*)

# The Image
# Lost and Found

Now a great sign appeared in heaven: a woman, adorned with the sun, standing on the moon, and with the twelve stars on her head for a crown.  —Revelation 12:1

The Image of Guadalupe is linked, in a surprising way, to the early icons brought to Constantinople by the Empress Eudokia and others in the fifth century. Two hundred years later, a young monk who had spent several years in Constantinople brought to Rome a statue of the Virgin that apparently resembled the one now in the Santuario d'Oropa near Turin. The young monk, who later became Pope Gregory the Great, carried the statue around Rome, using it as a palladium—a religious object to counteract evil. The statue was credited with ending a plague then rampant in Rome. Later, Pope Gregory gave the statue to Saint Leander of Spain.

The story picks up again more than six hundred years later in the fourteenth century. According to papers about the victory of the Christians over the Moors at Salido in 1340, Spanish King Alfonso XI credited his success to a discovery made in 1328. In that year the Virgin appeared before a shepherd named Gil Cordero and directed him to search a cave by a river on the plain of Estremadura. There he discovered a well-preserved statue of the Virgin.

The name of that river translates as "hidden channel." We know it better as Guadalupe.

A Hieronymite monastery was established there under the tutelage of Cardinal Don Pedro Barroso. Before long, the surrounding community grew into the town of Guadalupe. In it could be found the richest museum in all Europe. The museum was visited and generously supported by royalty and such notables of the fifteenth and sixteenth centuries as Columbus, Cortes, and Admiral Andrea Doria. The opulent monastery and museum of Guadalupe reigned supreme for more than a century, being gradually eclipsed only by the Escorial of Philip II in the late sixteenth century.

According to manuscripts in the monastery archives and the Spanish National Library, the statue of Our Lady of Guadalupe-Estremadura was twice hidden. It was first put away for safekeeping in Byzantine times. It was hidden again in A.D. 711 at the beginning of the Muslim invasion of Spain. Historian Jacques LaFaye has quoted from an anonymous codex dated A.D. 1440. It tells of

> that time when all the Christians fled from Seville. Among them were some saintly priests who took with them a statue of Our Lady, Holy Mary, . . . and in these mountains the priests dug a cave that they surrounded with large gravestones; inside they placed the statue of Our Lady, Holy Mary, together with a small bell and a reliquary containing a writing which told how this statue of Holy Mary had been offered at Rome to the archbishop of Seville, Saint Leander, by the doctor of the church, Saint Gregory.[1]

LaFaye noted that another part of the codex describes how an apparition of Mary led the shepherd to discover the statue. Although LaFaye believed that the statue is genuinely ancient, he doubted its association with Gregory and Leander and completely dismissed the idea of any apparition. Instead, he pointed out the many similarities in the legends concerning discovery of the Estremaduran statue and the Mexican Image. He believed that the story of the Virgin's apparition before Juan Diego and the miraculous appearance of her image on his tilma is an echo of the story surrounding the discovery of the Virgin's statue in Spain two hundred years earlier.

Yet the similarities in the two legends can just as easily be used to support the actuality of apparitions at the two sites and, incidentally, to confirm the early existence of a shrine on Tepeyac hill. This is indicated indirectly by the testimony given at the official inquiry in 1556, when Alonso de Santiago, one of those objecting to Archbishop Montufar's approval of the Virgin of Guadalupe, said:

---

[1] Jacques LaFaye, *Quetzalcoatl and Guadalupe*, trans. Benjamin Keen (Chicago: University of Chicago Press, 1976).

The Archbishop ought to have sent orders to that hermitage, that it not bear the name "Our Lady of Guadalupe" but "of Tepeaca or Tepeaquilla"; because, if in Spain Our Lady of Guadalupe was so named, it was because there was there a town so named.[2]

Not only does Santiago's testimony provide independent confirmation of the existence of a shrine on Tepeyac hill, but it gives a clue as to why the name given to the shrine should be so controversial. For ostensibly there was no reason why the Mexican shrine should not have been named the Virgin of Tepeyac. The fact that it was instead named the Virgin of Guadalupe indicates that something surely out of the ordinary happened on that Mexican hilltop.

The story of the discovery of the Spanish statue was well known at the time. Is there any reason to doubt that Guadalupe was chosen as the name for the Mexican shrine on Tepeyac because, as in Spain, an apparition of the Virgin occurred there?

Whatever the skepticism expressed by historians over the centuries, people continued to believe in the miraculous nature of the Image of Guadalupe. One evidence of that belief, and the growing acceptance of the Image by Church authorities, is the widespread dispersal of copies of the Image throughout the western hemisphere and even in Europe.

I have already mentioned the copy of the Virgin of Guadalupe that was carried on Giovanni Andrea Doria's ship in the famous sea battle of Lepanto. That copy, believed to be the earliest copy still extant, now hangs on the wall high above the altar of Santo Stefano d'Aveto in Italy, where I viewed it in the spring of 1981.

It is not quite three feet in height—about half the size of the original, which measures 66 by 41 inches. I climbed a twenty-five-foot ladder to get within four feet of the picture. Close study revealed that every detail of the copy is identical with the original, including the number of stars on the mantle—forty-six—the number of years it took to build the temple

---

[2]*Informacion de 1556*, in F. de Jesus Chavet, *El ulto Guadalupano del Tepeyac*, 6th ed. (Mexico City: Centro de Estudios Bernardo de Sahagun, 1978) 241 (folio 15b): "*Senor Arzobispo quisiese* que, por devocion se fuese a aquella hermita, habia de mandar que no se nombrarse de Nuestra Senora de Guadalupe sino de Tepeaca o Tepeaquilla; porque si en Espana, Nuesta Senora de Guadalupe tenia aquel nombre, era porque el mesmo pueblo se decia asi de Guadalupe."

in Jerusalem. The only exception is the shape and color of the gold rays and crown. These appear more skillfully done in the copy than in the original. In any case, it is now recognized that the gold rays were added to the original at a later date and may have been retouched at various times in the past.

In 1580, only a few years after this famous copy of the Image was carried to Europe, a portrait of the Virgin was painted by an anonymous artist on the wall of an adobe hut belonging to an Indian. Although the hut was inundated by the terrible floods that submerged much of Mexico City for five years, the picture on the wall survived. A shrine to hold the wall painting was built in 1595 and named Our Lady of the Angels. The painting and the shrine still exist.

Our Lady of the Angels resembles the Image of Guadalupe in several striking details. The mantle is identical—blue with a gold border and sprinkled with gold stars. The only difference is that in Our Lady of the Angels the mantle does not cover the Virgin's head. The robe, too, is the same—rose-colored with gold fleur-de-lis. And the clasp at the throat is identical—oval-shaped with a cross in the middle. In both paintings gold rays surround the figure and under the Virgin's feet is a crescent moon.

Many of these motifs are derived from a description of the Virgin given in the Bible in Revelation 12. Besides the golden rays of the sun surrounding her and the moon at her feet, the Bible passage mentions a golden crown, present in both pictures. Such Revelation 12 motifs appear in portraits of the Virgin dating back at least to the eighth century. But it is important to remember that these particular details—the gold rays, stars and border of the mantle, the gold crown, the gold fleur-de-lis on the robe, and the moon and angel at her feet—are now known to be later additions. Comparison of the Image with Our Lady of the Angels would seem to indicate that those additions were made in the late sixteenth century.

Since then, millions of copies of the Virgin of Guadalupe have found their way into the hands of worshipers. One of the earliest pieces of printed matter in the New World is a small line engraving of the Image that must have been distributed by the thousands to the newly converted Christians of Central America. In the eighteenth century famous painters like Miguel Cabrera produced a series of paintings illustrating the legend of Juan Diego, and these too were copied widely. In the nineteenth century the Emperor Maximilian, a devoted believer in the Virgin of Guada-

lupe, had her image stamped on banners that hung in his court and en-
graved on medals he presented to his soldiers.

The historian Northrop was struck by the overwhelming influence of
the Image on Christian worship in Mexico. He wrote:

> Anyone who has walked among the mountains or through the fields and
> orchards of Switzerland, Bavaria, Austria, or France will recall coming
> again and again, even in out-of-the-way places, upon little shrines of the
> Christ hanging upon the Cross. This is the true and natural symbol for
> orthodox Catholicism. Nevertheless, in journeys covering hundreds of
> miles radiating from Mexico City in every direction, no such image of
> the Christ located in the countryside or even made very conspicuous in
> the churches comes to mind; in its place appears instead the Madonna
> of Guadalupe.[3]

Despite religious and, later, political opposition to the Virgin of
Guadalupe, the ardent devotion the painting inspired has never flagged.
In the earliest history of New Spain, written by a Franciscan who was
beloved by the Indians and became known by the name "Motolinia"
("threadbare"), there is a remarkable account of the unprecedented conversion of the Aztecs to Christianity, which seems to have occurred a few
years after 1531, the date historians now pinpoint as the year when the
miraculous Image of the Virgin appeared on Juan Diego's tilma.

Motolinia claimed five million conversions, and described the method
he used to arrive at this estimate:

> I calculate the number of the baptized in two ways: first, by the towns
> and provinces which have been baptized; and second, by the number of
> priests who have administered the Sacrament. . . . From the sixty who
> are here in this year 1536, I subtract twenty who have not baptized be-
> cause they are new to the country and do not know the language. For
> the forty who remain, I would estimate one hundred thousand baptisms
> each. . . . there must have been baptized up to the present day, nearly
> five million.[4]

---

[3]F. S. C. Northrop, *The Meeting of East and West* (New York: Macmillan,
1946) 28.

[4]Toribio de Motolinia, *History of the Indians of New Spain*, trans. Elizabeth
A. Foster (Westport CT: Greenwood Press, n.d.).

By 1541, Motolinia said that some nine million Aztecs had become Christians. Although he made no direct reference to an apparition of the Virgin at the site or to a shrine there containing the Image, he did make specific reference to Tepeyacac, an alternate spelling for Tepeyac:

> Since this report was copied, over [sic] five hundred thousand have been baptized. . . . in 1537 in the province of Tepeyacac alone, there have been baptized by actual count over [sic] sixty thousand souls.[5]

A popular theory among those who doubt the supernatural origin of the Image of Guadalupe is that the church commissioned the picture and invented the legend of Juan Diego to account for it, as a device to attract the Indians to Christianity. Yet the fierce jealousies between the Spanish religious authorities that are described in Motolinia's *History of the Indians of New Spain* would seem to argue against a church conspiracy to invent a useful "miracle." For such a conspiracy to be effective, a certain measure of unanimity on the part of the church fathers would have been essential. The dissension revealed by the official inquiry of 1556, with its open hostility to the growing devotion to Guadalupe, is a further reason to doubt that the church "concocted" the miracle on Tepeyac hill.

Motolinia's writings contain tantalizingly indirect references to that miracle. He referred to "strange things" that happened "ten or twelve years" before the publication of his history in 1541. In a later comment, he reported that he had heard of these strange occurrences, but he "neither confirms nor denies" their actuality.

The ambiguity of church authorities toward the Virgin of Guadalupe gradually gave way to wholehearted acceptance. By the time of the second official inquiry in 1666 attempts were already being made to give solid scientific support to the miraculous nature of the painting that had inspired such widespread devotion among the common people of Mexico. By the nineteenth century some of the highest figures in the Church and the government, including Emperor Maximilian, were enthusiastic Guadalupans. The foundation for historical research on the Image was laid in the eighteenth century by Sigüenza y Góngora, a professor of mathematics who had been given an entire portfolio of ancient Indian documents

---

[5]Ibid.

by his royal friend Don Juan de Alva Ixtlilcóatl. Alva Ixtlilcóatl, who had received these treasures from his father, lacked an heir, so he bequeathed the portfolio to Sigüenza, to be given in turn to the Mexican National Library.

Although there is disagreement as to how much mention was made of the painting in this portfolio, there undoubtedly was some reference to it, for the Ixtlilcóatls, father and son, believed implicitly in the miracle on Tepeyac hill. When Sigüenza, who had been skeptical, read through the documents, he also was convinced.

In 1746, the historian Beneducci Boturini published a history of New Spain based on a wealth of sixteenth-century documents that he had gathered together. Of all those who have tried to trace the history of the Image, Boturini was one of the most enthusiastic:

> Hardly had I arrived in Mexico when I felt myself driven by an invariable attraction to undertake research into the prodigious miracles of Our Lady of Guadalupe. I discovered that its history was based on a *single tradition*, and that it was not known where or into whose hands the written proof of such a great prodigy had fallen.[6]

Even scholars who dismiss the idea of a supernatural origin for the painting are grateful for Boturini's surviving documents. Among them is a list of thirty-one sources confirming the apparition of Mary in 1531. Boturini spent seven years compiling the list, which he published in his *Prologo Galeato*. His famous *Museum Catalogue of Indian History* also contains significant Guadalupan references. He was later expelled from Mexico for political reasons, and it is tragic that he was already dead when the king of Spain appointed him "Historiographer of the Indies."

Despair almost surely was a major factor in Boturini's death. His thirty-one Guadalupan sources were confiscated by the Mexican authorities, along with the rest of his scholarship, at the time of his expulsion. The documents survived, but their elaboration, known only to Boturini, did not.

---

[6]Benaducci Boturini, *Idea de una Nueva Historia Generalde la America Septentrional* (Madrid, 1746) in LaFaye, *Quetzalcoatl and Guadalupe*, 248, 262, 268.

Eventually, the Boturini collection came into the hands of a Mexican aristocrat, M. F. de Echeverría Vetia. Echeverría was not originally a believer in the miraculous Image. What caused him to reassess his views? It must have been some document he found in the Boturini collection, for immediately after receiving it he became a vigorous proponent. The Boturini collection now rests in the archives of the Basilica of Our Lady of Guadalupe in Mexico City.

At the beginning of Mexico's war of independence from Spain in 1810, the revolutionary leader Father Hidalgo, a parish priest who had been greatly influenced by the French Revolution, adopted the Image of Guadalupe as the revolutionary symbol. The treaty that ended the Mexican-American War of 1846–1848 was named the Treaty of Guadalupe-Hidalgo, further evidence of the overwhelming devotion accorded the Image in the nineteenth century.

In the twentieth century scholars have brought. to light several additional documents that—while of course not proving that an apparition of the Virgin actually occurred in 1531—support the view that belief in the apparition was widely held at the time. Church leaders today no longer quarrel with that belief, which has persisted with undiminished strength through the centuries. On October 12, 1945, Pope Pius XII specifically confirmed the Church's view by stating,

> On the shores of Lake Texcoco flowered the Miracle, and on the cloak
> . . . was painted a most lovely Portrait, by brushes not of this earth.

In 1976 a new basilica was built in Mexico City to house the sacred painting. There, in January 1979, the newly designated pope, John Paul II, made clear that the Roman Catholic Church now accepted, without reservation, the importance of the Virgin of Guadalupe in the spiritual life of Mexico.

# The Right Eye
# of the Virgin

The highest that man can attain in these matters is wonder.
—Goethe

In addition to my search for historical evidence that would confirm or
deny the truth of the Guadalupan legend, I wanted to find out what kinds
of scientific research had already been done on the painting. One of the
most tantalizing leads I followed concerned what seems to be the
reflected image of a man's head in the right eye of the Virgin.

Anyone who examines the painting at close range can see this apparent reflection. How long it has been part of the painting no one knows,
for documented note of it did not occur until 1929 (Marcué), and it was
not until after 1951, when the anomaly was rediscovered by Carlos
Salinas, that scientific examination was made.

One of those called in to examine this particular aspect of the sacred
Image was an ophthalmologist, Dr. Javier Torroella-Bueno. What he
discovered is truly amazing.

Before I could understand the importance of his discovery I needed
to learn something about the physiology of the human eye, in particular
what is called the Purkinje-Sanson principle (named after Jan Evangelista
Purkinje, Czech physiologist, and Louis Joseph Sanson, French physician,
who independently described the phenomenon). In its simplest form, the
Purkinje-Sanson law states that whenever we see any object, the object
is reflected in each eye, not once but in three different places. This
threefold reflection is caused by the curvature of the eye's cornea. Two
of the reflections are always right side up and one is always upside
down. Depending on the angle at which the object is seen, the three
reflections occur on different parts of the eye because of the differing

angles of curvature of the cornea. The curvature also causes the reflected images to be distorted in a varying degree.

On May 26, 1956, Dr. Torroella sent the following letter to Carlos Salinas:

> If we take a light source and put it in front of the eye . . . we see the cornea, the only part of the eye which can reflect an image in three places [the Purkinje-Sanson principle]: the front surface of the cornea, and both front and rear of the lens surfaces, immediately behind. . . . The image of the Virgin of Guadalupe, which has been given to me for study, contains in the cornea these reflections. . . . In the images in question, there is a perfect collocation in agreement with this [principle], the distortion of the figures even concurring with the predicted curvature of the cornea.[1]

In other words, in the eyes of the Virgin of Guadalupe, not only is the cornea curved, but it is curved exactly the way the human eye is curved. The location of the reflections, which can be seen in both eyes, conforms exactly to what occurs in the living eye. One is near the temple, while in the other eye the reflection is near the bridge of the nose. Thus the images are said to be "collocated," that is, they are located where they would have to be in order to conform to the way our eyes work together to perceive a single object.

Dr. Torroella's findings were corroborated by another ophthalmologist, Dr. Rafael Torija-Lavoignet, who invited me to his office to discuss the conclusions he had reached on examining the painting twenty-five years earlier. He gave me a copy of a statement published in 1956:

> When the ophthalmoscopic light is directed to the pupil of the human eye, one sees a luminous reflection. . . . Lighting the pupil of the eye of the Image of the Virgin, there appears the same luminous reflection . . . impossible to obtain on a simple surface, and moreover on one opaque, as [is] the said painting. . . . I, with the aid of the ophthalmoscope, proceeded to examine the eyes of diverse paintings, even of

---

[1]Carlos Salinas and Manuel de la Mora, *Descubrimiento de un Busto Humano en las Ojos de la Virgen Maria de Guadalupe* (Mexico City: Editorial Tradicion, 1976) 52.

photographs (in each case, they were of readily distinguishable persons), but found no reflections in any of them.[2]

Two years later, on September 20, 1958, Dr. Torija published his study of the Purkinje-Sanson effect as exhibited in the Guadalupan painting. The "human bust" reflection is readily visible, he states, especially in the cornea of the right eye. "In addition to the human bust," continues Torija, "there are two luminous reflections, which together correspond to the three images of Purkinje-Sanson."[3]

Dr. Torija explained to me that it is necessary to dilate the eye in order to perceive the second of three reflections. After examining reflections he saw in the right eye of the Virgin, Dr. Torija concluded that they conformed perfectly with the requirements of the Purkinje-Sanson law.

In his published study Torija stated that the "human bust . . . image reflected in the cornea is not an optical illusion caused by some irregularity in the [cactus cloth]." Dr. Torija made clear to me that he was not denying that there might be an imperfection in the weave, but that the way the eye of the painting reflected the light of the ophthalmoscope was exactly the way the curvature of the human eye reflected the same light. In other words, whether the reflections in the eye of the Virgin are caused by an irregularity in the coarse cactus cloth is essentially irrelevant.[4]

Dr. Torija is convinced that the Image of Guadalupe is a miraculous creation. Additional research in the intervening years has buttressed this conviction.

In 1975 the glass was removed so the picture could be examined by another ophthalmologist, Dr. Enrique Graue. Dr. Graue reported: "The total sensation is that of seeing a 'living eye' and, really, it cannot be thought less than something supernatural."[5]

Another discovery was made two years ago by Dr. José Aste-Tonsmann, who received his doctorate in systems engineering from Cornell

---

[2]Personal interview with author; also reprinted in ibid., 54.
[3]Ibid., 56-58.
[4]Ibid.
[5]Ibid., 8-9.

University. Using a photograph of the Image marked off in one-millimeter squares and using computer amplification to magnify each square 2,500 times, he not only confirmed the existence of the Purkinje-Sanson images in the right eye, but also found that in the iris of the left eye could be seen at least four figures, one of which appeared to be an Indian peasant with his hands lifted in prayer. Dr. Aste wrote: "I believe, without fear of error that this person is Juan Diego." The other figures, he theorized, represent those present when Juan Diego opened his tilma, including Bishop Zumárraga and the translator Juan Gonzalez, and were imprinted in the left eye (of the picture) at the time it made its miraculous appearance on the tilma. They are what the Virgin herself saw at that moment, according to Dr. Aste.

Aste generously shared with me his procedures and conclusions during my January 1982 visit to his Mexico City home, and the account here is drawn from his work. While it is beyond the scope of this book to give comprehensive details of his research, a brief survey of his achievements and the astonishing results would make a most interesting addition to this study.[6]

Assisted by computer, Aste has reproduced an enhanced view of the tilma, with attention being given to sections as small as $6/_{1,000,000}$ meter. He has concentrated attention on the eyes of the Virgin through a process called digitalization. This photo-enhancing involved the assignment of numerical equivalents to originally qualitative values. The resulting printout allows us to evaluate images too small to interpret visually.

Aste studied both eyes in microscopic detail. Before this research, although the suggestion of human shapes has been discerned in the left eye (that is, the eye of the painting that appears to the observer's right), most attention had been focused on the right eye. This right eye (which is the one farthest away from the vertical seam) contains an obvious "human bust" observed before the advent of computer technology. Now, computer enhancement has revealed unanticipated detail in both eyes. Because of this research, weight has been added to the argument that the shapes are genuine images. If what is captured in the eyes is some kind

[6]José Aste-Tonsmann, "Analisis por Computadora en los Ojos de la Virgen de Guadalupe," in *Milciades* (Mexico City: Editorial Circulo Farmaceutico, S.A., 1981) 1:68-71.

of visual record, it is logical to assume that the "human bust" in the right eye, located slightly off center, will have a corresponding image in the left eye.

The bust in the left eye (as seen from the position of observer) is in the upper right quadrant of the pupil. Aste located that image in the left eye in the "upper center." This slight difference in position would be acceptable because of the lateral distance between the eyes in the human face.

Perhaps the most startling thing revealed in this contemporary analysis is the questioning of the identity of the dominant feature in the eye as being the Aztec Juan Diego. Almost immediately after its announced discovery, this anomaly was made to correspond in popular piety with the recipient of the apparition. Indicative of his high scholarship generally, Aste disclaims the connection by noting that the Aztecs of the time are known to have been clean-shaven. The dominant "human bust" clearly provides the shape of a full beard extending down from the chin.

From the larger perspective, however, the issue is not who the face-shape represents, but what it is, and consequently what it implies about the Image of Guadalupe.

It will be remembered that the work of Dr. Torija placeed great emphasis on the apparent conformity of the bust in the right eye with the Purkinje-Sanson optical law, that Torija perceived not one but three images of the same figure in geometrically proper alignment. Does Dr. Aste's work substantiate this original hypothesis, which is also supported by other ophthalmologists? Yes, it does. Pointing to a page in his own book for added emphasis, Aste told me that the computer "had corroborated with sufficient clarity" these internal images, just where they would have to be if explicable as Purkinje-Sanson phenomenon.

The computer engineer calls the dominant image not "Juan Diego," but "a bearded Spaniard." He readily notes that the image is much clearer in the Virgin's right than in her left eye, but computer analysis adds something in terms of the enhancement that was invisible to the naked eye of all previous observers. Connected to the beard is a shape that could be the right hand of the figure.

The new examination finds an image of the "bearded Spaniard" in both eyes of the painting, and has indicated the proper locations. We have spoken in some detail about the right-eye image. Why was the image of the left-eye "Spaniard" not seen for what it was at an earlier date? Aste's modest answer is also reiteration of the significance of the remarkable, new study: it is a matter of magnification. A vague figure was observed but not discerned with sufficient clarity before the computer was employed simultaneously to amplify and enhance the shape.

Other quasi-human shapes are reported by Aste to exist in both eyes of the painting. Of particular note is a face in the pupils that, like the "bearded Spaniard," is reminiscent of human anatomy. It reveals a bearded, balding head, and the beard is longer and is pointed, both features different from that of the former human torso. Having obtained a painting by the famous Guadalupan artist Miguel Cabrera that portrays Bishop Juan de Zumárraga, Aste is able to show a certain resemblance between the two profiles. Cabrera lived some two centuries after the event in question, but it is quite possible that he had as a model an accurate portrait of the Mexican prelate.

This "Zumárraga-form" is perceivable, as was said, in the pupils of both eyes. The difference is that the indistinct shape is in the right eye, rather than the left—just the opposite as with the "bearded Spaniard"—so the description provided by Aste for the left eye is necessarily more nearly complete than for the right. But if, as he has said, the discovery of the major bust in the left eye was a function of the greater magnification or resolution available via computer, one might wonder why the same would not apply to the "Zumárraga" in the eye so intensely studied in the past.

The contribution of the computer is not exclusively one of literal enhancement or even clarification through magnification. As important as these functions can be, there is another property of this process that in the present research is of far greater importance. It is associated with a technique known to mathematicians as "statistical mapping." What is ultimately impressive about Aste's work is not the perceivable facial shapes; if it were, we might well be studying a modified Rorschach ink blot. Rather, what impresses the qualified observer is that these several anomalies are positioned in the corresponding eyes in such a fashion as to be virtually beyond coincidence.

A random accumulation of pigments, some of which seem to resemble known forms, is interesting but subject always to wide interpretation. In the case at hand, the shapes are positioned at *precisely* the locations demanded by optical geometry and therefore cannot be dismissed by a disinterested student.

Nor are these the only shapes discovered and aligned in the respective eyes in the new research. Also discerned is a cluster of forms (Aste calls this the "indigenous family group") and, even further, a seated figure. This figure appears to have the right leg drawn up in front and the left leg extended parallel to the floor. This "seated Indian" appears to be female, and quite large, the image having a total length of more than four millimeters. The Peruvian researcher observed in our conversation that though he had studied this one shape for more than a year, he was still learning new things about it, and about other mysterious forms by means of comparison to it.

Is Dr. Aste's discovery no more substantial than the human shapes we see in the clouds, the result of what Father Harold J. Rahm once termed a "pious imagination"?[7] Or is it one more of the mysteries surrounding the Virgin of Guadalupe that our science, no matter how advanced, has as yet been unable to solve?

I am not sure. Do the "human figures" perhaps constitute no more than one would be led to expect statistically? Are they but *chance* phenomena? Or is there a still more fundamental objection, that there are no figures in the eyes, nothing to "explain" in the first place? By this view, the putative shapes are projections only. The phenomena would then constitute a sort of Rorschach test.

But if, on the other hand, the figures are indeed genuine, then we are presented with a double helping of mystery. The Image of Guadalupe itself is said to be *acheiropoietos*—not made with human hands—and if the shapes in the eye are real, then they provide additional valuable evidence of such natural phenomena, for they too are "not made with human hands."

We know too little about the nature of perception itself to provide a conclusive answer. I find myself troubled with the idea of "reflections" of supposedly near subjects requiring microscopic enlargement even to be seen. Yet the possibility of Purkinje-Sanson phenomena remains intriguing.

---

[7]Harold J. Rahm, *Am I Not Here?* (Washington NJ: Ave Maria Institute, 1962) 144.

It should be noted that hesitation about the eye figures (the study of which is a legitimate part of any history of the Guadalupe mystery) is not antireligious. The projection view is, in fact, the opinion of Father Rahm (mentioned above). The church has always avoided falling into the hands of its enemies precisely by choosing to conduct its own rigorous tests of possible miracles well in advance of endorsing them.

# SCIENCE
## ANd THE MIRACulOUS

> You yourself are even another little world, and have within you the sun
> and moon and also the stars.        —Origen, *Liviticum Homiliae*

The basilica authorities had agreed to allow us to make a close examination of the painting with the glass removed and to take infrared
photographs of it, provided we avoided exposing the picture to excessive
heat. The hot lights usually required for photography were out of the
question. We arranged to use special lights that threw off little heat and
to monitor the surface of the painting continuously with a thermometer.

The date was set for our long-awaited rendezvous—May 4, 1979.
Philip Callahan and I arrived in Mexico City four days earlier to make
sure the cameras, lights, and other equipment necessary for our examination of the Guadalupe Image had been assembled and were in perfect
working order.

What would we find when we viewed the Image close at hand? What
would infrared reveal? Would new scientific techniques solve the
mysteries surrounding the painting?

Callahan and I had little sleep the night before. In the early evening
of Friday, May 4, after the basilica had been closed to worshipers, we
arrived at the administration building behind the new basilica as arranged,
only to meet with disappointment. The Image could not be viewed that
evening because only two people knew the combination to the huge vault
where the painting was kept when it was not on display. One of the two,
Abbot D. Guillermo Schulenburg-Prado, was out of the country on
business. The other, the new church's architect, refused to open the vault
without the direct authorization of the abbot. We would have to wait until
the abbot returned on Monday.

To fill the time and satisfy our curiosity we decided to visit the chapel on top of Mount Tepeyac. Scenes of Juan Diego's life were depicted on the wall panels in the foyer to the main sanctuary. On examining one panel, Dr. Callahan exclaimed: "Now I'm sure they'll let us photograph the Guadalupe on Monday."

Puzzled, I asked him the reason for his certainty. He pointed to three birds in the panel. "I owned birds of the same species when I was a boy," he explained, "and not a one is a native to the Americas. It has to be a good omen."

He was right. On Monday morning the young woman who had been assigned to be our translator informed us that the glass cover would be removed from the painting late that same evening. She cautioned: "But don't tell anyone—anyone at all—what is going to happen to-night—especially not the press. If the newspapers find out, some one might put 'Yanquis Invade National Symbol' in three inch letters on the front page."

Unfortunately, on the preceding Friday we had given a preliminary interview to a reporter from United Press International. We worried the entire day that news of this interview would reach the basilica authorities.

At eight o'clock on Monday, May 7, 1979, Callahan and I arrived at the entrance to the basilica's administration building, where we were to meet our translator. Except for an occasional late pilgrim or night watchman, we were alone.

The minutes crawled by. Callahan stopped pacing and sat down on the steps of the old basilica, where the Guadalupe had reposed until 1976 when the church was closed because it had become structurally unsafe. Equally distraught, I climbed the wide steps leading up to another sanctuary, high on a hill behind the administrative building. I looked out over the city, then returned to the plaza below where Callahan sat with his head buried in his hands.

Abruptly, out of the darkness of an underground garage, our trans-lator appeared, along with several men in clerical garb. I had met most of them. As Callahan and I were guided to the upper floor of the church, several of our escorts expressed their apprehension concerning what was soon to occur.

Callahan, a Catholic, asked to receive Communion, and his wish was granted. Then we watched while a huge stainless-steel door was opened and twelve men slid the life-sized painting backward into the vault.

In the old basilica, which was built in the early eighteenth century, there was no vault behind the Image.

Anyone who wished to examine the picture at close range was required to stand on a specially constructed platform, a cumbersome arrangement at best.

It took the twelve-man crew two hours to remove the heavily bejeweled outer frame, then the bulletproof glass, and finally the inner frame. While some of the men carefully placed the silver-backed picture in the position we had requested, others washed the glass, and still others bustled around looking for electric outlets for our lights. We loaded our three cameras and rehearsed procedure. Then we set to work.

We first examined the picture with the naked eye and with a hand lens, from less than a half inch from the surface. We saw immediately that the gold rays surrounding the figure were badly chipped and scarred. The clerics must have noted the disappointment on our faces, for they quickly reminded us that the metallic gold rays were a later addition to the painting and were in no way indicative of the substance of the original Image.

In the pages that follow I am going to quote extensively from Dr. Callahan's report (with his kind permission). Regarding the gold used in the painting, he reported:

> The gold paint of the sun-rays is metallic gold, opaque to near infrared rays, and that of the stars, et cetera, of the mantle are of an unknown pigment (probably alumina hydrate natural earth ocher). . . . these details were added by human hands long after the original painting was formed. . . . the sun-rays, stars, and mantle trim will . . . deteriorate with time.

Our close examination with the naked eye confirmed the remarkable state of preservation of the original. There is no evidence whatsoever of cracking. Yet paintings less than half the age of the Guadalupe Image commonly show a web of hairline cracks across their entire surface, caused by drying of the paint. Any moisture left in the paint, or whatever was used to color the Image, would surely have evaporated in the four hundred and fifty years of its existence.

Another remarkable aspect of the painting that is immediately apparent to the naked eye is the way it seems to change both size and coloration when viewed from various distances. Callahan, who has studied the phenomenon of iridescence on bird feathers and insect scales, explained that this strange effect is caused by the diffraction of light from the surface:

> Beyond six or seven feet . . . the skin tone becomes what might best be termed an olive-green, an "Indian Olive," or gray-green tone. It would seem that somehow the gray and "caked"-looking white pigment of face and hands combines with the rough surface of unsized tilma to "collect" light, and diffract from afar the olive-skinned hue. . . . Such a technique would seem to be impossible to accomplish by human hands; however, it occurs often in nature. In the coloring of bird feathers and butterfly scales, and on the electra of brightly colored beetles. Such colors are physically diffracted . . . do not depend on absorption and reflection from molecular pigments, but rather on the "surface-sculpturing" of the feather or the butterfly scales.

Callahan concluded:

> The same physical effect is quite evident in the face of the *Imagen*, and is easily observed by slowly backing away from the painting until the details of the imperfections of the tilma fabric are no longer visible. At a distance where the pigment and the "surface-sculpturing" blend together, the overwhelming beauty of the olive-colored Madonna emerges. . . . the expression suddenly appears reverent yet joyous, Indian, yet European, olive-skinned, yet white of hue. . . . It is a face that intermingles the Christianity of Byzantine Europe with the overpowering naturalism of the New World Indian.

After our preliminary examination, we took seventy-five photographs, forty of which were on specially prepared infrared film. We made both full-figure shots and extreme close-ups of the face, hands, and other areas of special interest. Working past midnight into the early morning, we spent four hours with the mysterious painting—probably a longer stretch of time for uninterrupted examination than was available even to the artists and physicians who took part in the official inquiry of 1666.

At dawn I was on a plane bound for my home in Pensacola; Callahan was in another plane, his destination Gainesville, also in Florida but several hundred miles southeast. He carried with him the precious film, and he had taken special precautions to ensure its safety during preboarding and customs inspections. A few days later, I was present in his darkroom when he developed the film. Our excitement grew as we waited for the developer to work.

The first picture developed appeared fuzzy. Callahan assured me that such fuzziness is normal with infrared film. It was a good indication, he said, that the rest of the processing would be successful. As it turned out, every one of our seventy-five photographs developed perfectly.

Not before he had spent two months in examination and evaluation of the photographs was my friend ready to release his preliminary findings to the press, cautioning even then that final conclusions might well be years in the future. The most important of his findings—and the primary aim of our examination—regarded the presence or absence of an undersketch. While the absence of a preliminary sketch is not incontrovertible proof that the Guadalupe is miraculous, the *presence* of such a sketch would have proved once and for all that an earthly artist had created the painting.

The practice of making a rough sketch before proceeding with a portrait traces back to antiquity. The apocryphal *Acts of John*, which dates to the second century A.D., relates that the artist took two days to paint a portrait of the apostle, spending the entire first day drawing a true-to-life sketch. Only when he was satisfied with the sketch did the artist take up his paintbrush.

Callahan examined the close-up infrared photographs of the fold shadows of the robe and mantle. He wrote:

> The fold shadows of the robe may, under cursory examination, appear to be thin sketch lines. However, closeups of both the robe and the mantle show them to be broad, and also *blended with the paint* and, therefore, uncharacteristic of undersketching. . . . as in the case of the blue mantle, the shadowing of the pink robe is blended into the paint layer, and no drawing or sketch is evident under the pink pigment.

The closeups of the Virgin's hands led to the same conclusion:

> The . . . hands, as in the case of the robe and mantle, show no undersketching whatever. The shading, coloring, and pigments of the . . . hands are inexplicable.

Some thin black lines, which outline the entire left side of the figure from the shoulder downward, might seem to provide a sort of preliminary guide for the painting of the body, but the fact is that these black lines were applied some time after the blue of the mantle but before the gold border had been added. Callahan reported:

> Since the black is opaque to the infrared, the border outline shows up much better than it does in the visible [light] photographs where it is obscured by the shadows of the mantle. . . . the added black outline trim often misses the incorporated shadow . . . around the edge . . . a careless job of outlining for Gothic emphasis.

Finally Callahan examined closeups of the beautifully serene face of the Virgin. Of all parts of the painting, undersketching was most likely to be found here. He found none.

> The eyes and shadows around the nose are simple dark lines that are not underdrawn but are, rather, part of the face pigment. . . . [The face] does not show an underdrawing or sizing of any type. . . . These are characteristics which, of themselves, render the painting fantastic.

Examination of the infrared photographs led to another important discovery—the existence of four fold-marks at about the level of the Virgin's hands. Callahan observed that

> Two of the four fold-lines are easily visible across the Virgin's body. The two top fold-lines cross the entire body but end *at the edge of the mantle*. They also cross the tassel at the center of the painting. They do not appear at all across the sunburst, or any part of the background surrounding the body of the figure.

From his unprecedented discovery of these folds, Callahan was able to deduce when certain additions had been made on the original painting. In particular, he observed that "the background was added after the rest of the painting was formed." Had the background been present when the Image was folded (perhaps when the painting was moved during the floods that inundated Mexico City about 1630), it would have shown the same fold-marks as are apparent on the body of the Virgin.

In addition to the background, the black outlines, the gold rays, and the gold trim and stars on the mantle and fleur-de-lis on the robe, it is generally agreed that the entire bottom third of the painting is a later addition.

Callahan recorded that

> The cherub is at best a mediocre drawing. The arms are clumsy and out of proportion and obviously added to support the Virgin Mary. The face is lifelike, but has none of the beauty or genius of technique shown in the elegant face of the Virgin. . . . the hair is probably black oxide of iron. It overlaps the moon, as shown by the drawn line which circumscribes it. . . . the red of the angel's robe, unlike the Virgin's delicately colored . . . is laid on thickly and is completely opaque to the infrared, indicating that it is in all probability red oxide, an extremely permanent pigment, yet chipping at its outer edges.

The wing feathers seem to be of the same form of red. The wing blue is "badly cracked [and] probably a form of copper-oxide 'Mayan Blue,' laid on, like the thick black of the moon, so thickly as to be subject to heavy cracking."

Callahan concluded that "the angel was added after the moon" because of the overlapping of the angel's hair. This part of the painting, he notes, is "in an extremely bad state of repair." The same black appears on the brooch at the neck of the Virgin and in the hair of the cherub. Like the gold leaf of the rays, it is completely opaque to the infrared, indicating a metallic base to the paint.

Callahan suggested three possibilities for the black: silver nitrate, carbon black, and iron oxide (slate black is too gray). Carbon black, which did not come into use until 1884, must be excluded, since these additions to the original painting had clearly been made, on the evidence of extant copies, by the seventeenth century. While Callahan emphasized that no certain identification of color composition could be made without chemical analysis, he favored iron oxide:

> chemically ferric oxide . . . known by painters as Mars Black. . . . [It] is a dense, opaque, permanent color almost brownish in undertones.

It also has a tendency to turn browner with time. "Since it is a heavy pigment," Callahan added, "it would be expected to crack away with age if not properly bound to the canvas." Both browning and cracks are visible in the bottom third of the painting.

The bottom fold of the Virgin's robe presents another curiosity. The same distinctive fold can be seen in other famous paintings of the sixteenth century in America, in particular, The *Tribute Roll of Montezuma*. In that picture hundreds of Aztecs are shown presenting their tilmas as gifts to their new rulers, the Spaniards. Brightly colored, each of the tilmas shows a consistent and apparently symbolic shape. It involves the folds placed diagonally on opposing corners of the tilma. Callahan calls it the "Aztec tilma-fold."

Whatever its earlier purpose for the Aztecs, the fold was added to the bottom of the Guadalupe. "The simple fact," Callahan noted, "is that some artist (not a very good one) went to great pains to duplicate the tilma-fold at the base of the Virgin's robe." The tilma-fold was, like the cherub, put on after the crescent moon because "between the line and the visible black moon, one half of the moon was brushed over by the lower edge of the robe and (in infrared) shows through it." He observed that there is a

> poorly drawn black line which folds into the base of the robe. The same black line was drawn above the foot, but is cracked off. Both the posterior portion of the foot and the moon lie *under* this segment of the . . . fold, but barely show through because of the opaque (even to infrared) paint of the moon.

Every artist who has ever examined the painting, from 1666 onward, has remarked on the apparent absence of sizing. But they are only partially correct. The bottom third of the Image we found to be clearly if clumsily sized by the layered application of paint to the originally unsized surface. Moreover, brushstrokes, so notably absent from the original parts of the painting, can readily be seen here. The blue color of the tilma-fold is not of the same composition as the blue of the Virgin's mantle. The turquoise blue of the mantle is radically and strangely different.

Callahan reported that the mantle

> is of a dark turquoise blue, more toward a blue than a green hue. It does not appear to be what artists call turquoise green (cobalt oxides mixed

with chromium and aluminum). It is also unlikely to be Bremen or Lime blue (mixtures of copper hydroxide carbonate). [Although] Bremen/Lime Blue can be mixed to a great number of shades of blue or greenish blue. . . . the mantle shade is [closer] to the hue seen on early Mayan wall-paintings, or on the cured animal-skin 'books' of the Mixtecs. These colors are likely to have been made from copper oxide.

Callahan also observed, however, that "this presents an inexplicable phenomenon, because all such blues are semipermanent and known to be subject to considerable fading with time, especially in hot climates." The blue of the mantle, Callahan noted, "is of even density and not faded . . . of unknown, semitransparent blue pigment . . . bright enough to have been laid on last week."

Callahan found the rose-pink of the robe even more mysterious than the blue of the mantle.

The robe is highly reflective of visible radiation, yet transparent to the infrared rays. . . . of all the pigments studied, the rose is by far the most transparent. . . . it is unlikely to be either cinnabar or hematite, both of which are Indian red pigments; nor is it orange mineral (too yellowish), as all of these minerals are opaque and not transparent to infrared rays. Red lead may be excluded for the same reason. Red oxide is an absolutely permanent pigment. . . . it would be a likely candidate, except that it is also extremely opaque to infrared rays. This leaves little but the modern aniline reds. [Yet] there is no evidence anywhere in this painting of any modern aniline colors. . . . it appears to be inexplicable.

Also inexplicable is the black of the Virgin's hair. Unlike the brownish black of later additions to the painting, "The black of the hair," Callahan writes, "cannot be iron oxide or any other pigment that turns brown with age, for the paint is neither cracked nor faded."

But perhaps the most amazing part of the painting is the Virgin's face. I have already mentioned how the color of the face seems to change when viewed from different angles because of the diffraction of light from the rough threads of the cactus cloth. This explanation aside, the question of what substances were used to color the face has been a source of unending controversy. Coley Taylor, New York journalist who coauthored a book entitled *The Dark Virgin*, published in 1956, believed some sort of "wash" or "dye" was used. Dr. Eduardo Turati, an ophthalmologist who studied the "human bust" phenomenon of the eyes in 1975, suggested that the colors were in the original threads before they were woven into cloth. Others have theorized that the original parts of the painting, including the face, were stamped on the cloth in some fashion.

When Callahan and I viewed the face through a magnifying glass on the night of May 7, 1979, we realized that no one explanation could cover all its mysterious properties. Later, on examining the infrared close-up photographs of the face, Callahan found the cheek highlight of special interest. It was produced, he reported, from "an unknown white pigment that is practically 'caked on' the coarse fabric. At first glance, it would seem to be blurred in infrared, and then semitransparent to that radiation." Furthermore, he observed:

> If the cheek layers were lime or gypsum, it is almost certain that such an extremely thick application would have cracked over the centuries.
>
> The pensive-meditative expression, beautiful to behold, is formed by the simple, narrow dark lines that make up the eyebrow, ridge of nose, and mouth. . . . close-up, the face appears almost devoid of depth, . . . but from afar, there is an elegant depth of expression. . . .
>
> The painting . . . takes advantage of the unsized tilma to give it depth and render it lifelike. This is particularly evident in the mouth . . . where a coarse fiber of the fabric is raised above the level of the rest of the weave, and follows perfectly the ridge of the top of the lip. The same rough imperfections occur below the highlighted area on the left cheek, and below the right eye.

Dr. Callahan concluded:

> I would consider it impossible that any human painter could select a tilma with imperfections of weave positioned as to accentuate the shadows and highlights so to impart such realism. The possibility of coincidence is even more unlikely!

In this he agrees with the many millions who, over the centuries, have decided that the beautiful face of the Virgin is a miraculous achievement. No amount of scientific analysis can account for the overall effect of the painting. Both Callahan and I were forced to admit that, in some irreducible way, the Image of Guadalupe is indeed a miracle.

# The Living Image

I started to wonder what would happen if the cosmos were not a desert and its beauty not a mask or deception. . . . then suddenly a wonderful encounter took place in Dresden. . . . The eyes of the Heavenly Queen . . . pierced my soul. I cried joyful and yet bitter tears, and with them the ice melted from my soul.
—Sergius Bulgakov, on first seeing Raphael's *Sistine Madonna*

Some may find it ironic that in our skeptical age the tools of science have been used not to disprove but in some degree to authenticate miracles of the past. Our discovery of the absence of undersketching in the Guadalupe Image and our inability to account for the remarkable state of preservation of the unsized cactus cloth as well as the unfading brightness of the paints or dyes used in the original parts of the painting placed Callahan and me firmly among those who believe the Image was created supernaturally.

What we saw during those four hours that we examined and photographed the picture in the vault of the basilica on May 7/8, 1979 also strengthened our conviction that further research, while incapable of solving the essential mystery of the Guadalupe's creation, could greatly enrich our understanding of the picture. Accordingly, on June 17, 1979, I wrote Monseñor Enrique Salazar as follows:

We would like to request of you, Mons. Salazar and Abbot Schulenburg-Prado, the opportunity to do some further photography and other research, i.e., ultraviolet photos; far-infrared spectroscopy and Fourier analysis—this requires computerization, but can in many cases determine the atomic and molecular nature of the colors. . . . It is precisely this area which is the basis of Dr. Callahan's worldwide reputation. Further studies (. . . over a period of years more probably than months) would be radiographic and/or X-ray fluorescence analysis; perhaps carbon-14 analysis; ideally, ion microprobe analysis, sometime.

What we propose as a logical next step, *because the infrared photos show neither an underdrawing of the face of the Sacred Image, nor any evidence of brushstrokes thereon,* is to try to ascertain the chemical nature of the colors. One or more of the methods outlined above may be able to do this. (italics added)

After further correspondence with Monseñor Salazar and with Monseñor Maurilio Montemayor, secretary of the Center for Guadalupan Studies, whose encouragement of our research was invaluable, we were able to arrange a second viewing of the painting in April 1981. At that time we were able to accomplish some of the eight specific procedures requested by Callahan that I had described to Abbot Schulenburg in a letter the preceding September:

(1) twenty near-ultraviolet photographs of the front of the tilma
(2) twenty near-ultraviolet photographs of the back of the tilma
(3) twenty near-infrared closeup photographs of the back of the tilma
(4) twenty (computer enhancement type) black-and-white photographs of the tilma back
(5) twenty additional near-infrared closeup photographs of the "fold-lines" of the obverse side of the tilma, including any pulled threads or such areas where the cloth might seem to be damaged.
(6) One small fiber from the edge of the tilma, for a laboratory (spectroscopic; electron-microscopic) study of its composition.
(7) The collection of any particulate matter (probably microscopic) from the *Imagen,* by means of the gentle application of specially prepared small cloths.
(8) The coincident exposure of special-grade color photographs of the two sides of the *Imagen.*

Ultimately, only spectrophotometry of the painting was allowed. This was done by Donald J. Lynn, a scientist with the Jet Propulsion Laboratory in Pasadena, California. He reported no unusual lines in the spectrum he obtained.

Meanwhile, based on what we have discovered so far and on my own continuing historical studies, certain suspicions regarding the painting can be laid firmly to rest. For example, to account for the brightness of the colors in mantle and robe, many people have suspected that artists employed by the church have, from time to time, in the last four hundred and fifty years, been called in to retouch the portrait. But in the original portions of the painting there is absolutely no sign of retouching—no brushstrokes, no cracked or chipped pigment, no layering of paint. In short, the unfading brightness of the turquoise and rose colors remains inexplicable.

Did a human artist paint the portrait—perhaps the Aztec artist Marcos Cipac mentioned in the first official inquiry of 1556? If he did, he did so without making a preliminary sketch—in itself a near-miraculous procedure. My own theory is that Cipac may well have had a hand in painting the Image, but only in painting the additions, such as the angel and moon at the Virgin's feet. How the life-sized figure of the Virgin was imposed on the rough cloth of a peasant's cape in the first place remains mysterious. Even if at some later date we determine that it was stamped, dyed, or woven into the fabric, we are scarcely closer to solving the mystery.

❑

In the years since I first stood in a line of worshipers to view the Virgin of Guadalupe, my fascination with the miraculous Image has steadily deepened. While scientific and historical research will answer some of the many remaining questions surrounding the painting, the *essential mystery* of Guadalupe—her hold on millions of Christians over the centuries—is irreducible.

# Epilogue

Earlier in this book I noted that "efforts to confirm or repudiate the supernatural origin of the Image of Guadalupe are not new" (13). Now, after having presented the original account and, along the way, as many alternatives with which I am familiar, it seems appropriate to bring into focus my own personal viewpoint. As the sensitive reader will have previously noted, an existentialistic stream runs concurrently beneath the obvious landscape of objective "scientific" examination.

My pursuit of the Guadalupan phenomenon was from the beginnings a pilgrimage—an act of thanksgiving. The idea to propose scientific study was, however important, an afterthought. In the fall of 1978, I was aware of an impending scientific study of the Shroud of Turin, but the travel plans I made to go to Italy were not motivated by science; rather, they were of the domain of an anxiety-ridden search. As Teresa Levine has put it so well:

> Someday you will find that they are no longer simmering but have suddenly burst into flame. These questions may be thought of in another way—as figures who are standing offstage in the darkness of the wings. But there are times whey they will come to the center of the stage and shout and scream at you. They will scream their importance on the center of the stage when you have a personal crisis or when your whole society is in crisis and a revolution seems about to break out. Some-times . . . when you suddenly feel that you have lost your bearings, that you don't know what you believe, that you have no convictions, and that you have a sense of vast inner emptiness, a sense of nothingness, these questions will then thunder loudly in that emptiness within you.[1]

I was in Mexico City rather than Turin because of a sea-change in my consciousness. That change was not defined by but was itself previ-

---

[1]Teresa Z. Levine, *From Socrates to Sartre: The Philosophic Journey.*

ous to the question that launched my Guadalupe odyssey. Refreshed and thankful, the question then formulated itself in my mind: Is there anywhere any object comparable to the Shroud of Turin?

So, when I first stood outside the Basilica of Guadalupe, it was—as you now can better understand—an offering of thanks. The scientific proposals, however exciting in themselves, were never foremost. If this account foreshadows my own conclusions regarding the Image, so much the better. But I do not think that a few words concerning the scientific rationale itself will pose any obstacle of distraction.

Precisely *what* did I expect to find, if I was actually given the unlikely permission to undertake scientific study of the Image of Guadalupe? This is a question of no small moment.

I expected to find whether or not there was anything unexplainable about the painting, when subjected to the sophisticated regimen of contemporary scientific examination. Was there any part, any feature of the Image itself that was *not* reducible to the familiar and the mundane? And what, if some defiant aspect was really discovered, would that imply? Something truly alien would mean *what*?

Here is the critical point: I assumed—apparently paralleling the assumptions of the Turin Shroud researchers—that the alien would be equivalent to the supernatural, so that belief in the supernatural would be returned to intellectual respectability. But this apparently critical point turned out to be the wrong point.

For a large number of persons, the Guadalupe Image is supernatural by virtue of its classic characteristics: the unfaded brightness of its colors, together with the absence of disintegration of its rough-hewn and unsized "canvas" base.[2] It was the mystery of the colors to which emphasis was given by the Physicians Royal in the *Información* of 1666 and the

---

[2]Well after the publication of my first book on the Guadalupe story (*The Image of Guadalupe*, Doubleday, 1983) I located an interesting statement in a nineteenth-century history of Mexico: *Conquest of Mexico* by William Hickling Prescott. Prescott reports that the sixteenth-century Aztecs knew of and applied a type of preservation to *ayate*. If this is correct, then it would appear that at least one of the putatively miraculous characteristics of the *Imagen* is subject to reduction. Ironic indeed is the observation of a reviewer that Prescott's *Conquest* glows with the colors of a fine painting.

apparent impossibility of multimedia composition by Cabrera in the next century. Through the work of an initially skeptical intellectual—who was himself a priest—named Ignacio Bartolache late in that same eighteenth century, the problem of the unrotting *ayate* was readdressed. Bartolache, it will be recalled, turned from doubt to belief when he found a painstakingly recomposed duplicate picture set nearby the original, both discolored and decomposed.

If we prudently forego making any categorical decision on the basis of the putative "reflections" in the eyes of the Madonna, what other protocol should in the interest of explanation be pursued? The answer, of course, lies in terms of the classic features of mystery reviewed immediately above: Could we but obtain a piece of the Gualdalup cloth, we could quickly and surely analyze and determine "once and for all" whether or not there was any "unsolved mystery" inherent.

This is precisely what, over a period that stretched into years, I tried to do. But, I was unable ever to obtain a piece of the fabric possessing the necessary characteristics to perform any conclusive study. Through the kindness of a Mexican gentleman, Dr. Palacios-Bermudez, I was lent an encased filament of the cactus cloth held to be from the original Image. I obtained low-magnification photographs of this fiber (see the illustration), and even talked at length with an electron microscopist who himself examined the tiny string. But all of this came to naught when it was reconfirmed that what the unaided eye saw was correct—there was no coloring of any kind or amount remaining on it.

Before any of these efforts took place, I had learned that just such an obvious chemical analysis as I wished to do had already, many decades earlier, been carried out. Indeed, it was as a matter of reconfirmation of the incredible results claimed for that test that I so wished to do another. The same source that had introduced me to the claims of "a human bust in the eye of the Virgin of Guadalupe" also made a claim that was at least startling if not sensational. The source claimed that in the mid-1930s a highly respected Mexican organic chemist, Dr. Ernesto Sodi-Pallares, had obtained and then hand-carried to Germany two Guadalupan fibers *uno rojo y uno amarillo* ("one red and one yellow"). Once in Germany, so the story went, he submitted these fibers to test by his own mentor, a world-famous chemist who was later awarded the Nobel Prize. Accordingly, not only in my printed source but also according to multiple witnesses whom I personally interviewed, this test or tests proved unable to determine anything whatsoever about the nature of the red or the yellow coloring agent.

If this were true, the implications were *estupendoso*: we would have our long-imagined alien/supernatural artifact, and with it, long-sought relief from an oppressive worldview, materialism.

Try to imagine the quality of my disappointment when, the efforts of the Nobel laureate's son notwithstanding, I could find no independent record of the supposed experiment.

Of itself, this is no basis for concluding against the reality of the experiment nor its prima facie incredible results. I have always remembered a fine young Mexican physician, a Guadalupan believer, who showed me an article in a periodical from the early 1940s. This short article did appear to provide restatement of both test and test results. It may have been the source for Salina's mid-1970s claim. But, however interesting in itself, what was needed was something unequivocal—and that meant primary source data—especially for such unprecedented conclusions as those reported. After many apparently fruitless efforts to locate in Mexico any primary documentation, I recall that I began to feel as if I were on some sort of UFO or "flying saucer" search. A long-standing interest of mine, I had many years before become disillusioned with the search to ascertain whether or not flying saucers were "real." Always only dead ends and bickering rather than perceptible progress ever seemed in the offing.

One tack—and it is not an inherently unfair one—taken by many frustrated with an apparently insolvable problem is honestly to give up. Whether reserving final judgement or (especially) reaffirming the sufficiency of the way of faith, such a position is very easy to adopt. It appears not only the *obvious* but also the *only* alternative.

Yet, I have come to believe that nothing could be farther from the truth. Does my own failure to find what I scientifically sought in the Guadalupe painting leave me with the dilemma of reluctant dismissal on the one horn or a return to inchoate faith on the other? No. There is a third way.

In logic, one is advised (and then, immediately and to his peril, forgets) that the truth of a conclusion is only as good as the truth of its antecedents, its premises. A false premise always and necessarily produces a false conclusion. I believe that is what befell me. I presumed that to pursue the Guadalupan mystery by way of the scientific method *as currently understood* was the route to wisdom.

I still believe the tantalizing hints might *also* eventually produce some sort of empirical and readily "objective" truth to some super- or paranormal aspect to the Guadalupan phenomenon. But, if so, it will be a both/and rather than an either/or matter.

What I have in mind as the "other" rather than the "either/or" is not easy to explain, because we are all used to thinking in rather black or white categories. By "other" I mean to suggest that the Guadalupan mystery exists *whether or not* it happened in historical time. Objectively, I believe it *may* have happened in historical time. As for the question of its happening in other than historical time, I would say yes, whether in the former category or not.

My incorrect presupposition when I began researching the Guadalupe Image fifteen years ago carried with it a subtle and dangerous reciprocal: if the supernatural nature of the painting were not validated by virtue of some unexplainable quality, then *because no such quality existed*, the painting was not, by definition, supernatural.

This means that I was incorrect to suppose that I could only reassert the supernatural by discovering the supernatural, *and* I was also incorrect in supposing that if I did not, then the painting must be entirely mundane. After achieving the proper frame of mind, I came to recognize what now seems obvious, that should a supernatural being so decide, it could with equal facility employ materials for a painting native to this our realm with at least as much ease as that of some other realm. Consequently, down crashes the whole idea that to localize the unexplainable is the exclusive way of reaffirming its reality.

The other or third way to which I made reference at the beginning of this chapter, the subject for its own volume, involves what Jung called synchronicity. In the chapter above entitled "Impossible Coincidences" a brief account was presented of the strange unpredictable conjunction of events occurring from the (conscious) beginnings of my Guadalupan pursuits. There were and have been many more such "coincidences." I shall mention some of the more memorable as a means of expressing my personal view of the Guadalupan stature as—fundamentally—the *Hodegetria*, "Pointer of the Way" (chapter 4).

Soon after my return from my first sight of the Guadalupe Image, the purpose of the pilgrimage complete and with the bonus of cordial invitation to pursue research, I became unexpectedly able to move into a house in a quaint old neighborhood. The house, itself quite old, is situated directly across from a park. One afternoon, a television interviewer seeking directions to my home exclaimed that his wall map gave him a sure guide, "right across from Estramadura Park." I replied that I was indeed across the street from a park and was pleased to know not only that the unmarked park had a name but also that it was appropriately Spanish. Many street names in my city, Pensacola, Florida, are Spanish-named, and so I thought no more of the information. The interview went quite well, as did the broadcast.

Only much later that night, with the lights extinguished and while walking down the hallway to the bedroom did I suddenly recognize where I had heard the name *Estramadura* before: it is the name of the region of Spain from which the first statue named *Guadalupe* originated. To someone outside the gestalt, this (like most synchronicities) may appear "merely coincidental." Nor do I make any attempt to persuade anyone that such incidents have extraordinary value. But to myself this "happenstance" left me with fifteen minutes worth of chills as I continued back and forth in the hall before being able to retire. I had not been looking to purchase a house, much less one across from an unmarked park named only on the largest of maps, in an unmistakably Guadalupan perspective.

The indigenous Mexican Indians came to the hill of the apparition, Tepyac, for many decades before Cortes invaded in 1519. Perhaps this sacred site is some sort of Mesoamerican "power hill," the junction of what some claim exists from an antique Europe as "Ley Lines." I have it on excellent authority that the noted Sanskritist Evans-Wentz had concluded that some kind of veridical apparition did really occur.[3]

---

[3]If I understand correctly his thesis, Jacques LaFaye denies that there is any need to *ever* evoke a supernatural explanation to account sufficiently for *any* putatively historical occurrence. I disagree with such a viewpoint precisely because I think it no less dogmatic than the alternative presumption that such a supernatural explanation is necessarily demanded. As an example of this difference I briefly cite a heretofore apparently unappreciated defense of the position that is specifically contrary to that of LaFaye. The story of the Guadalupe apparition was *not* (as LaFaye suggests) an invention of the mid-seventeenth-century Mexican pronationalists; it was instead what it always claimed to be—indigenous

Whether it beckons to other apparitions[4] (*Zeitoun* being the most pronounced probability), to other *acheiropoietos* Marian paintings, or to both, I do not doubt that the Guadalupe Mystery is a catalyst to increased horizons for humankind.

---

to the prior, mid-sixteenth century; further, it was—at least it was sincerely and at that early time *believed* to be—accountable only in terms of supernatural proportion.

There is no argument that, if one invariably applies Occam's razor, all possibly supernatural phenomena will be reduced to the mundane; the "unknown" will become merely the conditionally or technically unknown, waiting only for the development of a more refined historiography. Thus, Juan Diego's cape full of fresh roses is reduced to Saint Isabel's or Saint Elizabeth's apron carrying bread for the poor.

To the promised citation: the interested reader is directed to the work of Brother Gregorio Lopez. Lopez, who lived until the late sixteenth century, had come to Mexico sometime—apparently early—in the reign of Philip II of Spain. The Guadalupan connection is that the impetus for his move was his favorable impression, already received back in Spain, that something most unusual had been believed to have occurred in New Spain.

[4] I take the idea of a veridical apparition seriously. As this present work goes to press, I am nearing completion of a book on this subject. The nucleus of my theory is in bare outline as follows: hyperdimensional spaces are a reality. The mystery of perception involves in a fundamental way what B. B. Mandelbrot named *fractals*. Specific to the perception of apparitions as fractal images is the recognition that inherent in the nature of the fractal is the "nesting" nexus, i.e., that every individual image contains within itself an apparently infinite number of progressively smaller but otherwise identical *representations*. What appears to human consciousness as an apparitional entity is the three-dimensional, fractalar *reduction* of a four-dimensional (or higher) form. As a three-dimensional object casts a shadow in two dimensions, analogously a four-space dimensional referent casts a shadow into (our) length-width-depth locus of the continuum. One name for this, given a hypercube as cause, is "supercube" (George Gamow). The "supercube" as seen by ourselves is a *cube-within-a-cube*. There are recorded instances, in modern times, reporting just such phenomena in glass, especially glass shards. These would be "frozen" instantiations or "apparitions-in-stasis."

**16.** The new Basilica of Our Lady of Guadalupe, dedicated on October 12, 1976. *(Photo courtesy Milton M. Knight)*

**17.** The high altar of the new basilica. The Image of Guadalupe is encased in a glass frame on the wall behind the altar. *(Photo courtesy Milton M. Knight)*

18. The author and Dr. Philip Callahan. (*Photo courtesy of author*)

**19.** Setting up the camera in front of the Image to take the infrared photographs for spectroscopic examination. *(Author's photo)*

**20.** Dr. José Aste-Tonsmann, author of a recent book on the eyes of the Image of Guadalupe, at a symposium examining his work with computer enhancements of the Image. *(Author's photo)*

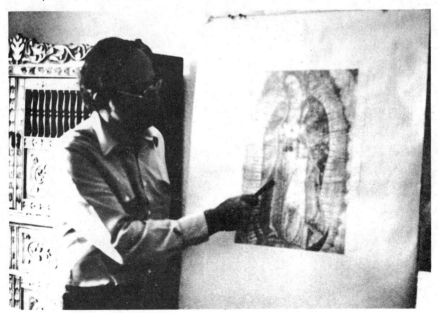

21. Closeup of the face of the Image of Guadalupe. While the eyes have always been especially striking, they have received extraordinary attention during the last fifty years because of the image of a human bust that appears in them upon close examination.

These photographs show computer-enhanced printouts and enlargements of the eyes of the Image. The computer only enlarges what is already there. It does not add anything that is not already present in the subject.

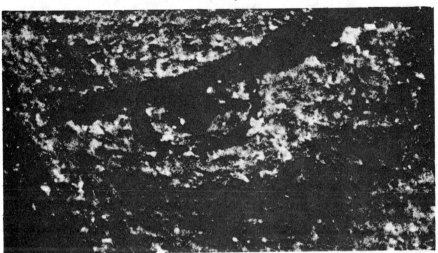

**22.** The right eye of the Image, magnified ten times. (*Author s pnoto*)

**23.** The location of the figures in the right eye, outlined in white for clarity. Note particularly the large figure at the right, the "human bust" that was discovered over fifty years ago. (*Photo courtesy Dr. José Aste-Tonsmann*)

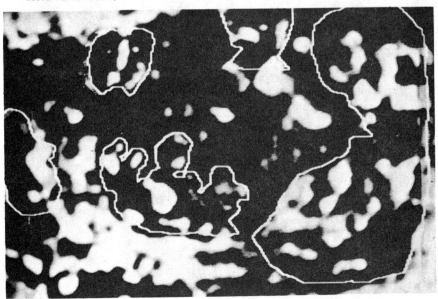

24. Rare photograph of a thread from the original tilma.
*(Photo by author)*

25. The Coptic church at Zeitoun, near Cairo, the scene of a modern-day sighting of the Virgin Mary. (*Photo courtesy of author*)

**26.** The first visit outside the Vatican by the new Pope, John Paul II, was to Puebla, Mexico, to a meeting of Central and South American bishops, and then to the Basilica of Our Lady of Guadalupe in January 1979. In late 1981, this larger-than-life statue was installed in the old basilica courtyard in memory of his stay. (*Author's photo*)

# Appendixes

## The Nican Mopohua[1]

*The* nican mopohua—*Nahuatl for "an account"—is the most complete and probably the earliest written record of the legend of Juan Diego. For many years it was believed to be a concoction of the seventeenth century when interest in anything relating to the Virgin of Guadalupe was at an all-time high. The name of the reputed Indian author, Antonio Valeriano, was thought to be a pseudonym, since no independent proof of Valeriano's existence could be found.*

*The recent discovery of a sixteenth-century civil paper signed by "Antonio Valeriano" confirmed his existence and also helped to establish that the* nican mopohua *was composed in the sixteenth century. Most scholars now believe the* nican mopohua *was written sometime between 1551 and 1561, twenty to thirty years after the miraculous happenings on Tepeyac hill and in the bishop's palace.*

Ten years after the seizure of the city of Mexico war came to an end, and there was peace amongst the people; in this manner faith started to bud, the understanding of the true God, for whom we live. At that time, in the year fifteen hundred thirty-one, in the early days of the month of December, it happened that there lived a poor Indian, named Juan Diego, said being a native of Cuautitlán. Of all things spiritually he belonged to Tlatilolco. On a Saturday just before dawn, he was on his way to pursue divine worship and to engage in his own errands. (As he reached) the

---

[1]Translated from the Spanish by Cleofas Callero, M.F.A., in Harold J. Rahm, S.J., *Am I Not Here?* (Washington NJ: Ave Maria Institute, 1961); reprinted by permission of AMI International Press. This work is originally from P. Mariano Cuevas, S.J., *Album Historico Guadalupano del IV Centenario* (Mexico D.F.: Escuela Tipografica Salesiana, 1930) Tercera Decada—1551–1561.

base of the hill known as Tepeyacac, came the break of day, and he heard singing atop the hill, resembling singing of varied beautiful birds. Occasionally the voices of the songsters would cease, and it appeared as if the mount responded. The song, very mellow and delightful, excelled that of the *coyoltotl* and the *tzinizcan*[2] and of other pretty singing birds. Juan Diego stopped to look and said to himself: "By fortune, am I worthy of what I hear? Maybe I dream? Am I awakening? Where am I? Perhaps I am in the terrestrial paradise which our elders had told us about? Perhaps I am now in heaven?" He was looking toward the east, on top of the mound, from whence came the precious celestial chant; and then it suddenly ceased and there was silence. He then heard a voice from above the mound saying to him: "Juanito, Juan Dieguito."

Then he ventured and went to where he was called. He was not frightened in the least; on the contrary, overjoyed. Then he climbed the hill, to see from where he was being called. When he reached the summit, he saw a Lady, who was standing there and who told him to come hither. Approaching her presence, he marveled greatly at her superhuman grandeur; her garments were shining like the sun; the cliff where she rested her feet, pierced with glitter, resembling an anklet of precious stones, and the earth sparkled like the rainbow. The mezquites, nopales, and other different weeds, which grow there, appeared like emeralds, their foliage like turquoise, and their branches and thorns glistened like gold. He bowed before her and heard her word, tender and courteous, like someone who charms and esteems you highly. She said: "Juanito, the most humble of my sons, where are you going?"

He replied: "My Lady and Child, I have to reach your church in Mexico, Tlatilolco, to pursue things divine, taught and given to us by our priests, delegates, and Our Lord."

She then spoke to him, revealing her holy will. She told him: "Know and understand well, you the most humble of my sons, that I am the ever-virgin Holy Mary, Mother of the True God for whom we live, of the Creator of all things, Lord of heaven and the earth. I wish that a temple be erected here quickly, so I may therein exhibit and give all my love, compassion, help, and protection, because I am your merciful mother, to

---

[2]These are Nahuatl names for two varieties of birds for which there are no English equivalents.

you, and to all the inhabitants on this earth and all the rest who love me, invoke, and confide in me; listen there to their lamentations, and remedy all their miseries, afflictions, and sorrows. And to accomplish what my clemency pretends, go to the palace of the bishop of Mexico, and you will say to him that I manifest my great desire, that here on the plain a temple be built to me; you will accurately relate all you have seen and admired, and what you have heard. Be assured that I will be most grateful and will reward you, because I will make you happy and worthy of recompense for the effort and fatigue in what you will obtain of what I have entrusted. Behold, you have heard my mandate, my humble son; go and put forth all your effort."

At this point he bowed before her and said: "My Lady, I am going to comply with your mandate; now I must part from you, I, your humble servant."

Then he descended to go to comply with the errand, and went by the avenue which runs directly into Mexico City.

Having entered the city, and without delay, he went straight to the bishop's palace, who was the recently arrived prelate named Father Juan de Zumárraga, a Franciscan religious. On arrival, he endeavored to see him; he pleaded with the servants to announce him; and after a long wait, he was called and advised that the bishop had ordered his admission. As he entered, he bowed, and on bended knees before him, he then delivered the message from the Lady from heaven; he also told him all he had admired, seen, and heard.

After having heard his chat and message, it appeared incredible; then he told him: "You will return, my son, and I will hear you at my pleasure. I will review it from the beginning and will give thought to the wishes and desires for which you have come."

He left and he seemed sad, because his message had not been realized in any of its forms.

He returned on the same day. He came directly to the top of the hill, met the Lady from heaven, who was awaiting him, in the same spot where he saw her the first time. Seeing her, prostrated before her, he said: "Lady, the least of my daughters, my Child, I went where you sent me to comply with your command. With difficulty I entered the prelate's study. I saw him and exposed your message, just as you instructed me. He received me benevolently and listened attentively, but when he replied, it appeared that he did not believe me. He said: 'You will return;

I will hear you at my pleasure. I will review from the beginning the wish and desire which you have brought.' I perfectly understood by the manner he replied that he believes it to be an invention of mine that you wish that a temple be built here to you, and that it is not your order; for which I exceedingly beg, Lady and my Child, that you entrust the delivery of your message to someone of importance, well known, respected, and esteemed, so that they may believe in him; because I am a nobody, I am a small rope, a tiny ladder, the tail end, a leaf, and you, my Child, the least of my children, my Lady, you send me to a place where I never visit nor repose. Please excuse the great unpleasantness and let not fretfulness befall, my Lady and my All."

The Blessed Virgin answered: "Hark, my son the least, you must understand that I have many servants and messengers, to whom I can entrust the delivery of my message, and carry my wish, but it is of precise detail that you yourself solicit and assist and that through your mediation my wish be complied. I earnestly implore, my son the least, and with sternness I command that you again go tomorrow and see the bishop. You go in my name, and make known my wish in its entirety that he has to start the erection of a temple which I ask of him. And again tell him that I, in person, the ever-virgin Holy Mary, Mother of God, send you."

Juan Diego replied: "Lady, my Child, let me not cause you affliction. Gladly and willingly I will go to comply your mandate. Under no condition will I fail to do it, for not even the way is distressing. I will go to do your wish, but perhaps I will not be heard with liking, or if I am heard I might not be believed. Tomorrow afternoon, at sunset, I will come to bring you the result of your message with the prelate's reply. I now take leave, my Child, the least, my Child and Lady. Rest in the meantime."

He then left to rest in his home.

The next day, Sunday, before dawn, he left home on his way to Tlatilolco, to be instructed in things divine, and to be present for roll call, following which he had to see the prelate. Nearly at ten, and swiftly, after hearing Mass and being counted and the crowd had dispersed, he went. On the hour Juan Diego left for the palace of the bishop. Hardly had he arrived, he eagerly tried to see him. Again with much difficulty he was able to see him. He kneeled before his feet. He saddened and cried as he expounded the mandate of the Lady from heaven, which God grant he would believe his message, and the wish of the Immaculate, to erect her temple where she willed it to be.

The bishop, to assure himself, asked many things, where he had seen her and how she looked; and he described everything perfectly to the bishop. Notwithstanding his precise explanation of her figure and all that he had seen and admired, which in itself reflected her as being the ever-virgin Holy Mother of the Savior, Our Lord Jesus Christ, nevertheless, he did not give credence and said that not only for his request he had to do what he had asked; that, in addition, a sign was very necessary, so that he could be believed that he was sent by the true Lady from heaven. Therefore, he was heard, said Juan Diego to the bishop: "My lord, hark, what must be the sign that you ask? For I will go to ask the Lady from heaven who sent me here."

The bishop, seeing that he ratified everything without doubt and (was) not retracting anything, dismissed him. Immediately he ordered some persons of his household, in whom he could trust, to go and watch where he went and whom he saw and to whom he spoke. So it was done. Juan Diego went straight to the avenue. Those that followed him, as they crossed the ravine, near the bridge to Tepeyacac, lost sight of him. They searched everywhere, but he could not be seen. Thus they returned, not only because they were disgusted, but also because they were hindered in their intent, causing them anger. And that is what they informed the bishop, influencing him not to believe Juan Diego; they told him that he was being deceived; that Juan Diego was only forging what he was saying, or that he was simply dreaming what he said and asked. They finally schemed that if he ever returned, they would hold and punish him harshly, so that he would never lie or deceive again.

In the meantime, Juan Diego was with the Blessed Virgin, relating the answer he was bringing from his lordship, the bishop. The Lady, having heard, told him: "Well and good, my little dear, you will return here tomorrow, so you may take to the bishop the sign he has requested. With this he will believe you, and in this regard he will not doubt you nor will he be suspicious of you; and know, my little dear, that I will reward your solicitude and effort and fatigue spent on my behalf. Lo! go now. I will await you here tomorrow."

On the following day, Monday, when Juan Diego was to carry a sign so he could be believed, he failed to return, because, when he reached his home, his uncle, named Juan Bernardino, had become sick, and was gravely ill. First he summoned a doctor who aided him; but it was too late, he was gravely ill. By nightfall, his uncle requested that by break of day he go to Tlatilolco and summon a priest, to prepare him and hear his confession, because he was certain that it was time for him to die, and that he would not arise or get well.

On Tuesday, before dawn, Juan Diego came from his home to Tlatilolco to summon a priest; and as he approached the road that joins the slope to Tepeyacac hilltop, toward the west, where he was accustomed to cross, said: "If I proceed forward, the Lady is bound to see me, and I may be detained, so I may take the sign to the prelate, as prearranged; that our first affliction must let us go hurriedly to call a priest, as my poor uncle certainly awaits him."

Then he rounded the hill, going around, so he could not be seen by her who sees well everywhere. He saw her descend from the top of the hill and was looking toward where they previously met. She approached him at the side of the hill and said to him: "What's there, my son the least? Where are you going?"

Was he grieved, or ashamed, or scared? He bowed before her. He saluted, saying: "My Child, the most tender of my daughters, Lady, God grant you are content. How are you this morning? Is your health good, Lady and my Child? I am going to cause you grief. Know, my Child, that a servant of yours is very sick, my uncle. He has contracted the plague, and is near death. I am hurrying to your house in Mexico to call one of your priests, beloved by our Lord, to hear his confession and absolve him, because, since we were born, we came to guard the work of our death. But if I go, I shall return here soon, so I may go to deliver your message. Lady and my Child, forgive me, be patient with me for the time being. I will not deceive you, the least of my daughters. Tomorrow I will come in all haste."

After hearing Juan Diego's chat, the Most Holy Virgin answered: "Hear me and understand well, my son the least, that nothing should frighten or grieve you. Let not your heart be disturbed. Do not fear that sickness, nor any other sickness or anguish. Am I not here, who is your Mother? Are you not under my protection? Am I not your health? Are you not happily within my fold? What else do you wish? Do not grieve nor be disturbed by anything. Do not be afflicted by the illness of your uncle, who will not die now of it. Be assured that he is now cured."

(And then his uncle was cured, as it was later learned.)

When Juan Diego heard these words from the Lady from heaven, he was greatly consoled. He was happy. He begged to be excused to be off to see the bishop, to take him the sign or proof, so that he might be believed. The Lady from heaven ordered him to climb to the top of the hill, where they previously met. She told him: "Climb, my son the least, to the top of the hill; there where you saw me and I gave you orders, you will find different flowers. Cut them, gather them, assemble them, then come and bring them before my presence."

Immediately Juan Diego climbed the hill, and as he reached the summit, he was amazed that so many varieties of exquisite *rosas de Castilla* were blooming, long before the time when they are to bud, because, being out of season, they would freeze. They were very fragrant and covered with dewdrops of the night, which resembled precious pearls. Immediately he started cutting them. He gathered them all and placed them in his tilma. The hilltop was no place for any kind of flowers to grow, because it had many crags, thistles, thorns, *nopales*, and *mezquites*. Occasionally weeds would grow, but it was then the month of December, in which all vegetation is destroyed by freezing. He immediately went down the hill and brought the different roses which he had cut to the Lady from heaven, who, as she saw them, took them with her hand and again placed them back in the tilma, saying: "My son the least, this diversity of roses is the proof and the sign which you will take to the bishop. You will tell him in my name that he will see in them my wish and that he will have to comply to it. You are my ambassador, most worthy of all confidence. Rigorously I command you that only before the presence of the bishop will you unfold your mantle and disclose what you are carrying. You will relate all and well; you will tell that I ordered you to climb to the hilltop, to go and cut flowers; and all that you saw and admired, so you can induce the prelate to give his support, with the aim that a temple be built and erected as I have asked."

After the Lady from heaven had given her advice, he was on his way by the avenue that goes directly to Mexico; being happy and assured of success, carrying with great care what he bore in his tilma, being careful that nothing would slip from his hands, and enjoying the fragrance of the variety of the beautiful flowers.

(When he reached) the bishop's palace, there came to meet him the majordomo and other servants of the prelate. He begged them to tell (the bishop) that he wished to see him, but none were willing, pretending not to hear him, probably because it was too early, or because they already knew him as being of the molesting type, because he was pestering them; and, moreover, they had been advised by their coworkers that they had lost sight of him, when they had followed him. He waited a long time. When they saw that he had been there a long time, standing, crestfallen, doing nothing, waiting to be called, and appearing like he had something that he carried in his tilma, they came near him, to see what he had and to satisfy themselves. Juan Diego, seeing that he could not hide what he had, and on account of that he would be molested, pushed or mauled, uncovered his tilma a little, and there were the flowers; and upon seeing that they were all different *rosas de Castilla*, and out of season, they were thoroughly amazed, also because they were so fresh and in full bloom, so fragrant and so beautiful. They tried to seize and pull some out, but they were not successful the three times they dared to take them. They were not lucky because when they tried to get them, they were unable to see real flowers. Instead, they appeared painted or stamped or sewn on the cloth.

Then they went to tell the bishop what they had seen and that the Indian who had come so many times wished to see him, and that he had reason enough to wait so long anxiously eager to see him. Upon hearing, the bishop realized that what he carried was the proof, to confirm and comply with what the Indian requested. Immediately he ordered his admission. As he entered, Juan Diego knelt before him, as he was accustomed to do, and again related what he had seen and admired, also the message. He said: "Sir, I did what you ordered, to go forth and tell my *Ama*, the Lady from heaven, Holy Mary, precious Mother of God, that you asked for a sign so that you might believe me that you should build a temple where she asked it to be erected; also, I told her that I had given you my word that I would bring some sign and proof, which you requested, of her wish. She condescended to your request and graciously granted your request, some sign and proof to complement her wish. Early today she again sent me to see you; I asked for the sign so you might believe me, as she had said that she would give it, and she complied.

She sent me to the top of the hill, where I was accustomed to see her, and to cut a variety of *rosas de Castilla*. After I had cut them, I brought them, she took them with her hand and placed them in my cloth, so that I bring them to you and deliver them to you in person. Even though I knew that the hilltop was no place where flowers would grow, because there are many crags, thistles, thorns, *nopales*, and *mezquites*, I still had my doubts. As I approached the top of the hill, I saw that I was in paradise, where there was a great variety of exquisite *rosas de Castilla*, in brilliant dew, which I immediately cut. She had told me that I should bring them to you, and so I do it, so that you may see in them the sign which you asked of me and comply with her wish; also, to make clear the veracity of my word and my message. Behold.

Receive them." He then unfolded his white cloth, where he had the flowers; and when they scattered on the floor, all the different varieties of *rosas de Castilla*, suddenly there appeared the drawing of the precious Image of the ever-virgin Holy Mary, Mother of God, in the manner as she is today kept in the temple at Tepeyacac, which is named Guadalupe. When the bishop saw the image, he and all who were present fell to their knees. She was greatly admired. They arose to see her; they shuddered and, with sorrow, they demonstrated that they contemplated her with their hearts and minds. The bishop, with sorrowful tears, prayed and begged forgiveness for not having attended her wish and request. When he rose to his feet, he untied from Juan Diego's neck the cloth on which appeared the Image of the Lady from heaven. Then he took it to be placed in his chapel. Juan Diego remained one more day in the bishop's house, at his request.

The following day he told him: "Well, show us where the Lady from heaven wished her temple be erected."

Immediately, he invited all those present to go. As Juan Diego pointed out the spot where the Lady from heaven wanted her temple built, he begged to be excused. He wished to go home to see his uncle Juan Bernardino, who was gravely ill when he left him to go to Tlatilolco to summon a priest, to hear his confession and absolve him. The Lady from heaven had told him that he had been cured. But they did not let him go alone, and accompanied him to his home. As they arrived, they saw that his uncle was very happy and nothing ailed him. He was greatly amazed to see his nephew arrive so accompanied and honored, asking the reason of such honors conferred upon him. His nephew answered that when he left to summon a priest to hear his confession and to absolve him, the Lady from heaven appeared to him at Tepeyacac, telling him not to be afflicted, that his uncle was well, for which he was greatly consoled, and she sent him to Mexico, to see the bishop, to build her a house in Tepeyacac. Then the uncle manifested that it was true that on that occasion he became well and that he had seen her in the same manner as she had appeared to his nephew, knowing through her that she had sent him to Mexico to see the bishop. Also, the Lady told him that when he would go to see the bishop, to reveal to him what he had seen and to explain the miraculous manner in which she had cured him, and that she would properly be named, and known as the blessed Image, the ever-virgin Holy Mary of Guadalupe. Juan Bernardino was brought before the presence of the bishop to inform and testify before him. Both he and his nephew were the guests of the bishop in his home for some days, until the temple dedicated to the Queen of Tepeyacac was erected where Juan Diego had seen her. The bishop transferred the sacred Image of the lovely Lady from heaven to the main church, taking her from his private chapel where it was, so that the people would see and admire her blessed Image. The entire city was aroused; they came to see and admire the devout Image, and to pray. They marveled at the fact that she appeared as did her divine miracle, because no living person of this world had painted her precious Image.

# THE PRimiTivE RElATioN[1]

*Nahuatl scholar Ángel M. Garibay-Kintana is credited as discoverer of the "Primitive Relation" in the Mexican National Library Archives in the early 1950s. He believed it was written about 1573 by the historian Juan de Tovar, who transcribed it from an earlier source. According to Garibay, the earlier source was Juan González, credited by most Guadalupan historians with being the man who translated Juan Diego's startling message into Spanish in Bishop Zumárraga's palace in 1531. Much fine work has been done since 1976 by Father Mario Rojas of the Center for Guadalupan Studies. The English translation that follows is by James A. Guest.*

## Our Lady of Guadalupana

1. This is a great marvel that our Lord God did by means of the forever-virgin Holy Mary.
2. Here it is.
3. That which you should notice, that which you should hear in what miraculous manner it was desired that a house should be erected, that a dwelling should be established, that would be called Queen Holy Mary in Tepeyac.
4. This is what occurred: a poor man of the village, a *macehual* of great piety,
5. said to be laborer (poor creature, poor yokel) there in Tepeyac, was going by walking along the peaks
6. (to see if by chance a little root might have broken through the ground ), struggling to earn his living.
7. There he saw the beloved Mother of God, who called him and said to him:
8. "MY LITTLE SON, GO TO THE EMPEROR OF THE GREAT CITY MEXICO,

---

[1] The following English translation is by James A. Guest and is reprinted here by permission of Dr. Guest.

9. SAY TO HIM WHO THERE IS GOVERNING THAT WHICH IS SPIRITUAL, TO THE ARCHBISHOP

10. THAT I WISH WITH A GREAT DESIRE THAT HERE IN TEPEYAC THEY MAKE ME A DWELLING, THAT THEY RAISE UP TO ME MY HOUSE

11. SO THAT HERE THEY COME TO KNOW ME WELL, THAT THE FAITHFUL CHRISTIANS MAY COME HERE TO PRAY TO ME

12. HERE I WILL CONVERT TO ME (IN IT) WHEN THEY MAKE ME THEIR INTERCESSOR."

13. Then that poor little man went to present himself before the great governing priest archbishop, and said to him:

14. "My Lord, I am not going to importune you but, behold, Our Lady of the Heavens has sent me,

15. told me that I should come to say how much she desires that there in Tepeyac should be made, should be erected for her, a house in order that there Christians may supplicate her.

16. She also said to me that something very close to her (in her riches) that there might be converted when they will go there to invoke her."

17. But the archbishop gave him no credit but rather said to him:

18. "What are you saying, my son? Perhaps you have dreamed this or perhaps you have been drunk!

19. If, in truth, it is certain that which you say, (say) to her, to this Lady that said it to you, that she give you any sign

20. In order that we may believe that it is really true what you are saying."

21. He returned, our poor little man, became extremely sad, and there appeared to him again the Queen

22. And when our little man saw her he said to her:

23. "My child, I went whither you did send me, but my lord did not believe me,

24. Moreover he said to me that perhaps I dreamed it or perhaps I had been drunk .

25. And he said to me that in order to believe it you should give me a sign in order that it be carried out."

26. And when Our Lady the Queen, the beloved Mother of God, then said to him:

27. "DON'T BE SAD, MY YOUNG ONE, GO TO GATHER, GO CUT SOME LITTLE FLOWERS WHERE THEY ARE BLOOMING."

28. These flowers only by miracle were growing there, because at that season the earth was very dry, nowhere were flowers opening.

29. When our little man cut them he put them in the hollow of his cloak.

30. From there he went to Mexico to tell the bishop:

31. "My Lord, here I bring the flowers that Our Celestial Lady gave me in order that you may believe her word is true, her will, that I have come to tell you, that it is certain that which she said to me."

32. And when he opened his cloak, in order to show the flowers to the archbishop, there also was seen in the cloak of our little man,

33. There was painted, there was converted into a signal portrait the Virgin Queen in prodigious form so that the archbishop believed.

34. On seeing it, they knelt and admired her.

35. And, in truth, the very Image of the Virgin Queen is here only by miracle; in the cloak of the poor man (it) was painted as a portrait, where now (it) is placed as a light for the whole universe.

36. There come to know her those who supplicate her

37. And she, with her pious maternity ( with her maternal affection), there helps them, gives them what they ask.

38. And, in truth, if someone fully recognizes by her intermission, and totally gives himself to her, loving her for her intercession, the beloved Mother of God will convert him.

39. In truth, it will help much to her, it will show to her, to whom they esteem (that) they have begun to put themselves under her shadow, under her care.

# THE CODEX SAVILLE:
# AMERICA'S OLDEST BOOK[1]

*The Codex Saville, a pictorial calendar, was found in Tetlapalco in Peru in 1924 by the anthropologist M. H. Saville. It now resides in the collection of the Museum of the American Indian in New York City. The following commentary on the Codex Saville was written by Mariano Cuevas, S.J., who translated the calendar in 1929.*

It has always been the privilege of the first reader of or commentator on a codex to give it an appropriate name. It is therefore a pleasant duty for me to designate as the "Codex Saville" this pre-Columbian Mexican historical paper, now for the first time published under the auspices of the United States Catholic Historical Society. This is the honor due to Doctor Marshall H. Saville whose merits and brilliant successes during the last forty-five years are so much appreciated by students of the ancient history of the Latin-American countries. The Codex Saville was recently secured in Lima, Peru, by the Heye Foundation of New York City. It is still only provisionally catalogued.

Its size is fifty-seven by five inches (1.45 m. x .126 m.). It is made of the native *maguey* or *agave* American fibre, conglutinated by a vegetable pulp called in the Nahuatl, or Mexican language, *zazalic*. Some linen finished marks that appear here and there on the codex are only surface clothprints from outside pressure on the paste used at some recent date to put together the separate fragments.

Three small patches of the same native paper were pasted on the main part of the codex in order, it seems, to correct some dates or data. One of these little papers pasted about the year 1453 is of the utmost importance. Notice that the upper part of it was afterwards scratched out, thus again making visible the original painted sign of the cycle:

---

[1]The following commentary on the Codex Saville is by Mariano Cuevas, S.J. It appeared as "The Codex Saville: America's Oldest Book" in *Historical Records and Studies* 19 (September 1929): 7-20; copyright 1929 by the U.S. Catholic Historical Society, it is reprinted by permission.

similar to the one at the right side of the year 1455.

Originally the codex was not in colors. These were poorly applied by the painter of the upper and later part sometime about 1531.

There were at least two *tecuilos* (painters) of this *amatl* (painting paper). The first one started before 1454, possibly in 1440. The last one, of the remaining part of the codex, was at his work about 1557. Names of different Mexican towns, most of them now illegible, were written all through the codex in Spanish characters but with the typical Indian handwriting of the middle of the sixteenth century.

Before commenting on the codex, I believe that some previous interpretations will be welcomed by readers who are not very familiar either with codex-reading or with Mexican history and technique.

1. *The Chronology*. The elaborated Chronology of the Nahuatl peoples, inherited, it is said, from the Toltecs, has has been fully illustrated and published by many first-class authors.[2] For the reading of this codex it is sufficient to bear in mind the main divisions of the time among the old Mexicans.

Their cycle was not of one hundred years but the natural one of fifty-two years. This cycle was divided into four groups each of thirteen years (4x13 = 52). Four different signs were employed to name the years: *tochtli* (rabbit), *acatl* (reed), *tecatl* (stone), and *calli* (house). Each year was named by a combination of one of the above four signs mentioned in succession, together with the corresponding number of the group of thirteen years, named also successively, e.g.: (1) *tochtli*, (2) *acatl;* (3) *tecatl,* (4) *calli,* (5) *tochtli*—and so on until after a period of fifty-two years the same sign occurred again with the same number. The same combination could not occur in the same cycle. A picture of the Codex Aubin shows the cycle as conceived by the Aztecs.

2. *History*. Sometime about the end of the twelfth century of our Christian era, seven branches of the race Nahuatl (meaning "clear-talking people") were wandering at a very slow rate all through the present Mexican republic. The origin of these people still remains a mystery. With the exception of their very last "pilgrimage" in the region south of the grand plateau of Mexico, the rest of their wanderings are almost lost

---

[2]See Orozco y Berra, *Historia Antigua y de la Conquista* (Mexico, 1870) vol. 2, chap. 3.

in the mist of prehistory. One of these seven branches called *Tenochca* (and later on *Mexicas*) made a final halt (1318) in the center of the present city "because there they saw the unequivocal sign given by the gods: an eagle upon a cactus, devouring a snake."[3]

An older tradition, given as such by a reliable source,[4] was that the sign to stop the wandering of the Mexicans should be a white oak in the middle of the lake.

As soon as the Mexicans made up their minds to remain there, they began to build dikes to protect their swamp dwellings or Venice-like town from the main surrounding lake. Hence the very first name of the city, which, according to Tezozomoc, was not Mexico but *Atlitc* meaning "water surrounded by a wall." The finishing of that dike six years later in 1324 was a very good reason to give that date as the foundation of the city. The Codex Saville by giving the date 9 *acatl* (1319) for the end of the pilgrimage, or the beginning of the foundation and *Itecpal* (1324) for the end of the foundation, is the best solution to the endless dispute about that important event.

After the death of the leading warrior Tenoch, of which the date is uncertain, the Mexicans began to "elect" their absolute rulers. Following the Spanish conquerors, although most improperly, we still call them "kings." Their names, as well as the hieroglyphs to represent them, and the corresponding English translations, appear below (p. 99).

Progress and existence itself became impossible for the *Mexicas*, confined as they were, within their diked swamp town. Meanwhile the Tecpanecas, owners of the lands surrounding the lake, would not give the Mexicans any chance for expansion. The latter, furthermore, were continuously insulted by the powerful Tecpaneca king, Maxtla. He used to call them effeminate people. Mexican ambassadors sent to him by

---

[3]If it is not a real tradition, it is at least the hieroglyphic of what really took place, namely, that the *Mexicas* stopped where their explorers bade them. Now, the name of the first explorer was Cuahu-cóatl meaning precisely "eagle with snake" and he was sent and supported by Tenoch meaning precisely "cactus upon the rock." An eagle upon a cactus therefore could be in Mexico a unique and unequivocal sign.

[4]*Cronica de Mexico* by Don Fernando Alvarado Tezozomoc, written in the 16th century (printed in Mexico in 1898) chap. 9.

King Itzcóatl in 1432 were again insulted, dressed in women's robes, and forced to return in this guise from Tlatilolco to Mexico. No one was thus more humiliated than the head of the embassy, the famous Moctezuma, afterwards emperor of Mexico.

Acamaphichtli
Handful of reeds—female snake.

Huitzihuitl
Hummingbird.

Chimalpopoca
Smoking shield.

Itzcóatl
Sword-back serpent

Moctezuma I.
Wrathy-lord

Axayacatl
Face in the water.

Tizoc
Wounded leg.

Ahuizotl
Water rat.

Moctezuma II.
Wrathy-lord

The rage provoked in the Mexicans by such an outrage finally led to the ferocious war lasting five years. The courage of the Mexicans and the skill of their leaders, Moctezuma and King Itzcóatl, defeated the Tecpanecas. They became masters of the whole grand valley of Mexico and thus had an open door for the conquest of the rest of the country.

King Itzcóatl, after celebrating his triumph, initiated a series of substantial reformations in the politics of his much-enlarged kingdom. A very useful one was the institution of a council of state to be formed by four prominent noblemen. Out of them the successor for the throne would be selected, thus avoiding the probability of much dangerous competition. This council was instituted during the fifth year of Itzcóatl, A.D. 1437.

Moctezuma I, the next following ruler, carried a victorious war against his powerful Huaxteca neighbors, thus becoming master of the eastern seashore and proving once more that the women's dresses given him by Maxtla were not appropriate.

King Axayacatl was shamefully defeated by Tarasco's brave warriors. Old Mexican historians prefer to forget this unexpected misfortune. They rather turn their eyes to the pompous allies of Mexico, the rich and learned kings of Tezcoco. The most conspicuous of them was Netzahualpilli meaning "hungry child." He was crowned in 1471.

After the short reign of Tizoc, the Mexican throne was occupied by Ahuitzotl, one of the most cruel monsters ever seen. The dedication he made of the double temple in 1487, sacrificing more than 20,000 innocent men, stained the history of Mexico forever.

Moctezuma II, crowned in 1502, saw the punishment of such a "civilization" when in 1519 "white men, bearded, silver plated" riding on "big hornless deers" with sword and fire overthrew the Mexican empire.

Nothing from that time could be recorded with pleasure by the *tlacuilos* or history painters. All was pain and disgrace, until in 1526 the Franciscan missionaries, who had arrived two years before, mastered the Nahuatl language and began the evangelization of the country. They started this work by erecting an enormous wooden cross that could be seen for several leagues around in their churchyard.

*Another fact of great importance in Mexican history and Mexican life was the apparition of the Madonna that, according to tradition and reliable documents, took place in 1531 a few miles north of Mexico City, where the present national shrine stands.*[5]

All this data about the old Mexican history has been known and printed by many reliable authors long ago and independently of this codex. In fact, no one seems to have even quoted it, lost as it was from the middle of the sixteenth century.

It is fascinating to find records of these main lines in a five-centuries-old document that is a firsthand work taken directly from contemporary life by eyewitness historians. Of course this book, like all of its age and like many of our own age, would be meaningless without some commentaries. They could only be memoranda, to be commented on by specially trained men called *amoxoaque*, meaning "men explaining the old paintings."

---

[5]Italics added.

### The Reading of the Codex

The series of historical events herein contained was methodically coordinated and framed in the vertical ruler-line, at the right as you read. There we find, reading up, eight groups, each consisting of a corpse and a living man; both are connected with one of the discs in the line indicating a certain year.

From the characteristic hieroglyphic placed upon the heads of most of the figures, we can recognize with certainty the series of the Mexican rulers from 1422 to 1520. It is a most valuable chronological series and may give the last word to what seemed an endless controversy about the dates of the Mexican rulers prior to 1468.

According to the *tlacuilo* of the Codex Saville the succession of the Mexican "kings" was as follows:

| | |
|---|---|
| Acamapichtli | undated |
| Huitzihuitlt | died 1422 |
| Chimalpopoca | 1422–1432 |
| Itzcohuatl or Itzcóatl | 1432–1445 |
| Moctezuma I | 1455–1467 |
| Axayacatl | 1468–1481 |
| Tizoc | 1481–1486 |
| Ahuitzotl | 1486–1502 |
| Moctezuma II | 1502–undated |

Cuitlahuac and Cuauhtemoc, emperors for a short time after the conquest, are, of course, not mentioned.

The inside column of the codex, which is also to be read upwards, has to be divided into two parts: the pre-Cortésian and the post-Cortésian.

### Part 1. Pre-Cortésian

A.  In the first one we have to consider several groups. The arrival of the Mexicans at their final resting place is clearly shown by the natural sign of footprints. Some archaeologists maintain that footprints of both feet, in hieroglyphic language, mean the act of climbing. If this statement is well founded, so much the better for the reading of this part, since the last steps of the Mexicans were nothing but climbing to the two-thousand-feet-high Mexican plateau, where they settled.

B. The majestically enthroned king, as far as we can tell, is Acama-pichtli. The nickname for the position he held at the time of his election was *chuacóatl* (female serpent). Now the woman's face in an initiated snake body is just his hieroglyphic expressed in this very way by the Codex Mendocino.

C. The oldest name of the City of Mexico. Its white oak as well as the "water surrounded by a dike" cannot be better indicated than they are in the group nearer to the sign "one stone."
   The period of five years (1318–1324) that it took the Mexicans to settle and build their city is just the one given in our codex. The date "nine reed" (1318) coincides with the last footprint or final stop of the Mexicans, and the year "one stone" (1324) is in front of the white oak. These are precisely the two dates given by the best authorities for the foundation of Mexico. The signs, though, are to be related to the fourteenth century. The Indian painter living in the fifteenth century was as sure as we are now, by tradition, that the two dates were "nine *acatl*" and "one *tecpal*," but he could not be more precise, for, as is pointed out, their method for discriminating the cycles, although theoretically good, was out of practice in the fifteenth century.

D. A period of twenty-two years was covered by the big war and con-quest of the Tecpaneca lands. The two dominant figures, Itzcohuatl and Moctezuma, are embraced in it. The main political event, the Council of the Four selected noblemen, is clearly shown. Here they are seen having a regular session just in front of the date of their appointment, on the fifth year of Itzcohuatl's reign, ten *acatl* (1437).

E. The final triumph of Moctezuma I, the "woman-like" ambassador, over his powerful enemies, is beautifully synthesized by the picture of a king in full robe and with hair, dressed like a woman but, nevertheless, on top of another much bigger king whose crown has been taken away. This symbolizes the events of the year 1457 as it appears in our codex.

F.  The names of towns written in Spanish characters, although of much later date, make us think that the second painter, or a commentarist, in the sixteenth century had some vague idea that these three groups give a comprehensive view of that important war period. The towns whose names are still legible certainly were very closely connected with that war. Such are, for instance, Azcaputzalco, Talalpam (*sic*), Heichilputi (*sic*; now Churubuzco), Tacahuaya (for Atlicuahuayan, now Tacubaya), Miahuatlan, and Tepeapulco.

G.  The next prominent figure in the codex is King Netzahualpilli, just facing the year 1471 of his election. His hieroglyph is most appropriately that of a "hungry child." The reasons for his appearance among the Mexican rulers were given in our previous notes in this commentary.

H.  Ahuitzotl, the monster of cruelty and his indelible national crime, the dedication of the two parallel shrines (Huitzilopochtli and Tezcalipuca), are plainly painted in the direction of the sign "ten rabbits" (1486), the very date given by the best historians.

 On the upper part of the sheet, under the date "two houses" (1533) a similar figure is given. Evidently it is not a picture of any actual event, as no Mexican ruler, no temples, no sacrifices had lasted after 1526. It is the picture of something that the *tecuilo* took for granted that was going to happen, if the ritual laws were to be kept, and, we add, if the Spanish conquest, which the *tecuilo* could not foresee, had not come to stop that rite or sacrifice. This anonymous king with his two temples ready for the sacrifice is possibly the expression of the sacrifice prescribed for the middle of the cycle, which had to occur precisely in 1533, or (1507–1526) as it comes in our codex.

### Part 2. Post-Cortésian

A.  A Spanish *conquistador* of the beginning of the sixteenth century, riding on a horse (designed here by a hand probably used to painting only deer) is attacking, sword in hand, a Mexican Indian. Such a group by the year one *acatl* (1519), the very year of the arrival and first attacks of Cortés, can only be the expression of the conquest of

Mexico. Nothing justifies the importance here given to Tetlapulco, a village no more in existence and where no battle took place. Is the conqueror here painted, a portrait of Hernán Cortés? There can be nothing but conjecture on this particular point.

B.  It is certain that the famous Franciscan cross was first erected in 1526 when the missionaries moved from their little chapel (at the corner of what is now Argentina and Guatemala streets) to the site of their new church of San Francisco. The new possessor of the codex believed (and he was right) that such an important fact ought to be recorded. But he was not familiar with the old Aztec way of recording nor with the Spanish way either. So in a very childish but typical Indian fashion he expressed his date with the few figures he happened to know: (4 4 4 4 4 2 = 26). This probable explanation is confirmed by the similar method used to record the foundation of the town of San Marcos (4 4 4 4 4 4 4 = 32).

C.  The painting of the holy evangelist with his symbolic lion would not be of itself the sign of a town. But the *marqués* crown on top of the saint has no meaning unless it refers to the Marquis of Salinas to whom this very town (some twenty miles north of the City of Mexico) had to pay annual tribute, as we know from the *Anales Franciscanos*. The only reason we find for including the San Marcos foundation among the main events of the country is that such a town was of great importance—for the painter.

D.  *A Virgin with her hands folded near her heart, her head bent towards her right shoulder, dressed in a salmon-colored tunic and a greenish-blue mantilla (see the unique design as in the original) is the Virgin of Guadalupe as venerated in Tepeyac, four miles north of the City of Mexico, and some six miles south of San Marcos. By painting it a little lower than the year 1532 it is well indicated that her year was 1531.*

E.  The bell seen as a part of the lost portion of the codex is not of sufficient moment to warrant any serious consideration.

### The Age of the Codex

The Codex Saville is especially important on account of its extra-ordinary antiquity among the historical codices of America. At first glance, any expert archaeologist would refer the pre-Cortésian part of the codex to the middle of the fifteenth century. A closer analysis leads us to the same conclusion for several good reasons:

1. Its paper is of the most primitive and rough material. There is nothing on it like the coating used by all the *tecuilos* of the end of the fifteenth century to smooth the surface and make it ready for painting.

2. The designs are genuinely simple, very different from those colorful and standardized gods and warriors so much in vogue in the time of Ahuitzotl 1486. The pre-Hispanic part, the only one we are referring to now, was designed with deep black vegetable ink. The colors are of that much later date when the *tecuilo* had lost even the notion that the *colpilli* or royal crowns were golden instead of blue.

3. Another sign of antiquity is the absence of more progressive writing. The oldest kind of script is called figurative. Its meaning is simply the painted design. At most it uses some natural sign, like footprints for human walking. The second kind is called ideographic, which expresses abstract ideas or verbs, as, for instance, a conventional curve line before the lips meaning to ask and another one to answer. The third kind of script, called phonetic, paints two or more things whose names, if articulated, sound like the name of a very different object. The last two ways of writing are not used at all in this codex because they are of later invention. The codex therefore is much older than the end of the fifteenth century.

4. The system of dating in our codex is the same substantially as the one used by the other historical codices, but it is handled in a very awkward way. It only gives you the initial sign for every *tlalpilli* (period of thirteen years ) leaving you the task of counting when you try to find a year. Now the need for improvements in chronology, as

well as in any other human invention, is a natural proof of antiquity just as an oil lamp must be much older than an electric light.

5. We can go still further to a more certain and definite statement: The codex was started before the year A.D. 1454. It is admitted by first-class historians that the "Feast of the Sacred Fire" that, according to civil and religious law, was to be celebrated at the end of the cycle, namely, in the year "thirteen houses" (1453), was transferred, owing to the necessary correction of the calendar, to the first morning of "two reeds" (1455). Now the original writer, when making his para-digm or frame of the vertical dating line with its circles, signs, and symbols, painted the symbol of the cycle just in front of 1453.

When, later on, he knew of the correction of the calendar (that news must have been given sometime before the end of 1453), he had to correct his time line as he did, by painting the symbol (a bundle of reeds) in 1455 and pasting a small piece of paper on top of the old one.[6] If he had to correct the mistake he made in 1453, he certainly made the mistake and did the painting in or before 1453. It is hard to tell how long before 1454 the codex was in existence. From the details, though, of the Itzcóatl period, which seem to be taken from personal impressions of the *tecuilo*, one feels inclined to believe that he began to write sometime around 1440. After we have reached this conclusion it is quite natural to ask: "Do you think this is the oldest book in America?" It is hard to tell; but if we take the word "book" in its formal sense, I think we can give an affirmative answer.

---

[6]Orozco y Berra, *Historia antigua y de la Conquista de México*, 2:44, 45, 90, maintains that the correction took place in 1299. He tried to prove his statement by quoting the Tellerian and Vatican Codices, but the fact is, they do not show the original sign of the correction for that remote year. Perfect reproductions of them are in all good libraries.

# THE Apologia of MiER

December 12, 1794, was a fateful day in the life of Servando Teresa de Mier, aristocrat and Dominican priest. On that feast day of Our Lady of Guadalupe he delivered a sermon at the Tepeyac sanctuary on his theory of the origin of the mysterious painting. Mier's view so inflamed the bishop of Mexico City that the bishop had him expelled from the country.

The exiled intellectual went first to Spain, then to England where he remained for many years. The irony of his expulsion lies in the fact that he actually believed the cloak painting to be miraculous.

Mier's idea appears at once fascinating and preposterous. The Image of Guadalupe was a miracle, he said, but not one primarily involving an apparition to a sixteenth-century Christianized Aztec. The miraculous painting in his view, was far older, having been originally the property of an evangelizing Christian, whom the Indians called Quetzalcóatl!

The mythic legend of Quetzalcóatl is well known. The story involves a bearded white man visiting the Indians, only to depart abruptly. The loss is lessened somewhat by Quetzalcóatl's promise to return someday. Scholarship has revealed that a real person, a Toltec king who called himself Topiltzin-Quetzalcóatl, did live at the end of the tenth century. But how did Topiltzin acquire the name Quetzalcóatl?

According to the thesis of Mier, which was borrowed from the historian Ignacio Borunda, a bearded Jew had made his way from distant India to Mexico about the sixth century. This man, who was large by Indian standards, spoke of himself as "Saint Thomas."

It was under the auspices of this "Saint Thomas" that somehow (the details are lost) the Guadalupe cloak first presented itself. If this account is true, the painting is a thousand years older than generally thought and has no more than a peripheral relationship with the Tepeyac locale.

The tradition of associating Juan Diego's apparition of Mary with the Madonna cloak is primarily Indian. The Spanish colonists do not seem to have embraced the belief for several decades, and, in any event, codices reproduced elsewhere in this book clearly originate as Indian memory. If by Mier's time the church was able to exile one of its own for merely expressing public doubt concerning the received tradition, this only illustrates the extent to which the creoles now considered themselves specifically Mexican, not European.

As the Guadalupan tradition was becoming stronger, another Indian legend was becoming favorably, if somewhat reluctantly, assessed. This was the legend relating Quetzalcóatl and Saint Thomas, which Mier had resurrected in 1794. The idea had never gained wide notice, but as Mier himself indicated, it had been carefully preserved by the intelligentsia:

> I was not surprised by this preaching, for I had heard about it from infancy from the mouth of my learned father. All that I have since learned has confirmed its existence, and I do not believe a single cultured American does not know about it or doubts it.[1]

Mier's claim is confirmed in the testimony of several different regions and strengthened in that the different titles suggest independent traditions: "Bochica" is cited in Colombia, "Viracocha" in Peru, "Zume" in Brazil and in Paraguay. In nearer Mayan Yucatan, the name given was "Kulkulkan."[2] The historian Juan de Tovar (see "The Primitive Relation," above) was instrumental in the initiation of the Quetzalcóatl-Saint Thomas tradition. Written in the last quarter of the sixteenth century, de Tovar's *History* calls specific attention to it. "For a better understanding of this," he observed, "it must be remembered that long ago there lived in this country a man who, according to tradition, was a great saint and came to this land to announce the Holy Gospel."[3]

Some one hundred years later, following the inquiry of 1666 in which he had testified (see chapter 2, above), mathematics professor Bercero Tanco appears to have been the first person directly linking the Saint Thomas legend with that of the Virgin of Guadalupe. His testimony was later published in book form, selections from which follow, below ("Proofs of the Apparition").

---

[1]Servando Teresa de Mier, *Memorias* (Mexico City, 1946) 1:5; cited in Jacques Lafaye, *Quetzalcóatl and Guadalupe*, trans. Benjamin Keen (Chicago: University of Chicago Press, 1976).

[2]Lafaye, *Quetzalcóatl*, 190.

[3]Juan de Tovar, *Manuscrit Tovar: Origines et croyances des Indiens du Mexique*, ed. Jacques Lafaye (Graz, Austria: Akademische Druck- und Verlagsanstalt, for UNESCO, 1972) 69; cited in Lafaye, *Quetzalcóatl*, 163.

Just a few years later, Manuel Duarte payed tribute to Tanco's work with the Saint Thomas legend:

> In order that you may know that he was in New Spain, read *The Apparition of the Virgin of Guadalupe*, printed in Mexico in 1675. . . . There you will see that Saint Thomas was in Tula [a town north of Guadalupe], as is clearly shown by the Bachiller Bercero, professor of the Mexican language, who read about it in the Indian histories which tell of the prodigious works and the doctrines taught by this Ketzalcohuatl. . . . In 1680, when I returned to the Philippines, I left a manuscript notebook of more than fifty-two sheets, containing information relative to the teaching of the apostle Saint Thomas in New Spain with the Bachiller Don Carlos de Sigüenza, professor of mathematics.[4]

These were the sheets that the historian Lorenzo Beneducci Boturini laboriously reassembled about 1745:

> Moreover, I have historical notes on the preaching of the glorious apostle Saint Thomas in America. These are contained in thirty-four China papers which, I suppose, were used by Don Carlos de Sigüenza in writing a book on the same subject.[5]

This returns us to the general period in which Borunda and Mier were beginning to work, the late eighteenth century. Recalling with Tovar that "those who found a tan skin in a village on the Gulf coast" were the same people who believed that Saint Thomas made an early mission to Mexico,[6] we must recognize the fact that almost every Indian record had been destroyed by the conquistadors. These ancient codices bore no words, for the early Indians employed picture writing exclusively. To

---

[4]Manuel Duarte, in Nicolas León, *Bibliografía Mexicana del siglo XVIII* (Mexico City, 1902–1908) "Pulma rica," 500-14; cited in Lafaye, *Quetzalcóatl*, 187, 191.

[5]Lorenzo Benaducci Boturini, *Catálogo del Museo Indiano del Cavallero Boturini* (Mexico City: Library of the Basilica de Guadalupe, n.d.); see also Lafaye, *Quetzalcóatl*.

[6]"It was a very ancient tanned skin on which were shown in Mexican hieroglyphics all the mysteries of our faith, though mixed with many errors." Tovar, *Manuscrit Tovar*, 73; cited in Lafaye, *Quetzalcóatl*, 163-64.

European eyes, the codices were caricatures if not blasphemies. Bishop Zumárraga himself consigned great numbers to public bonfires.

Borunda believed these Mexican hieroglyphs contained information quite unknown to anyone but the former inhabitants of Anahuac, the ancient Aztec empire. I reproduce from his rare work *General Key to American Hieroglyphics*:

> Concerning the figurative writings of really unknown signs . . . the eagle of the Church, an African bishop . . . flourishing at the end of the fourth century [Saint Augustine of Hippo?] . . . warned that the great remedy is the remedy of language . . . encouraging [us] to penetrate symbols or signals—an animal, by its smell, fire by its smoke, and so forth—as many things giving more delight to truth when discovered through images and symbols.
>
> [The Indians] invented significant hieroglyphics . . . not with letters but with sculptured figures. . . . [the result is that] hieroglyphics [are arranged] in such a manner that sacred ideas can be expressed, whereas . . . languages of various nations would be unable to adequately convey these intentions.
>
> When we are thus soaked in our own literal writing, gradually perfected in the space of many centuries, we do not remember the particular style that would make understandable concepts which were familiar in terms of ancient characters. We are left today with paradoxical inconsistencies, such as one equal to ten; another to fifteen; and yet another, to twenty.
>
> And in this regard we are aware that it is declared, in the Orient, without the likelihood of explanation in terms of communication from Europe or the Occident and still much less from New Spain, that it was initially in Mylapore [modern Madras] ancient capital of the coast of Coromandel and the Gulf of Bengal, the same named by the Portuguese "Canamina," in allusion to the canes [hieroglyphics] . . . that they were using in ancient times.[7]

Here we have the basis for Mier's alternative history of the Guadalupe Image. The Dominican believed that Saint Thomas of Mylapore, India, introduced the gospel into ancient Mexico. The Incarnation

---

[7]Ignacio Borunda, *Clave general de jeroglíficos americanos* (Mexico City: Library of the Basilica de Guadalupe, n.d.) 14-19.

necessarily involves the sacred relationship between Mother and Son. The cloak bearing the Madonna portrait belonged to Saint Thomas and had remained generation after generation a signal reminder of the Christian who had come from the East.

The connection of the Image on the cloak with Tepeyac hill, Mier asserted, had little to do with a baptized Aztec renamed "Juan Diego." For centuries this hill had been a favorite place of worship for the Indians—a fact that almost certainly lay at the root of sixteenth-century Franciscan opposition to the Guadalupan cultus, such as that of the missionary-historian Bernardino de Sahagún. He wrote,

> Now that the church of Our Lady of Guadalupe has been built the Indians also call her 'Tonantzin,' on the pretext that the preachers call Our Lady, the Mother of God, 'Tonantzin.' . . . This is an abuse that should be stopped, for the true name of the Mother of God is not Tonantzin, but Dios-Nantzin, 'God' and 'nantzin.' . . . The Indians today, as in the old days, come from afar to visit this Tonantzin.[8]

Sahagún's objection was that the unique status of Mother of God (Dios-nantzin) was diluted by being equated with Our Mother (Tonantzin, the Aztec goddess).

Mier, on the other hand, believed the connection between Tonantzin and the Virgin Mary to have been made centuries before—in the times of Saint Thomas-Quetzalcóatl. If that was so, there was no abuse by the sixteenth-century Indians. They were only doing what they had always done. Although they did not remember her as such, Tonantzin, gradually distorted through the centuries, was originally both Our Lady and the Mother of God, Mary:

> Who then was this Tonantzin or Tzenteotenantzin whom Quetzalcóatl taught the Indians to know and who from those remote times had been venerated on Tepeyac hill, also named Tonantzin? She was a virgin, consecrated to God, in the service of the Temple, who by the will of

---

[8]Bernardino de Sahagún, *Historia general de las cosas de la Nueva España*, 4 vols. (Mexico City: Porrúa, 1958) 3:352; cited in Lafaye, *Quetzalcóatl*, 216.

heaven conceived and bore . . . the Lord with the Crown of Thorns, Teohuitznahuac, who partook of both human and divine natures.[9]

Mier was pointing to the several corresponding roles shared by Tonantzin and by the Virgin Mary. His ultimate point, however, was that the Guadalupe cloak dates from a period far earlier than the sixteenth century. Given the legends of ancient voyagers to America—Irish Saint Brendan in the sixth century and Welsh Prince Madoc in the twelfth century—plus parallel tradition by the Mormons, decoding any symbolic significance abiding in the Image of Guadalupe becomes desirable. Perhaps carbon-14 testing on the tilma to determine its age, until now considered unnecessary on the presumption that the tilma could not be more than four hundred and fifty years old, would shed some light on the matter.

---

[9]Servando Teresa de Mier, "Manifesto apologético," in *Escritos inéeitos* (Mexico City: Library of the Basilica de Guadalupe, n.d.); see also Mier, "Apología del Dr. Mier," in *Memorias,* 1:37-38, cited in Lafaye, *Quetzalcóatl.*

# BERCERO TANCO:
# PROOFS of the APPARITION[1]

The news that exists in this city about the apparition of Our Lady, and of the origin of her miraculous Image, which is called Guadalupe, remains more vividly in the memory of the Indians because it was to Indians that she first manifested herself. Consequently, they recorded it and kept it as a memorable event in all their papers and writings. As among other traditions of their ancestors, so it is also necessary to establish first of all the level of belief that should be given to all their writings and memoirs.

The natives (particularly the Mexicans) had two systems of preserving their history, laws, judicial matters, and traditions of their elders, in [much] the same way the Western world does it.

One was by means of paintings of the events that depict them. These paintings were made quite vividly on coarse paper (*papel de estraza*), deer or other animal skins, which were tanned and prepared for this purpose in the manner of a soft papironn, and on each of these "canvases" they painted on the top, the bottom, on the sides, the signs of the years of each one of their centuries; which consisted of fifty-two solar years, and each year of three hundred sixty-five days. The natural months consisted of one full moon to the next (as the Hebrew has), and so they have only one name, which is Metzli.

But for all their rites, ceremonies, and sacrifices to their false deities, as well as for their festivities, the year had eighteen months of twenty days each, which amounts to three hundred sixty days, and after these had passed they added five, which they called "intercals," in the manner of our leap years, which did not belong to any month in the year.

They also painted the characters corresponding to the months and years in which the events took place, as well as the characters and figures corresponding to the king and the lords under which such events took place.

---

[1]Bercero Tanco, *Felicity of Mexico* (Mexico City: private edition for D. Felipe de Zuñiga y Ontiveros, 1780) 36-52; here translated by Jody Smith.

These paintings were and still are as authentic as the writings of our scribes, because they did not trust ignorant people, but only the priests and historians whose authority and credibility was greatly appreciated during the times of the Gentilísimo; so there is little doubt about the veracity of such recordings (characters and paintings); since they have to be exposed to the viewing of everybody in each century, inaccuracy would result in a loss of prestige by the priests. So if we discard the superstitions and rites in honor of their false deities, to whom they attributed some happy or unhappy events, the historical part is authentic and true.

The second form that the natives used to preserve their memorable events and to be transmitted from generation to generation was by means of songs (cantatas) composed by the same priests, with certain types of verse with some vocables added only to preserve a certain rhythm. These cantatas were taught to children who had certain abilities such as memory and musicality and who, upon reaching a certain age as well as proficiency, sang them during their festivities and celebrations, accompanied by *teponaztles*.

Through these songs, traditions and events older than five hundred and a thousand years passed from one generation to the next. Wars, victories, misfortunes, hungers, plague, and births and deaths of kings and lords were related; the beginning and end of different dynasties and all memorable events were thus described. From those maps, paintings, characters, and songs, the Rev. Fr. Juan de Torquemada drew his sources for the writing of the first volume of his *Indian Monarchy*, in which he relates the foundation of this city of Mexico, as well as many other older things, [such as] the life and death of all those who governed these kingdoms before the advent of the Spaniards.

The learned natives continued this same way of writing their history, even after they became subjects of the Crown of Castille, where they conform with the Spanish historians. And after the Indians learned to read and write the Spanish, many Indians continued writing in their Mexican language all the important events that were happening, as well as the old ones that they copied from their maps and paintings.

The members of Spanish religious orders used all these paintings to write about the history of the Land, believing them to be accurate and believing them to be true.

It is well known that the Franciscan monks founded in their convent in Santiago Tlaltelolco a college in which many Indian children learned to read and write the Spanish language, music, and Latin grammar and rhetoric, as well as other liberal arts subjects. These children became well-learned and worthy men, and they were the ones who showed the Spanish (our people) the way they should interpret their drawings and symbols, and also how to compute their centuries, years, months, days, and terms of numbers and figures.

From this we may infer that the Tlaxcaltecan and Acolhua Indians were the most intelligent and clever in the New World, although they also were the ones who were the most influenced by the rites and ceremonies with which they adored their false deities by means of the most cruel sacrifices.

Based on all these things I say and affirm, that among all the memorable events that the well-learned and wise Indians from the college of Santa Cruz (who were in the majority the sons of nobles and lords) painted in their former way for those who did not know how to read and write our alphabet; and with the letters of our alphabet for those who knew how to read them, [they affirmed] *the miraculous apparition of Our Lady of Guadalupe as well as the painting of her Sacred image* [italics added].

I certify to have seen and read an important and very old map written by the Indians with figures and signs in which they narrated events that took place more than three hundred year before the coming of the Spaniards to this land. This map with some lines added, with letters but in the Mexican language, was in the possession of D. Fernando de Alva [Ixtlilxochitl], who was the interpreter of the Indian Tribunal of the Viceroys, and who was a very wise and learned man who spoke and understood very well the Mexican language and who knew well the way of interpreting the paintings and characters of the Indians. Since he was a descendent, on his mother's side, of the king of Texcoco, he inherited and had in his possession many maps and historical papers in which all the events that took place during the lives of his ancestors were recorded.

And among the events related after the "pacification" of this kingdom and city of Mexico was the *one about the miraculous apparition of Our Blessed Lady of Guadalupe* [italics added].

He also had in his position a notebook written in Spanish and the Mexican language, by one of the most capable Indians from the College of Santa Cruz, in which *the four apparitions to Juan Diego were related, as well as the fifth one, to his uncle Juan Bernadino* [italics added].

The second way used by the Indians to preserve for posterity the important events were the cantatas, which I affirm and certify to have heard the old Indians [sing] during their religious ceremonies that they used to have before the flood of the Indian city and, when *they celebrated the ceremonies in honor of Our Lady of Guadalupe* in her sacred temple and in the plaza that faced the west side of the church cemetery. Many dancers formed a circle, and in the middle two old ones sung, accompanied by *teponaztli, the cantata that referred to the apparition of the Holy Virgin and which also told about the cloak or tilma, which was Juan Diego's cape, how it manifested itself to Juan de Zumárraga, first bishop of this city.* At the end of the contata, [were told] the miracles that the Lord had performed the day the Image of Our Lady was placed in her first hermitage and the joyful celebration by the Indians. And this was the end of the oldest and true tradition [italics added].

# ANNALS OF BARTOLACHE

*This is one of the annals found by Dr. José Ignacio Bartolache in 1787 in the library of the University of Mexico. These annals cover the period from 1454 to 1737. The references to Our Lady of Guadalupe are the following.*

13 caña, 1531: The Castilians walked the ground of Cuetlaxcoapan, city of Los Angeles (Puebla) and to Juan Diego was manifested the beloved Lady of Guadalupe in Mexico, at the place named Tepeyac. Pedernal year, 1548: Juan Diego, to whom appeared the Blessed Lady of Guadalupe, died.[1]

---

[1]P. Feliciano Veláquez, *The Apparition of St. Mary of Guadalupe* (Mexico City: Patricio Sanz, 1931); cited in *Guadalupan Documentation: 1531–1768* (Mexico City: Centro de Estudios Guadalupanos, 1980) 109.

# Reply of Fr. Juan de Tovar[1]

*This a reply by Fr. Juan de Tovar (ca. 1546–1626) to questions raised
by Fr. José de Acosta (ca. 1539–1600) concerning the degree of validity
assignable to the Indian codices in light of the Indians lack of an
alphabet. Tovar's answer illuminating on the entire matter of Guadalupan origins.*

Although I could have answered as soon as I received your letter and
could have given a solution to what you ask, nevertheless I was so
anxious for the history to find favor with you that I wanted to refresh my
memory more diligently. I communicated with some old Indian chiefs of
Tula, who are wise in these matters, very learned in this language, and
much like the old chiefs of Mexico and Texcoco, with whom I made the
history in this way. Viceroy Don Martin Enriquez, wishing to know these
peoples' antiquities exactly, ordered a collection of the libraries that they
had on these matters. The people of Mexico, Texcoco, and Tula brought
them, since these people were the historians and sages in these matters.
The Viceroy sent the papers and books to me with Dr. Portillo, formerly
vicar general of this Archbishopric, charging me to examine and study
them, and to make some relation to be sent to the king. . . . I looked over
all this history, the characters and hieroglyphs of which I did not
understand. Therefore it was necessary for the wise men of Mexico,
Texcoco, and Tula to meet with me, by order of the Viceroy. Talking
over and discussing the matter in detail with them, I made a thorough
history, which, when it was finished, was taken by Dr. Portillo, who
promised to make two copies with very fine pictures, one for the king
and one for us. At this juncture, it happened that he went to Spain, and
he was never able to make good his word, nor were we able to retrieve
the history. But as I had then investigated and discussed the matter at
great length, it remained strongly in my memory. In addition, I saw a
book made by a Dominican friar, a relative of mine, which was very

---

[1]George Kubler and Charles Gibson, *Tovar Calendar*, Memoirs of the
Connecticut Academy of Arts and Sciences 11 (New Haven CT: Yale University
Press 1951) 77-78.

similar to the ancient library that I had seen and which helped me to refresh my memory in making the history that you have now read. I put down what was most certain and omitted other dubious small matters, and this is the authority that it has, a great authority in my opinion, since in addition to what I saw in their own books, I discussed it, prior to the *cocolistle*, with all the old men whom I knew to have knowledge of it. No one disagreed, a rare circumstance among them. This is what I answer to your first question, concerning the authority of this history.

To the second question, how could the Indian retain so many things in their memory without writing, I repeat that they had figures and hieroglyphs with which they painted things in this way. Objects that could be represented *directly* were drawn in their own image. Whatever could not be represented directly was drawn with characters *representing* an image. In this way they [the Indians] drew what they wished. And as for their remembering the time in which each event took place, you have already read about the computation that these people used, how they made each fifty-two years a wheel, as I mentioned there. The wheel was like a century, and with these wheels they preserved the memory of the times in which the memorable events occurred, painting the events at the sides of the wheels with the characters mentioned. The wheels and circles of the years that I saw numbered four, since these people have no other count. From the time that they left the seven caves, mentioned at the beginning of that history, until the Spaniard came, three complete wheels had passed and the fourth was in progress. In these wheels all the events and memorable occurrences were indicated, as you will see in the wheel and the end of the calendar that goes with this. There they put the Spaniard with red coat and hat, as an indication of the time when the Spaniards entered this land. This was in the fourth wheel or age, during the sign that they call reed, which they painted in the form that you will see there. But it is to be noted that although they had different figures and characters with which they wrote, their method was less adequate than our writing, in which everyone knows verbatim what was written by the very words. They [the Indians] agreed only in the concepts.

But the words and forms of the orators' speeches and the many songs composed by the orators were known by all without any disagreement, even though they pictured them with their characters. In order to preserve the words that the orators and poets spoke, they held an exercise every day in the colleges of the young chiefs who were to succeed them, and with this continual repetition, the most famous orations of each time remained in their memory. This was a system for impressing the young with the fact that they were to be rhetoricians, and in this manner many orations were preserved verbatim from generation to generation until the Spaniards came. The Spaniards wrote in our letters many orations and songs that I saw, and thus they have been preserved. This is the answer to the last question, how it was possible to have this memory of the words, etc. And to add to what I have said here, I sent to you the orations of the Pater Noster, etc., and of the general confession, and other matters of our faith, as the ancients wrote and learned them by their characters, which were sent to me by the old men of Texcoco and Tula. And this will be enough to show in what manner the ancients wrote their histories and orations. Also I sent, besides the calendar of the Indians, another, very curious, in which their months and days and fiestas are equated with the fiestas and months and year of our ecclesiastical calendar. Certainly it excites admiration to see that these Indians achieved so much with their cleverness and skill, as you will see by these papers that I send.

# Pensacola and Guadalupe

I have recently discovered that the history of Pensacola, Florida (my hometown since 1970), and the history of the Guadalupe story are remarkably associated. The modern city of Pensacola (from *Panzacola*, an Indian name) appears to have been in pre-Columbian times known as *Ochuse* (sometimes *Achuse* and/or *Ichuse*[1]). The first European settler christened this particular location, along with the surrounding bay, *Santa Maria Filipina*. This was on August 14, 1559. So we begin with a correlation, since the *Guadalupe* Virgin is a type of, or name of, Saint Mary.

The area had been known to the Spanish as early as 1528, a bare ten years after Cortez invaded Mexico. This was also the year the first archbishop, the Franciscan Juan de Zumarrago, disembarked in New Spain, as the conquistadors referred to Mexico. In 1555, we find the second archbishop's Tepeyac sermon, in favor of the Guadalupan cultus. In the same year, the same Dominican, Alonso de Montufar, addressed the Spanish state in favor of a missionary enterprise to the northwest coast of Florida

> since we have it so near at hand, and know the numberless people which are lost therein from having none to preach them the Holy Gospel.[2]

This is an overtly religious conception of the expedition that eventually would be known as modern Pensacola. Moreover, exactly four years later, Montufar reiterated the crucial endorsement by personally pronouncing the solemn

---

[1]Herbert Ingram Priestley, *Tristan de Luna—Conquistador of the Old South* (Philadelphia: Porcupine Press, 1980; repr. of 1936 first edition by the Arthur H. Clark Co.) 104; see also 102-103.

[2]Ibid., 55-56.

blessing of the royal standard which the governor was to bear at the head of his forces. . . . the symbol of legal and regal authority.[3]

This came at the conclusion of the Viceroy Don Luis Velasco's commission to the expedition leader:

The sign of the Holy Cross upon which God our Lord redeemed mankind, in order that with these arms and with the evangelical preaching of the religious you may labor to bring these people into peace and into obedience to our holy mother church and into the dominion and overlordship of his majesty, without subjecting them to war, force, or bad treatment.[4]

These words were spoken to the "governor and captain-general,"[5] a man whose name has since become well known along the upper Gulf coast, Tristan de Luna y Arellano.

On June 11, 1559, thirteen ships, accommodating five hundred cavalry, one thousand colonists and servants, two hundred forty horses, plus all kinds of tools for both building and farming, put to sea. They sighted the Florida coast eight days later, but due to storms and unknown waters, it was almost two months before they arrived at their searched-for destination.

Immediately on disembarking, de Luna declared the name of his settlement *Santa Maria Filipina. Filipina* is easily attributable to Philip II, the king of Spain, but we are particularly interested in the assignment of the name "Saint Mary."

As far as I am aware, notwithstanding the favor already declared by Guadalupan Archbishop Montufar in Mexico, the immediate inducement for the name was the liturgical recognition for the date of August 15: called by de Luna "Lady Day of August," this particular date was already or recognized significance in the church calendar. Yet this is not a perfect match, for fifteen is not fourteen. The arrival was afternoon, and perhaps considered close enough to the Feast of the Assumption to make the dedicatin in its homage.

---

[3]Ibid., 84.
[4]Ibid.
[5]Ibid., 84-85.

## Part 2

De Luna's missionary settlement was soon ravaged by a hurricane and, with loss of provisions, reduced to virtual starvation. It was abandoned in 1561. Almost one hundred fifty years later the first permanent settlement by any nation was made in 1698.

If the Marian origin of de Luna y Arellano is interesting in its own right, the ultimate establishment in terms of honoring the Virgin Mary is even more remarkable. After much wrangling back in Spain, it was decided that the northwestern Florida coast would be thoroughly and scientifically surveyed. Memories of the original settlement had never entirely faded, especially the reputed excellence of the bay where de Luna had landed. This bay was rediscovered by Juan Jordan de Reiña. According to his records, on February 6, 1686, Jordan stated that he resighted the lost *bahía*:

> About eleven o'clock I saw a bay, the best I have ever seen in my life. We put into it, finding a depth of eight, nine, and ten fathoms at its entrance with is not very wide; after steering north and northeast inside this bay, I anchored in seven fathoms. Its opening lies almost north and south; the Indians call this bay *Panzacola*. . . .[6]

On March 25, 1693, the highly regarded scholar and scientist Don Carlos de Sigüenza y Gongora departed from Vera Cruz, bound for the recently rediscovered bay. Sigüenza had already gained high regard for his rational explanation of the nature of comets (e.g., that just named after Halley and which coincidentally had been visible in the year 1531) as well as his engineering and mapmaking prowess. Let us not also that March 25, the date of the Feast of the Immaculate Conception, may have been more than accidental in terms of propitious auspices for the departure from Mexico.

On April 7, the survey team arrived at their destination. While Andres de Pez was the formal leader of the survey team, it has been

---

[6]Irving A. Leonard, *The Spanish Approach to Pensacola, 1689–1693* (Alberquerque NM, 1939) in James R. McGovern, ed., *Colonial Pensacola*, the Pensacola Series Commemorating the American Revolution Bicentennial, 2nd ed. (Pensacola: Pensacola New Journal, Inc., 1976) 14.

noted that, de facto, it was Sigüenza who was the "leading spirit of the undertaking."[7] Delighted by the beauty of the area, Sigüenza tells us that he recalled the efforts of Tristan de Luna and specifically how the latter had first named the bay in honor of the Virgin and of Philip II:

> It seemed only right not to cheat it of such an honored name. . . . So, chanting the *Te Deum Laudamus* as best I knew how in the presence of her divine image because of the blessings we had received up to then from her holy land, I named it, in a prayer especially made for this occasion, the *Bahia de Santa Maria de Galve*.[8]

Admiral de Pez so authorized this name on all official charts and records.[9]

Sigüenza, in addition to being mathematics professor at the University of Mexico, possessed a widely inquiring mind. He became known, eventually, as owner of one of the most superb libraries in all New Spain. The centerpiece of this collection was a gift of Aztec royalty. So it is easy to recognize Sigüenza as the same already mentioned earlier in this book, and his royal friend is none other than Juan de Alva Ixlilcóatl, who had himself inherited the original codices from his own father by direct line of succession, including documentary source materials in Nahuatl, their language.

Both the Ixlilcóatls were strong believers in the reality of the Guadalupe Mystery.

Recall that Sigüenza's redecication of Tristan de Luna's name for both settlement and bay—Saint Mary's—occurred before "Her divine image." One naturally wonders what this means. The subject most popular in all Western art is the Virgin Mary. Is the adjective "divine" therefore metaphorically inclusive? No. Sigüenza, as a strictly orthodox Catholic, well knew that for all of the church's veneration of the Virgin, she was never considered to be divine. So, what prompted Sigüenza y Gongora to name the first permanent settlement Santa Maria "in the presence of *her divine image*"?

---

[7]Ibid., 23.
[8]Ibid.
[9]Ibid.

To answer this question adequately is only to reveal the name of the flagship on board which Admiral Andrez and Don Carlos were carried to *Santa Maria de Galve.*

Alongside the sloop *San José,* there floated in "the best bay I have ever seen in my life"[10] the frigate named *Nuestra Senora de Guadalupe.*[11]

---

[10]Ibid, 13.
[11]Ibid., 22, 52.

# Bibliography

Augustine of Hippo, *On the Trinity*, in *Patrologia Latina*, 5.42.8.

Behrens, Helen, trans., *The Virgin and the Serpent-God* (Mexico City: Editorial Progreso, 1966).

Brownrigg, Robelt, *Who's Who in the New Testament* (New York: Pillar Books, 1971).

Bulgakov, Sergius, *A Bulgakov Anthology*, ed. and trans. James Pain and N. Zernov (Philadelphia: Westminster Press, 1976).

Cabrera, Miguel, *Maravilla Americana–y conjunto de Raras Marvillas Observadas con la dirección de las reglas del arte de la pintura en la prodigiosa Imagen de nuestra Sra. de Guadalupe de México* (Mexico City, 1756; 2nd ed., 1977).

Callahan, Philip Serna, and Jody Brant Smith, *The Virgin of Guadalupe: An Infrared Study* (Washington DC: CARA, 1981). Spanish trans. F. Faustino-Cervantes, 1981.

De Clari, Robert, *The Conquest of Constantinople*, trans. Edgar H. McNeal (New York: Columbia University Press, 1936).

Delaney, John J., ed., *A Woman Clothed with the Sun* (Garden City NY: Doubleday, 1961).

Demarest, Donald, and Coley B. Taylor, eds., *The Dark Virgin: The Book of Our Lady of Guadalupe* (New York: The Devin-Adair Co., 1956).

Eusebius, *Ecclesiastical History*, trans. Kirsopp Lake and J. E. L. Oulton, Loeb Classical Library (Cambridge MA: Harvard University Press, 1926 and 1932).

Forsyth, Ilene H., *The Throne of Wisdom: Wood Sculptures of the Madonna in Romanesque France* (Princeton NJ: Princeton University Press, 1972).

Gibbon, Edward, *Decline and Fall of the Roman Empire* (New York: The Modern Library, 1932) vol. 3.

Grabar, André, *Christian Iconography* (Princeton NJ: Princeton University Press, 1961).

Graves, Robert, and Raphael Patai, *Hebrew Myths: The Book of Genesis* (Garden City NY: Doubleday, 1964).

Gregorovius, Ferdinand, *Athen und Athenais* (Dresden, 1887).

Hawkins, Columbine, "Our Lady of Guadalupe–Miracle of Iconography" (Lafayette OR: Trappist Abbey of Our Lady of Guadalupe, 1951).

Irenaeus, *Against Heresies*, ed. and trans. Alexander Roberts and James Donaldson, *The Ante-Nicene Fathers* 2 (Grand Rapids MI: Eerdmans, 1962–1966).

Jameson, Anna Brownell, *Legends of the Madonna—as Represented in the Fine Arts* (London: Hutchinson & Company, 1852).

Jung, Carl Gustav, *The Structure and Dynamics of the Psyche*, Bollingen Series 20, vol. 8, 2nd ed. (Princeton NJ: Princeton University Press, 1969).

Lacurdhas, V., ed., *Photius Homiliai* (Thessalonika, Greece: Etaireia Makedhonikon Spoudhon, 1959).

Lafaye, Jacques, *Quetzalcóatl and Guadalupe*, trans. Benjamin Keen (Chicago: University of Chicago Press, 1976).

López-Beltrán, Lauro, "Guadalupanismo Internacional" in *La Virgen de Guadalupe*, special ed. of *Mexico Desconocido* (Mexico City: Editorial Novaro, 1981).

Micheller, James A., *Iberia* (New York: Random House, 1968).

Motolinia, Toribio de, *History of the Indians of New Spain*, trans. Elizabeth A. Foster (Westport CT: Greenwood Press, 1972).

F. de Jesús Chauvet, *El Culto Guadalupano del Tepeyac*, 6th ed. (Mexico City: Centro de Estudios Bernardino de Sahagún, 1978).

*New Catholic Encyclopedia*, 1965 ed., 6:821-22.

Northrop, F. S. C., *The Meeting of East and West* (New York: Macmillan Company, 1946).

Rahm, Harold J., *Am I Not Here?* (Washington NJ: Ave Maria Institute, 1962).

Rojas, Mario, *Segundo Encuentro Guadalupano* (Mexico City, 1978).

Salinas, Carlos, and De la Mora, Manuel, *Descubrimiento de un Busto Hurnano en las ojos de Icz Virgen Maria de Guadalulupe* (Mexico City: Editorial Tradicion, 1976).

Sánchez, Miguel, *Imagen de la Virgen María, Madre de Dios de Guadalupe milagrosamente aparecida en México* (Mexico City, 1648).

Tyler, R., trans., *The Monastery of Guadalupe* (Barcelona,1930) 4.

Ugarte, José Bravo, *Cuestiones Guadalupanas* (Mexico City, 1946) part 1.

Wilson, Ian, *The Shroud of Turin*, rev. ed. (Garden City NY: Doubleday, 1979).

# Index

# The Republic of Korea

## A Study of the Educational System of the Republic of Korea and a Guide to the Academic Placement of Students in Educational Institutions of the United States

**Philip J. Gannon**

*President*
*Lansing Community College*
*Lansing, Michigan*

1985

A Service of the International Education Activities Group of the
American Association of Collegiate Registrars and Admissions Officers

Placement Recommendations Approved
by the National Council on the Evaluation
of Foreign Educational Credentials

Gannon, Philip J.
    The Republic of Korea

    (World education series)
    "A service of the International Education Activities
Group of the American Association of Collegiate
Registrars and Admissions Officers."
    Bibliography: p.
    Includes index.
    1. Education—Korea (South) 2. Students, Foreign—
United States. 3. Students—Korea (South) 4. School
credits—United States. I. American Association of
Collegiate Registrars and Admissions Officers.
International Education Activities Group. II. Title.

III. Series.
LA1331.G36   1985      370'.9519'5      85-14210
ISBN 0-910054-81-9

Publication of the World Education Series is funded by grants from the Director-
ate for Educational and Cultural Affairs of the United States Information Agency.

# Contents

## Tables

# Documents

# Preface

While working and visiting with the people of the Republic of Korea, I have studied their educational system and society, and have come to admire their perseverance and their ability to develop a unique culture. As an educator, I was intrigued by the fact that United States citizens of Korean birth were able, in a very short period of time, to adapt to the culture of the United States and make major contributions as professionals and business people. Their children, on the average, have performed scholastically above the norm in our educational system.

I found it of professional interest to search for the underlying answers to this performance, and to hypothesize regarding the next generation. The pathway which provided insight was the study of Korean history and the cultural roots which were responsible for their commitment to education and learning. I believe the staff and faculties of educational institutions in the United States, working with Korean students, will be impressed with their seriousness of purpose and their commitment to their studies and to their country.

The chapters which follow will trace the history and development of the present Korean educational system from pre-school through tertiary education, with advice to admissions officers and placement recommendations for Korean students making application to educational institutions in the United States.

The national flower of Korea, the *mugunghwa*, symbolizes the culture and character of the Korean people. The flower of this plant blooms over an unusually long period of time, and only through careful observation does one ascertain its beauty, longevity, color and perfection. To those not diligent in their observation, this beautiful flower, like Korea, hides itself from view as it blends with its environment. Similarly, the Korean culture, over 4000 years in evolving, has had to be tenacious to develop its own beauty, longevity and uniqueness.

# Acknowledgements

In the preparation of this document many people were helpful, making materials available and evaluating the information. It was only through their help that this document could be completed; however, any errors are my responsibility.

I would like to particularly acknowledge the following people for their willingness to assist and for their many hours devoted to reviewing this document and sharing their insight. Dr. Tai Sung Kim, Director of International Programs, and Vice President Jacqueline D. Taylor, Lansing Community College, guided me in the development of this document. A special thanks goes to two fine editors, Henrianne K. Wakefield and Lucy McDermott, and to the editorial staff of the World Education Series; and to my wife, Lois, for her encouragement. In the Republic of Korea three scholars were key individuals in the development of this material: Dr. Byung Rim Koo, Council Specialist of the Korean Council for University Education; Dr. Jeoung-Keun Lee, Executive Director, Vocational Training Research Institute; and Dr. Mu Keun Lee, Associate Professor, Vocational and Technical Education, Department of Agricultural Education, Seoul National University.

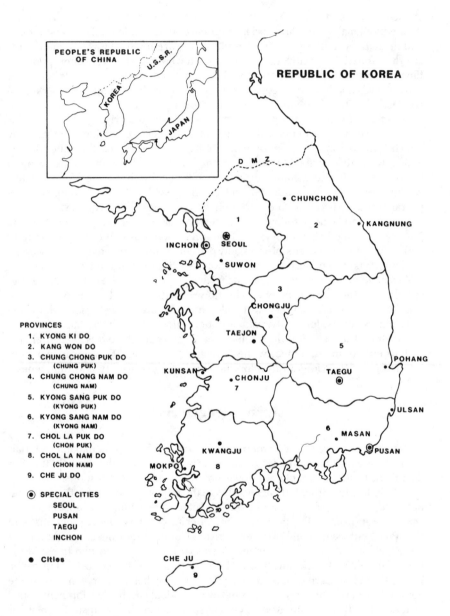

PEOPLE'S REPUBLIC
OF CHINA

U.S.S.R.

KOREA

JAPAN

REPUBLIC OF KOREA

D M Z

• CHUNCHON

1

2

• KANGNUNG

INCHON ◉ ⊕ SEOUL

• SUWON

3

CHONGJU •

4

TAEJON •

5

• POHANG

KUNSAN •

• CHONJU

7

TAEGU
◉

• ULSAN

6

• MASAN

MOKPO •

KWANGJU •

8

PUSAN

**PROVINCES**

1. KYONG KI DO
2. KANG WON DO
3. CHUNG CHONG PUK DO
   (CHUNG PUK)
4. CHUNG CHONG NAM DO
   (CHUNG NAM)
5. KYONG SANG PUK DO
   (KYONG PUK)
6. KYONG SANG NAM DO
   (KYONG NAM)
7. CHOL LA PUK DO
   (CHON PUK)
8. CHOL LA NAM DO
   (CHON NAM)
9. CHE JU DO

◉ SPECIAL CITIES
   SEOUL
   PUSAN
   TAEGU
   INCHON

• Cities

CHE JU
• 
9

# Chapter One

# Introduction

**Official name:** Republic of Korea.

**Location:** The Korean peninsula is located in northeast Asia and projects southeast from the People's Republic of China. Two rivers, the Tumen, flowing east, and the Yalu, flowing west, separate Korea from the People's Republic of China. A small area on the northeast is bordered by a maritime province of the Soviet Union. To the west, across the Yellow Sea, is the People's Republic of China. To the south lies the Pacific Ocean, and to the east lies the Japanese archipelago.

**Size:** The peninsula is approximately 600 miles long, between 125 to 200 miles wide, and consists of approximately 85,000 square miles. North of the demarcation line, 48,000 square miles comprise the People's Republic of Korea, and south of this line, 37,000 square miles comprise the Republic of Korea. The Korean peninsula corresponds in size to the state of Minnesota or the country of West Germany.

**Population:** Fifty-eight million (thirty-nine million in the Republic of Korea and nineteen million in North Korea). The population density of the Republic of Korea is among the five highest in the world. There are no racial minorities in Korea.

**Geography:** Approximately 70% of the country is mountainous. The mountains are found predominantly throughout the north and along the east coast of the peninsula. Consequently, most rivers flow to the west and south. Major mineral resources are found in the north. Agricultural resources are found in the southwest portion of the peninsula; rice is the predominant crop.

Korea is in the temperate monsoon zone with four distinct seasons: heavy rainfall in the summer; cool, pleasant weather in the spring and fall; and cold dry winters. Korea experiences few earthquakes or typhoons.

**Spoken language:** Korean.

**Written language:** *Hangul* with some use of Chinese characters. Transliteration makes it possible to use the Roman alphabet for sounds that are similar to Korean pronunciation. In the romanization of *Hangul*, the most commonly used approach is McCune-Reischauer, although the Republic of Korea in-

---

NOTE: Publications by the Ministry of Education of the Republic of Korea were used as primary sources of information throughout this book. These primary sources, along with others, will be listed in the "Useful References."

stituted a new system of romanization in 1984. McCune-Reischauer is the guideline used for this text.

**Major religions ranked by membership:** There is official separation of church and state. Freedom of religion prevails. Major religions in 1980 included: Buddhism (12,329,720); Protestantism (7,180,627); Confucianism (5,182,902); Catholicism (1,321,293); *Chondogyo*, the traditional Korean religion (1,153,677).

**Government:** The modern democratic political system in the Republic of Korea was introduced at the end of World War II. Throughout most of its history, the Korean political system centered on an absolute monarchy in which the king had ultimate decision-making power in administrative, legislative and judicial affairs. During the Japanese colonization, 1910 to 1945, control was centralized and absolute.

After World War II, from 1945 through 1948, the U.S. military served as the transition government. This introduced the Western democratic ideology and democratic political institutions to the Korean people. In 1948, when South Korea became independent, the first democratic constitution was promulgated. Since then, the Constitution has been amended five times. However, the fundamental democratic principles enumerated in the Constitution remain intact.

Korea has a presidential system of government with separation of powers and a system of checks and balances. The President is, nevertheless, the apex of the government, the head of state, the chief executive, the legislative leader, and the party leader and is elected by the presidential electoral college for no more than one seven-year term. The Constitution vests the President with the power to enact laws during national emergencies. Under the President, the Prime Minister heads the cabinet and other ministers.

The legislative power is vested in a unicameral National Assembly that is composed of 276 members. The members of the National Assembly are elected for four-year terms by popular universal election based on both territorial and proportional representations, two-thirds by direct popular vote and the remaining one-third by proportional representation. The one-third proportional representation allows a number of political parties to be represented in the National Assembly. This remaining one-third is selected by the party members on a proportional basis dependent upon the percentage of the popular vote each party receives. The chief function of the National Assembly is lawmaking. However, it also exercises power to oversee the administration. The members of the National Assembly serve for four years, but the President has the power to dissolve the Assembly in time of emergency.

Judicial power is vested in independent courts. Korea has three levels of courts: the Supreme Court, appellate courts, and district courts which include branch courts. The Supreme Court is empowered to make final judicial decisions and also has the power of judicial review. The Chief Justice is appointed by the President and serves for no more than one five-year term. All other justices are also appointed by the President and serve for five years; however, they can be reappointed.

Since Korea has a unitary system of government, the provincial and city governments are not fully independent from the national government. The governors and mayors are appointed by the President and serve at the pleasure of the President.

**Cities:** The government has designated the following four as special cities because of their size and complexity: Seoul, Pusan, Taegu, and Inchon. The population of the six largest cities in 1980 was as follows: 1) Seoul, the capital city (8,114,000); 2) Pusan (2,879,570); 3) Taegu (1,487,098); 4) Inchon (936,497); 5) Kwangju (694,646); 6) Taejon (508,574). As of 1980, Seoul and Pusan, the two largest cities, comprised approximately 30% of the population of the Republic of Korea.

The majority of the population of Korea is now located in urban areas. Consequently, the greatest number of students attend urban high schools. Over 90% of the college and university students are also found in urban areas.

**History:** Korea is located on a peninsula which, throughout its history, has been the crossroads of international spheres of influence and political interest. The country has been, and is, one of the catalytic ingredients between the cultures of China and Japan.

During its four thousand years of history, Korea, until the mid-1800s, was basically a fishing and agricultural country. Over the centuries, however, as Korean culture developed, a society evolved that was composed primarily of classes structured as follows: 1) royal families, 2) higher governmental officials, 3) educators and civil servants, 4) farmers, 5) craftsmen, and 6) merchants. Decision-making followed a hierarchical order centralized under a royal family until 1910 and under colonial rule until 1945, and provides some of the antecedents for a more centralized decision-making process and the highly developed social strata system found in modern Korea.

Confucianism and Buddhism have had a strong influence on the direction of and attitude toward the development of education throughout the history of Korea. Culture and education were also influenced predominantly by China through Manchuria and across the Yellow Sea. South Pacific influence is represented by the unique art and culture of the southern island province of Che Ju.

In its several thousand years of history Korea has been invaded by the Chinese, the Mongols, and the Japanese. Two major invasions by Japan in 1592 and 1597 devastated the country, and from 1910 to 1945, Korea was a colony of Japan. Near the end of the Yi Dynasty in the late eighteenth and early nineteenth centuries, Korea closed itself to outside influence. At that time Japanese power was on the ascendancy and, with the Japanese defeat of Russia in 1905, Japan's sphere of influence directly included the Korean peninsula.

In 1945, at the end of World War II, arrangements were made for the Soviet Union to accept the surrender of the Japanese in the northern part of Korea and for the United States to accept the surrender in the southern part. The United Nations requested that a national election for the Korean people take place across the total peninsula to establish a single government for the country. This election took place south of the 38th parallel under the direction of the United Nations; however, north of the parallel the Soviet Union did not allow an election to take place. Thus, the peninsula was divided along the 38th parallel, with the northern part under the influence of the Soviet Union and the southern part under the influence of the United States.

North Korea invaded South Korea on June 25, 1950. Sixteen nations fought in Korea under the flag of the United Nations to assist in the defense of the Republic of Korea. This conflict completely devastated the peninsula and ended in 1953 in a cease-fire with the country divided north and south by the demilitarized zone (DMZ) which generally follows the 38th parallel. Once more in its history Korea was divided. The desire for unification is a strongly held position by the Korean people.

**History of education:** The three early kingdoms of Korea—Koguryo, Paekche, and Shilla—laid the foundations for the beginnings of formal education, primarily for the upper class, with a curriculum centered in Buddhism and Confucianism. Table 1.1 presents the chronological chart of the kingdoms and governments of Korea.

During the Koryo period, movable metal type was invented 200 years before Gutenberg. Also during this period, a beautiful and highly refined craftsmanship evolved which was exemplified by Celadon porcelain. Korean craftsmen and scholars influenced the development of Japanese art and culture during visits to Japan.

The Yi dynasty, following the Koryo period, marked a time of great intellectual and artistic development that continued into the early twentieth century. During the reign of King Sejong (1418-1450), the *Hangul* alphabet was developed which allows the Korean language to be written phonetically and re-

**Table 1.1. Chronological Chart of Kingdoms and Governments of Korea**

| | 1st. Cent. | 7th Cent. | 10th Cent. | 14th Cent. | | 20th Cent. |
|---|---|---|---|---|---|---|
| Old Choson | Koguryo | Parhae | Koryo | Yi | Japanese rule | North Korea |
| | Paekche | Unified Shilla | | | | Republic of Korea |
| | Shilla | | | | | |
| B.C. | A.D. | | | | | |

Source: Adapted from Ministry of Education publication "Education in Korea," 1984.

placed the total dependency on Chinese characters. *Hangul*, a significant invention, thus made it possible to develop a system of reading and writing easily learned by all levels of Korean society.

Through the influence of Buddhism and Confucian studies and traditions from China, there evolved within the Korean culture: 1) a respect for and obedience to elders and superiors; 2) a tendency to adhere to tradition; 3) an unquestioning attitude towards the authority of the teacher and the presented subject matter; 4) the acceptance of a theoretical rather than an applied approach; 5) the acceptance of an emphasis on rote and memory rather than on inquiry and questioning; and 6) a willingness to endure suffering. These cultural influences have been both a strength and a concern. They have made it more difficult for Korea to evolve, in a short period of time, a system of education that fosters scientific inquiry in a modern sense, along with a research approach to library use, and an applied approach to vocational/ technical and engineering education.

In the 1880s, while Western education was first being introduced into Korea by Christian missionaries, Korean rulers adopted a closed-door policy against Western influence. Consequently, Korea did not develop the capability to deal with its neighbors nor share in their use of new technology.

During the Japanese occupation of Korea, the Japanese developed an industrial, business and educational infrastructure. Two educational systems were promoted, one for the Japanese and one for the Koreans. The Japanese system was open-ended and allowed for advanced training in science and managerial skills. The Korean system was very limited and primarily emphasized basic literacy. During the latter part of this colonial period, the study of Korean history and language was removed from the educational system. When Japan surrendered at the end of World War II, Korea was left with a work force that lacked experience or training in management, science, and technology. The educational system was predominantly staffed by Japanese at the administrative and supervisory levels. Consequently, when the Japanese occupation ended, Korea had very few trained teachers or administrators, and those that were trained had been brought up in a system of learning that was Japanese in its approach, emphasizing memorization and focusing on final examinations.

In 1950, with the economy in disarray and a newly established democratic form of government, the Korean War began. The conflict ended three years later, leaving the peninsula devastated with wide destruction of educational facilities, few trained educational personnel, and an influx of refugees from North Korea. The needs of the country and the desires of the people called for an open and available educational system.

In the mid-1950s, economic planning accepted two primary realities: few natural resources and an extremely weak economy. Plans were developed for an economy heavily directed towards international trade and heavy and light industry. To accomplish this goal, education became a central theme. Of major concern to Korea then and now is how to blend the uniqueness and cultural heritage of its past into a curriculum sensitive to science and technology. The dedication of the Korean people to education made it possible for Korea to move in the short period of forty years from an agricultural society to a developing country with a sophisticated educational system and a modern technological infrastructure.

Educationally, during this period the Republic of Korea faced a number of critical problems: 1) the development of a quality educational program during a period of rapid expansion; 2) the development of strategic planning that allowed for the amalgamation of educational ideas of other countries with the culture and needs of the Korean society; 3) the development of a philosophical base for education that would help Korea further develop its national identity; 4) the development of a structure and process that is sensitive to tradition and group identity and recognizes the needs of the individual; and 5) the development of an educational system sensitive to the worldwide scientific technological revolution and Korea's rapid urbanization.

## Overview of Education in Korea

The culture of Korea, as its history shows, has great respect for learning and education. In a period of approximately thirty years Korea has experienced an educational miracle. Illiteracy has virtually been eliminated in a span of twenty years. In the rapid expansion of its educational system, Korea searched the world for ideas and imported models and features of foreign educational systems.

The current educational system in Korea had its beginning in the late 1940s. At that time the Republic of Korea adopted to a great extent the educational structure used in the United States, based upon six years of primary education, three years of junior high school, three years of senior high school, and four years of college. Vocational/technical education was influenced by the German and Japanese systems of education.

The Constitution of the Republic of Korea requires universal education to be made available to all of its citizens.

Tables 1.2 and 1.3 describe the structure of the educational system of the Republic of Korea.

## Educational Governance

All education for pre-school through tertiary levels in both the public and private sectors is under the control of the Ministry of Education. The Ministry also has responsibility for supervision and implementation of educational policy. Consequently, it has authority for carrying out ministerial decrees, for the direction and supervision of the educational budget, for administration and personnel, and the authority regarding education as it relates to licensing and approval of the curriculum and its authorization in institutions. The Minister of Education is a member of the National Council of the President of Korea and participates in top level decisions within the central government. Also under the direction of the Ministry of Education are a number of institutes, associations, and educational research organizations.

Within each province and special city, a Board of Education, headed by the Superintendent, directs primary and secondary education, and manages other related educational programs in art, science, athletic and cultural activities in its local district. The Superintendent is a member of a local seven-member board. The Superintendent is recommended by the Board to the Ministry of Education, and then, upon the recommendation of the Ministry of Education, is appointed by the President.

Approximately 21% of the national budget is allocated for education. It is difficult to estimate the total budget for all of education inasmuch as this percentage does not include the funds raised and contributed for private education or non-formal education. A major share of the educational funding for primary and secondary education is financed by the central government. There is no tuition charge for primary school, but tuition is charged for both public and private kindergarten, junior high schools, senior high schools, and colleges and universities. National schools are sponsored by the central government. Public schools are sponsored by the city or provincial governments.

**Table 1.2. Educational System of the Republic of Korea, 1985**

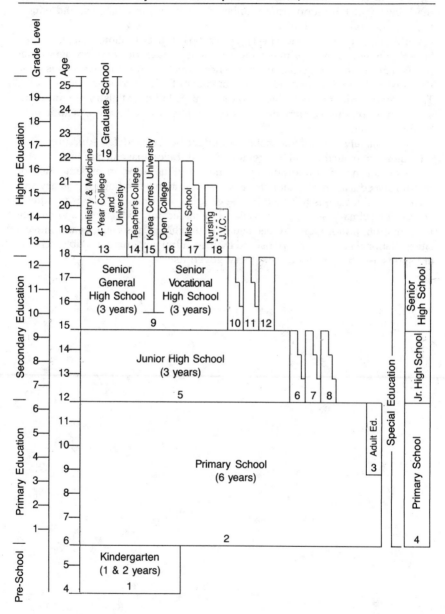

A. Pre-school Education (*Yua Kyoyuk*)
   1. Kindergarten (*Yuchiwon*)—Non-compulsory for 4-6 year olds.
B. Primary Education (*Chodung Kyoyuk*)
   2. Primary School (*Kukmin Hakkyo*)—Six years - compulsory.
   3. Civic School—Adult Education (*Kongmin Hakkyo—Sung-in Kyoyuk*)—Primary school for adult literacy training.
   4. Special Education (*Tuksu Hakkyo*)—Primary school, junior high school and senior high school for the blind, deaf and other handicapped.
C. Secondary Education—Junior High School (*Chungdung Kyoyuk*)
   5. Junior High School (*Chung Hakkyo*)—Three years - non-compulsory.
   6. Trade School (*Kisul Hakkyo*)—Junior high school concentrated on skills training with exit points at one, two and three years.
   7. Higher Civic School (*Kodung Kongmin Hakkyo*)—Junior high school for adult literacy training with exit points at one, two and three years.
   8. Specialized School (*Tuksu Hakkyo*)—Single purpose junior high school, e.g., music, dance, etc., with exit points at one, two and three years.
D. Secondary Education—Senior High School (*Kodung Kyoyuk*)
   9. Senior High School (*Kodung Hakkyo*)—Three years - non-compulsory. Senior high schools (general or vocational) lead to a senior high school diploma.
   10. Senior Higher Trade School (*Kodung Kisul Hakkyo*)—Trade school for students not in formal senior high school program. Exit points at one, two or three years.
   11. Specialized Senior High School (*Tuksu Hakkyo*)—Single purpose senior high school, e.g., music, science, physical education, etc., with exit points at one, two and three years. May lead to a senior high diploma.
   12. Air and Correspondence School—Senior High School Level (*Bangsong Tongshin Kodung Hakkyo*). Adult Education—Curriculum delivered predominantly via radio and correspondence leading to senior high school diploma.
E. Higher Education (*Taehak Kyoyuk*)
   13. College and University (*Taehak* and *Taehakkyo*) —Four-year programs leading to bachelor's degree. Medicine and dentistry are six-year programs. Colleges and universities may have graduate programs. To be considered a university, an institution must have three or more colleges.
   14. Teacher's College (*Kyoyuk Taehak* or *Sabom Taehak*)—National institutions with a four-year program for training primary and secondary teachers leading to a bachelor's degree. National primary teacher's colleges were formerly junior colleges.
   15. Korea Correspondence University (formerly Korea Correspondence College; also known as Korea Air and Correspondence College or University/*Bangsong Tongshin Taehak*)—Curriculum delivered predominantly via radio and correspondence leading to a bachelor's degree (for adult working students).
   16. Open College (*Kaebang Taehak*)—Two-year technical or commerce programs leading to diploma and four-year technical or commerce programs leading to a bachelor's degree.
   17. Miscellaneous School (*Kakchong Hakkyo*)—Postsecondary level. Single purpose school; predominantly theological seminaries. Instruction leads to a diploma. Not authorized by the Ministry of Education to grant bachelor's degree.
   18. Junior Vocational College (*Chonmun Taehak*)—Two-year vocational/technical programs; three-year nursing program.
   19. Graduate School (*Taehagwon*)—Two-year program for Master's Degree and a three-or-more-year program for Ph.D.

**Table 1.3. Student Flow Chart of the Republic of Korea, 1985**

| Pre-School | Primary School | Junior High School | Types of Senior High Schools Enrollment and % |
|---|---|---|---|
| Ⓚ—Ⓚ—▶ | ①—②—③—④—⑤—⑥▶ | A —⑦—⑧—⑨▶ | Exam B —⑩—⑪—⑫▶ |
| Kindergarten Students | Primary School Students | Junior High School Students | Senior High School Students |
| 254,432 | 5,040,958 | 2,735,625 | Gen. 1,200,448–(57%)<br>Voc. 891,953–(43%)<br>Total 2,092,401 |
| | | | Air & Correspondence Senior High School |
| | | | Exam B —⑩—⑪—⑫▶ |
| Non-Compulsory | Compulsory | Non-Compulsory | Non-Compulsory |
| Tuition<br>30% of age group attends | Free<br>99% of age group attends | Tuition<br>98% of primary school graduates attend | Tuition<br>88% of Junior High School graduates attend |

SOURCE: The information in this table was developed from reports and materials taken from publications furnished by the Ministry of Education and the Korean Council for University Education.

 No entrance examination for junior high school since 1969.

 A revised senior high school entrance system was implemented in 1974. This system requires a qualifying examination administered at the provincial or special city level to all applicants. Students are allocated within their districts by lottery, thus developing a more heterogeneous student body. Students who have passed the qualifying examination and who choose a senior vocational high school are screened for vocational aptitude. If selected, they may attend the senior vocational high school of their choice.

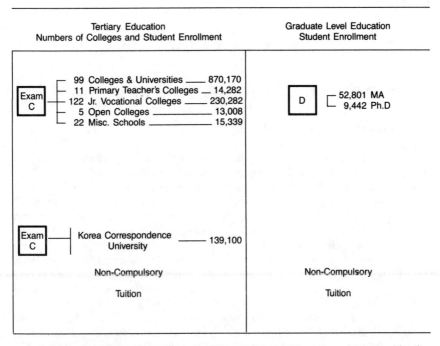

| Tertiary Education Numbers of Colleges and Student Enrollment | Graduate Level Education Student Enrollment |

All post-secondary institutions function under a quota system determined by the Ministry of Education.

Prior to 1981, all applicants took an M.O.E. qualifying examination based on the senior high school curriculum. Students who qualified through the M.O.E. examination then took an individual entrance examination administered by the college or university of their choice. If students met the standards of the institution of their choice, they were admitted to that college or university.

After 1981, the entrance examination administered by the individual college or university was discontinued, leaving only the M.O.E. Scholastic Achievement Examination for College Entrance (SAECE). The score on this examination and the senior high school academic record are submitted to the college or university of the student's choice. From this information the institution makes a decision regarding admission. The senior high school SAECE test is administered each November with test scores being received in December.

Graduate school students are selected by interview and on the quality of their baccalaureate academic record.

# Chapter Two

# Kindergarten, Primary and Secondary Education

## Kindergarten (*Yuchiwon*)

Kindergarten education in Korea goes back more than eighty years and has been predominantly in the hands of missionary groups and social groups. In 1962, the government established qualifications for kindergarten faculty and, in 1969, set curriculum standards. The government reinforced its emphasis on kindergarten education in 1977. Since that time kindergarten enrollment has increased; at present approximately 30% of the eligible age group (four and five-year-olds) is enrolled in kindergarten.

The kindergarten curriculum centers on the development of social skills, social adaptability, speech, reasoning skills, and physical and emotional health. Kindergarten places little emphasis on formal studies, but primarily focuses on the total development and the personality of the child.

The number of kindergartens in public programs in 1984 was approximately 3000, with 2000 in private programs. However, more students (137,000) were enrolled in the private kindergartens than in public programs (117,000).

## School Calendar for Primary and Secondary Education

The academic school year for Korean primary, junior high school, and senior high schools follows this schedule:

1) First semester—March through July with vacation from mid-July to mid-August,
2) Second semester — September through the end of December with vacation from the end of December through mid-February,
3) Test period and graduation—End of February.

Primary, junior high school, and senior high school students attend five-and-one-half days per week for two seventeen-week terms (approximately the equivalent of 220 days per year when taking into account the length of the day and testing period as compared to the academic year in the United States). Primary class periods are 40 minutes in length, junior high school class periods are 45 minutes in length, and senior high school class periods are 50 minutes in length.

## Primary School (*Kukmin Hakkyo*)

Although legislation dealing with compulsory primary education was initiated in 1948, the Korean War and the ensuing economic and political difficulties and social unrest hampered the development of universal compulsory education at the primary level. Full implementation of compulsory education therefore did not take place until 1962. During this time the major financial burden for this education fell upon parents, and they accepted this responsibility, often at great personal sacrifice.

Primary education for children six to twelve years of age is compulsory and free of charge. Its primary goal is to provide the basic skills and general education supportive of the Korean culture and essential to social living in a modern society. Nutrition is a key factor in primary education, with the government providing substantial support for this program. In excess of 99% of the eligible age group attends primary school. During the last several years the number of students in primary education has been decreasing due to a declining birthrate.

One of the highest priorities of the Korean government has been to improve the quality of primary education. The primary school curriculum is under constant review and revision. In 1978, textbooks became free of charge to the student and other forms of charges were eliminated. For many years the average size of elementary classes was 90 or more pupils. At times there were two or three daily shifts. Migration to urban areas has decreased the size of the school enrollment and class size in rural neighborhoods. In urban areas, however, although classes are not as large today as in the past, there are still many crowded schools with classes of over fifty pupils. This makes it difficult to individualize instruction and take care of individual student needs and requires a more rigid curriculum.

The goal of primary education is to develop an understanding of the Korean language and science concepts, and teach basic skills of living, including social and moral ethics, quantitative relationships, artistic appreciation and skills, and good health habits.

Table 2.1 indicates the distribution of student enrollment and the number of teachers in national, public, and private primary schools. Table 2.2 shows the primary school curriculum and time allocated to each subject.

**Table 2.1. The Status of Primary Schools, 1984**

| Classification | Schools | Students | Teachers |
|---|---|---|---|
| National | 16 | 16,627 | 362 |
| Public | 6,437 | 4,951,599 | 124,394 |
| Private | 75 | 72,572 | 1,471 |
| Totals | 6,528 | 5,040,958 | 126,227 |

Source: Adapted from the Ministry of Education publication "Education in Korea," 1984.

**Table 2.2.  Primary School Curriculum**

| Areas of Emphasis | Grade 1 | Grade 2 | Grade 3 | Grade 4 | Grade 5 | Grade 6 |
|---|---|---|---|---|---|---|
| Korean Language<br>Moral Ed.<br>Social Studies | 374(11) | 374(11) | 238  (7)<br>68  (2)<br>102  (3) | 204  (6)<br>68  (2)<br>102  (3) | 204  (6)<br>68  (2)<br>136  (4) | 204  (6)<br>68  (2)<br>136  (4) |
| Arithmetic<br>Science | 204(6) | 136(4)<br>68(2) | 136  (4)<br>102  (3) | 136  (4)<br>136  (4) | 170  (5)<br>136  (4) | 170  (5)<br>136  (4) |
| Fine Arts<br>Music<br>Phys. Ed. | 204(6) | 238(7) | 68  (2)<br>68  (2)<br>102  (3) | 68  (2)<br>68  (2)<br>102  (3) | 68  (2)<br>68  (2)<br>102  (3) | 68  (2)<br>68  (2)<br>102  (3) |
| Practical Arts | — | — | — | 68  (2) | 68  (2) | 68  (2) |
| Totals | 782(23) | 816(24) | 884  (26) | 952  (28) | 1020  (3) | 1020  (30) |
| Extracurricular Activities | — | — | 34+(1+) | 68+(2+) | 68+(2+) | 68+(2+) |
| Grand Totals | 782(23) | 816(24) | 918+(27+) | 1020+(30+) | 1088+(32+) | 1088+(32+) |

Source: Adapted from the Ministry of Education publication, "Education in Korea," 1984.

Note: The hours shown on this table represent the minimum time allocated per 34-week year. Figures in parentheses are hours taught per week.

## Secondary Education

### Junior High School (*Chung Hakkyo*)

Junior high school education for students 12 to 15 years of age is three years in length and non-compulsory. Tuition is charged. Over 98% of all primary school graduates move on to junior high school. It is hoped that by the late 1980s or early 1990s junior high school education will be compulsory and free. In 1969, entrance examinations for junior high school were discontinued and all applicants were accepted and allocated by lottery to the schools within their resident district.

Junior high school education builds upon the foundation laid in primary school and is a continuation of that program. The objectives of junior high school are to develop skills and attitudes essential for citizenship in a democratic society, respect for work, the proper code of conduct, initial skills for future occupations, self-discipline, critical thinking, good health habits, and physical fitness.

Junior high school subjects fall into one of the following twelve curricular areas: Classical Chinese, fine arts, foreign language, Korean history, Korean language, mathematics, moral education, music, physical education, sciences, social studies, and vocational skills and home economics. Extracurricular activities are also included as a part of the junior high school curriculum.

Junior high school students have very few electives. The principal and the staff may set the direction for the school by allocating the amount of time spent within Ministry of Education guidelines in compulsory as well as elective courses. This flexibility given to the principal and staff allows the school to emphasize certain areas such as science, mathematics or art. Most junior high schools do not provide a full offering of vocational curricula, so the selection in this program would normally be quite limited. Students who complete the three-year program are awarded a junior high school diploma (*Chung Hakkyo Chorupchung*).

Tables 2.3 and 2.4 describe the status of junior high schools and detail their curriculum and the number of hours taught.

**Table 2.3. The Status of Junior High Schools, 1984**

| Classification | Schools | Students | Teachers |
|---|---|---|---|
| National | 8 | 8,654 | 234 |
| Public | 1,579 | 1,820,065 | 45,023 |
| Private | 738 | 906,906 | 21,115 |
| Total | 2,325 | 2,735,625 | 66,372 |

Source: Adapted from the Ministry of Education publication, "Education in Korea," 1984.

**Table 2.4. Junior High School Curriculum**

| Areas of Emphasis | Grade 7 | Grade 8 | Grade 9 |
|---|---|---|---|
| Classical Chinese | 34 (1) | 34-68 (1-2) | 34-68 (1-2) |
| Fine Arts | 68 (2) | 68 (2) | 34 (2) |
| Foreign Language (English) | 136 (4) | 102-170 (3-5) | 102-170 (3-5) |
| Korean History | — | 68 (2) | 68 (2) |
| Korean Language | 136 (4) | 170 (5) | 170 (5) |
| Mathematics | 136 (4) | 102-136 (3-4) | 102-136 (3-4) |
| Moral Education | 68 (2) | 68 (2) | 68 (2) |
| Music | 68 (2) | 68 (2) | 34 (1) |
| Physical Education | 102 (3) | 102 (3) | 102 (3) |
| Science | 136 (4) | 102-136 (3-4) | 102-136 (3-4) |
| Social Studies | 102 (3) | 68-102 (2-3) | 68-102 (2-3) |
| *Vocational Skills & Home Economics:* | | | |
| Home Econ. (girls) Voc. Skills (boys) | 102 (3) | 136-204 (4-6) | — |

*(continued)*

| Agriculture, Commerce, Fisheries, Housekeeping, Technical | — | — | El. 1-2 170-238 (5-7) |
|---|---|---|---|
| Elective | 0-34 (0-1) | 0-34 (0-1) | 0-34 (0-1) |
| Total (Maximum/Minimum School Hours) | 1088-1122 (32-33) | 1088-1156 (32-34) | 1088-1156 (32-34) |
| Extracurricular Activities | 68- (2-) | 68- (2-) | 68- (2-) |
| Grand Total (Maximum/Minimum School Hours) | 1156-1190- (34-35)- | 1156-1244- (34-36)- | 1156-1224- (34-36)- |

Source: Adapted from the Ministry of Education publication, "Education in Korea," 1984.
Note: The hours shown on this table represent minimum school hours allocated per 34-week year. Figures in the parentheses are hours taught per week. El. = elective.

## Junior High School Entrance Examination (*Chung Hakkyo Iphak Sihom*)

The elimination of the entrance examination to junior high school resulted in a more academically heterogeneous student body. The primary school was no longer forced to teach to the examination, and the pupils, particularly those in graduating classes, were relieved of the pressure of competing for entrance into the more academically prestigious junior high schools in their district or within the country. Discontinuation of this test and the emphasis on preparing students for this examination allowed primary education to better develop its own purposes, objectives and curriculum. The rapid increase in numbers of primary school graduates going on to junior high school may be due to rising standards of living and increased social awareness of the need for education, as well as the elimination of the entrance examination.

## Senior High Schools (*Kodung Hakkyo*)

Senior high school for students ages 15 to 18 years is three years in length and non-compulsory. Tuition is charged. Approximately 88% of all junior high school graduates advance to senior high school. Senior high schools fall into two major divisions: general, with approximately 57% of the enrollment, and vocational with 43%. The curricula of both types of senior high schools expand and build on the junior high school program. Its major themes are centered around the following: 1) developing insight and judgment as they relate to Korea and its society, 2) the requirements for citizenship, 3) the sciences and liberal arts, 4) physical fitness, 5) background for future vocational choices, and 6) life direction.

A number of secondary schools are single-purpose in their curriculum. They include the core curricula of junior and senior high schools, but emphasize an area such as music or fine arts. Consideration today is being given for senior high schools that will emphasize science and mathematics. Students are recruited from across a special city, province or the country. Because of their special talent, students may receive a scholarship for tuition, room and board.

## Senior General High Schools (*Inmumgye Kodung Hakkyo*)

Senior general high schools have required subjects centering on 1) fine arts, 2) foreign language, 3) history, 4) Korean language, 5) mathematics, 6) military training, 7) physical education, 8) science, and 9) social studies. In the latter two years, in addition to required subjects, the students must choose among three major areas of emphasis: 1) humanities, 2) science, or 3) general vocational training.

## Senior Vocational/Technical High Schools (*Silopgye Kodung Hakkyo*)

Senior vocational/technical high schools offer the same required subjects as the senior general high school, but present them on more of an applied basis, rather than theoretical, and provide technical training, usually specializing by type. Forty-three percent of the senior high school students are in senior vocational/ technical high schools. Prior to graduation, senior vocational high schools require students to have experience in business or industry for a period of one to three months. The percentage of students attending the different types of senior vocational/technical high schools is presented in Table 2.5.

**Table 2.5. Types of Senior Vocational High Schools, 1984**

| | | % Attending | |
|---|---|---|---|
| Type | % Attending | Male | Female |
| Agricultural | 6.0% | 91.8% | 8.2% |
| Commercial | 43.0% | 28.0% | 72.0% |
| Fisheries/Marine | 1.0% | 98.0% | 2.0% |
| Vocational General | 5.0% | 24.0% | 76.0% |
| Vocational Comprehensive | 22.0% | 46.0% | 54.0% |
| Technical | 23.0% | 99.0% | 1.0% |

Source: Adapted from the Ministry of Education publication, "Education in Korea," 1984.

Some outstanding model senior technical high schools have been established by the national government in provinces and the special cities. These senior high schools receive special funding, have well equipped laboratories, select student bodies, and, in some cases, dormitories. Tuition, room and board are also made available to qualifying students. In some instances these institutions have a special affiliation with a business or industry. Because of the select student body and the quality of the program, many of these students are hired by businesses or industry on a preferential basis, or continue their education at the tertiary level in technical education.

The status of senior high schools including number of schools, students and teachers is described in Table 2.6.

**Table 2.6.  The Status of Senior High Schools, 1984**

| Classification | | Schools | Students | Teachers |
|---|---|---|---|---|
| General | National | 10 | 10,837 | 401 |
| High | Public | 436 | 476,914 | 15,730 |
| Schools | Private | 459 | 712,697 | 20,882 |
| | Total | 905 | 1,200,448 | 37,013 |
| Vocational | National | 4 | 8,468 | 336 |
| High | Public | 319 | 344,252 | 13,095 |
| Schools | Private | 321 | 539,233 | 15,834 |
| | Total | 644 | 891,953 | 29,265 |
| | Grand Total | 1,549 | 2,092,401 | 66,278 |

Source: Adapted from the Ministry of Education publication, "Education in Korea," 1984.

**Senior High School Entrance Examinations (*Kodung Hakkyo Iphak Sihom*)**

Prior to 1973, each senior high school developed and administered its own entrance test and selected students on the basis of their test results. This academically hierarchical approach determined which senior high school the student would be allowed to enter. Consequently, the makeup of a senior high school student body tended to be academically homogeneous. In 1973, this procedure was modified.

At the present time all students take a Ministry of Education or provincial qualifying examination for senior high school. The examination focuses predominantly on basic skills. Approximately 10% of the students who take this examination each year do not qualify for entrance to senior high school. Students who do not pass the admissions examination and those who fail to complete the senior high school program usually seek employment or may continue their education through adult education programs.

Individual students passing the qualifying examination for senior high school are assigned by lottery to the district senior high schools, an approach that results in a more heterogeneous student body comprised of students who range academically from mediocre through superior.

Urbanization and the development of new middle and upper class neighborhoods of high-rise apartments have led to middle and upper class families being concentrated in some areas. Therefore, senior high school student bodies in these neighborhoods are becoming both academically and by social class more homogeneous.

General, vocational/technical and specialized senior high schools are concentrated in the urban areas. In the rural and less populated areas of Korea, a senior general high school curriculum is normally followed with the possibility of a vocational option. Heavy and light industry and businesses are found predominantly in the urban areas or in communities that have a highly specialized industry. In these instances senior vocational/technical high schools may be found with their specialized curricula.

### Senior Vocational High School Selection Process
#### (*Silopgye Kodung Hakkyo Sunbal Chedo*)

During the last two decades, Korea's national policy has been to develop heavy and light industry committed to the use of new technology. Consequently, there has been heavy emphasis in the area of vocational/technical education. The Ministry of Education qualifying examination screens applicants both for senior vocational and senior general high schools.

Students selecting senior vocational high schools who have met at least the minimum standard in the Ministry of Education qualifying examination are then tested by individual senior vocational high schools, primarily for aptitude and dexterity, before the final selection is made for senior general high schools. Those who are not selected for senior vocational high schools do not lose their opportunity to continue their education. Their names are re-entered into the lottery system for selection to attend a senior general high school. This approach makes attendance at senior vocational high schools more attractive and has been somewhat successful. However, the majority of academically superior students still choose the senior general high school.

Of the number of both senior general high school and senior vocational high school students continuing their education beyond senior high school, approximately 15% is made up of students graduating from senior vocational high schools, with the predominant number of this percentage coming from senior commercial high schools. Senior commercial high schools, because of the employment opportunities they offer for women after graduation, attract academically above-average female students.

### Senior General and Senior Vocational High School Curricula
### (*Inmumgye* and *Silopgye Kodung Hakkyo Kyokwa Kwachong*)

The curricula of the senior general high schools and the senior vocational high schools are more structured than those in the United States. There are fewer electives and students are given less freedom to select an elective in Korea. Therefore, programs for students in both senior general and senior vocational high schools, depending upon their major, may include more concentrated study in the areas of mathematics, science and foreign language than programs for the average senior high school student in the United States. Senior vocational high schools may provide more concentration in vocational/technical subjects for students in Korea than their counterparts in the United States.

Students taking three years of mathematics in a science curriculum in a senior general high school will have taken mathematics courses which cover algebra, geometry, trigonometry, and some calculus. There is less concentration on mathematics in the humanities curriculum. This same approach is followed in the sciences where the students take biology, physics, chemistry, and some geology. For senior vocational/technical high school students, these courses are similar but more applied than theoretical and place less emphasis on foreign language. Test scores on the Scholastic Achievement Examination for College Entrance (SAECE) are usually lower for senior vocational/technical students than for senior general high school students. (See Chapter III, "Tertiary Education," for a discussion of the SAECE.)

Over the last decade Korean senior high schools have made tremendous progress from the standpoint of facilities and staffing; however senior high schools in the urban areas are crowded and have large class sizes. Except for model senior high schools in selected curricular areas, laboratories and equipment in the sciences or libraries are generally not adequate.

Because of a rapid increase in the size of the student body in secondary education, experienced faculty as well as vocational teachers with industrial and business experience have been limited in number. This problem, in combination with an educational approach that is more theoretical than applied in the areas of science and vocational education, has made it difficult for students to have an academic background that balances theory with inquiry and laboratory hands-on experience.

Tables 2.7 and 2.8 indicate, in general, the curriculum followed and subject unit allotment for the programs of the senior general high school and the senior vocational high school. Curriculum is under constant evaluation and change. Consequently, these tables should be used as guidelines.

### Examinations and Grading in Secondary Education

Over a period of ten years senior high school education in Korea has expanded very rapidly from 839,318 students in 1973 to 2,092,401 in 1984. This rapid

**Table 2.7. Recommended Ministry of Education Guidelines Senior, General High School Curriculum**

| Subjects | Required Subject † Units for 10th Grade | Students select one of three majors | | |
|---|---|---|---|---|
| | | Humanities Major 11th & 12th Grades | Sci. Major 11th & 12th Grades | Vocational Major for General High School 11th & 12th Grades |
| Moral education | 6 | — | — | — |
| Korean language (I, II) * | 14–16 | 14–18 | 8–10 | 2–8 |
| Korean history | 6 | — | — | — |
| Social studies | 4–6 | 4 | | |
| Geography (I, II) | 4–6 | 4 | — | Se. 1 2–6 |
| World history | 2 | 2 | | |
| Mathematics | 8–14 | 6–8 | 10–18 | 4–18 |
| Biology (I, II) | 4–6 | | 4 | |
| Chemistry (I, II) | 4–6 | | 4 | |
| Earth science (I, II) | 4–6 | — | 4 | Se. 1–2 4–12 |
| Physics (I, II) | 4–6 | | 4 | |
| Physical education | 6–8 | 8–10 | 8–10 | 4–8 |
| Military training | 12 | — | — | — |
| Music ─ or ─ Fine arts | 4–6 / 4–6 Se. 1 4–6 | Se. 1 4–6 | Se. 1 4–6 | Se. 1 2–6 |
| Classical Chinese (I, II) | — | 8–14 | 4–6 | 4–6 |
| English (I, II) or | 6–8 | 14–16 | 14–16 | 6–16 |
| Chinese, French, German, Japanese, Spanish | — | Se. 1 10–12 | Se. 1 10–12 | Se. 1 6–10 |
| Home economics / Industrial Arts | — | Se. 1 8–10 | Se. 1 8–10 | Se. 1 10–38 |
| Elective | — | 0–8 | 0–8 | 0–8 |
| Subtotal of units taught | ‡ 88–102 | 90–116 | 90–116 | 52–108 |
| Extracurricular activities | ──────── 12 ──────── | | | |
| Grand total | 204–216 | | | |

Source: Adapted from Ministry of Education publication "Education in Korea," 1984.

* (I) means required subjects.

  (II) means the elective subjects by course and program.

† 1 unit means a period of 50 minutes per week during one term (17 weeks). One week equals 5½ days.

‡ Individual high schools, by choice, may increase 10th grade required subject areas from 88 to 102 units. This allows, for example, a math, science or language emphasis. Same approach is allowed for 11th and 12th grades.

Se: Select

**Table 2.8. Recommended Ministry of Education Guidelines, Senior Vocational High School Curriculum**

| Subjects | Required † units for 10th grade | Elective units for 10th-11th & 12th grades | Agricultural | Voc. Comprehensive | Commercial | Technical | Fisheries/Marine | Vocational General |
|---|---|---|---|---|---|---|---|---|
| Moral education | 6 | — | | | | | | |
| Korean language (I, II)* | 14–16 | 2–8 | | | | | | |
| Korean history | 4 | — | | | | | | |
| Social studies | 2–6 ⎱ | ⎱ | | Required number of units in vocational/subjects | | | | |
| Geography (I, II) — or — | 2–6 ⎰ Se. 1 | Se. 1 2–6 | | | | | | |
| World history | 2 | | | | | | | |
| Mathematics | 8–14 | 4–18 | | | | | | |
| Biology (I, II) | 4–6 ⎱ | ⎱ | | | | | | |
| Chemistry (I, II) | 4–6 ⎰ Se. 2 | Se. 1–2 4–12 | | 82-122 Vocational/ technical units and courses assigned by senior vocational high school | | | | |
| Earth Science (I, II) | 4–6 | | | | | | | |
| Physics (I, II) | 4–6 ⎰ | ⎰ | | | | | | |
| Physical education | 6–8 | 4–8 | | | | | | |
| Military training | 12 | — | | | | | | |
| Music — or — | 4–6 ⎱ Se. 1 | Se. 1 2–6 | | | | | | |
| Fine arts | 4–6 ⎰ | | | | | | | |
| Classical Chinese (I, II) | — | 4–6 | | | | | | |
| English (I, II) or | 6–8 | 6–16 | | | | | | |
| Chinese, French, | | ⎱ | | | | | | |
| German, Japanese, | | Se. 1 6–10 | | | | | | |
| Spanish | | ⎰ | | | | | | |
| Subtotal of units taught | ‡ 72–84 | 10–38 | | 82–122 | | | | |
| Extracurricular activities | ———— 12 ———— | | | | | | | |
| Grand total | 204–216 | | | | | | | |

Source: Adapted from Ministry of Education publication "Education in Korea," 1984.
* (I) means required subjects.
 (II) means the elective subjects by course and program.
† 1 unit means a period of 50 minutes per week during one term (17 weeks). One week equals 5½ days.
‡ Individual high schools, by choice, may increase 10th grade required subject areas from 72 to 84 units. This allows, for example, a math, science or language emphasis. Same approach is allowed for 11th and 12th grades.
Se. = Select

growth has taken place at the same time that testing for entrance to colleges and universities has changed.

  Prior to the implementation of the lottery system in 1974, most cities and provinces had senior high schools that were ranked by the academic quality of the students, based upon their senior high school qualifying test scores and junior high school academic record. Consequently, there was extreme competition for entrance into the academically superior senior high schools. In these senior high schools with their select student bodies, faculty and staff concentrated on preparing their students for the senior high school academic achievement test and the individual college qualifying examination. This approach increased the opportunities for their students to qualify for the academically superior colleges and universities.

The present approach requires that students take only the Ministry of Education examination, the Scholastic Achievement Examination for College Entrance (SAECE). (See Chapter III, "Tertiary Education," for a discussion of the SAECE.) Previously, two tests were administered in the latter part of the senior high school year to determine eligibility for college and university admission. One was a general qualifying examination administered by the Ministry of Education across the country and the other was an individual test developed and administered by each college or university. The examination developed and given by each separate college and university has been discontinued. (See Table 1.3 in Chapter I, and "College and University Admissions Requirements" in Chapter III for more details.)

Senior high school graduation is not determined by SAECE test scores but by grades and completion of the senior high school program. The SAECE examination is administered to all senior general and vocational high school students in the latter part of their senior year. Those students who successfully complete the high school program are awarded a general high school diploma (*Inmumgye Kodung Hakkyo Chorup Chung*) or a vocational high school diploma (*Silopgye Kodung Hakkyo Chorup Chung* ). In determining the admissibility to colleges and universities the SAECE examination score and the senior high school academic record, accompanied by a recommendation from the senior high school principal, are required.

The grading system (*Chaechom Chedo*) predominantly used by secondary schools is A, B, C, D, and F, with D being the lowest passing grade. Most secondary school transcripts also include an explanation of the grading system, such as A = 90-100, B = 80-89, C = 70-79, D = 60-69, F = 59 or less.

## Secondary School Credentials (*Chung Kodung Hakkyo Songjokpyo*)

Secondary school credentials received for Korean students seeking admission into institutions in the United States are translated into English, and certified by the official seal of the secondary school and the signature of the secondary school official. Document 2.1 is an example of a junior high school transcript. Document 2.2 shows a high school transcript, including the grading system used, that has been signed by the Vice Principal and certified by an official seal.

Deed No _____          Date Mar.2, 1984

S A M S U N    J U N I O R    H I G H    S C H O O L

Name;
Date of birth;
Permanent address;

T R A N S L A T I O N    O F    S C H O L A S T I C    R E C O R D S

This is to certify that this student completed the whole three year
course with the below mentioned scholastic records.

| Subjects | 1st Year ( 7th grade ) | 2nd Year ( 8th grade ) | 3rd Year ( 9th grade ) |
|---|---|---|---|
| National Ethics | A | A | A |
| Korean Language | B | B | A |
| Civics | A | A | B |
| Korean History | ——— | B | A |
| Mathematics | A | A | A |
| Natural Science | A | A | A |
| Gymnastics | C | C | C |
| Music | B | B | A |
| Fine Arts | A | A | A |
| Chinese Composition | A | A | A |
| Technique | A | B | A |
| Technical Education | ——— | B | A |
| English | A | A | B |
| Ranks in school | 35 / 975 | 64 / 977 | 65 / 983 |

Grading System; A (90-100), B (80-89), C (70-79), D (60-69), E (50-59)

S T A T E M E N T    O F    C E R T I F I C A T I O N

I certify that this translation of scholastic records is correct.

Name of Notary _____    Signature _____

Address _____

Date _____

2.1.   Translation of Scholastic Records, Samsun Junior High School

# 信 一 中 高 等 學 校
# Shin-il Junior & Senior High School

Son 13, Miadong, Dobong-ku.

Seoul, Korea

☎989-4151~56

## TRANSCRIPT OF SCHOLASTIC RECORDS

Name:

Date: Feb. 29, 1984

Date of Birth:

Date of Enrollment: Mar. 2, 1979

Date of Graduation: Jan. 11, 1982

| | Subjects | 1st Year (10th Grade) 1st Sem. Units | Grade | 2nd Sem. Units | Grade | 2nd Year (11th Grade) 1st Sem. Units | Grade | 2nd Sem. Units | Grade | 3rd Year (12th Grade) 1st Sem. Units | Grade | 2nd Sem. Units | Grade |
|---|---|---|---|---|---|---|---|---|---|---|---|---|---|
| Compulsory | Bible | 1 | B | 1 | D | 1 | B | 1 | A- | 1 | B | 1 | C |
| | National Ethics | 1 | B | 1 | B | 1 | B | 1 | B | 1 | C | 1 | C |
| | Korean Language | 4 | C | 4 | C | 4 | C | 4 | C | 4 | C | 4 | C |
| | Civics | 2 | C | 2 | C | - | - | - | - | 2 | B | 2 | C |
| | Korean History | 2 | D | 2 | D | - | - | - | - | 1 | D | 1 | D |
| | World History | - | - | - | - | 2 | B | 2 | A- | - | - | - | - |
| | Geography | - | - | - | - | - | - | - | - | - | - | - | - |
| | Mathematics | 4 | C | 4 | C | 5 | B | 2 | C | 2 | B | 2 | C |
| | Biology | 2 | B | 2 | B | - | - | - | - | 2 | C | 2 | C |
| | Gymnastics | 3 | B | 3 | C | 3 | A | 3 | C | 3 | C | 3 | B |
| | Military Drill | 2 | C | 2 | B | 2 | B | 2 | A | 2 | B | 2 | B |
| | Music | 1 | B | 1 | A | 1 | C | 1 | C | - | - | - | - |
| | Fine Arts | 1 | B | 1 | B | 1 | A | 1 | C | - | - | - | - |
| Elective | Korean Grammar | 1 | C | 1 | C | - | - | - | - | - | - | - | - |
| | Korean Classics | - | - | - | - | - | - | - | - | - | - | - | - |
| | Physics | - | - | - | - | 3 | A | 3 | A | 2 | B | 2 | C |
| | Chemistry | - | - | - | - | 2 | B | 2 | B | 2 | C | 2 | C |
| | Geology | - | - | - | - | 2 | C | 2 | B | 2 | C | 2 | C |
| | Commerce | - | - | - | - | 2 | B | 2 | B | 2 | C | 2 | B |
| | English | 6 | D | 6 | B | 3 | D | 3 | B | 3 | D | 3 | D |
| | German | 2 | D | 2 | D | 2 | D | 2 | D | 2 | D | 2 | C |
| | Average Grade | | | | | | | | | | | | |

*Grading System; A (90-100), B (80-89), C (70-79), D (60-69), E (50-59); E is the lowest passing mark.

This is to certify that above records are true and correct.

Byung Taik Ahn
Principal (Vice-Principal)

Shin-Il Senior High School

2.2.　Transcript of Scholastic Records, Shin-il Junior and Senior High School

Document 2.3 is the accompanying Certificate of Graduation. Documents 2.4 and 2.5 attest that the translation of the transcript is true and correct according to the notary public. Document 2.4 would probably be included with other credentials (transcript, certificate of graduation) sent to the requesting institution to which the Korean student has applied. Many times the documents also have a certificate of notarization affirming the accuracy of the translation of the document. The notary public in Korea is certified by the government and does not merely authenticate the signature of the principal or student, but attests to the fact that the document is an accurate translation of the academic record. Normally, the documents include a certificate of graduation signed by a school official and a transcript of grades. Document 2.6 shows a Certificate of Graduation from Kyunggi High School and Document 2.7 the official transcript, including the student's numerical average and class rank for each year. Some credentials give class rank and the grading system employed but do not provide the SAECE test score or scores.

## Private Secondary Schools (*Sarip Chung Kodung Hakkyo*)

At the junior high and senior high school levels there are many private schools. Over the last several decades these schools have evolved predominantly from gifts from individuals, businesses or industries desiring to help in the education of Korean students. A number of these private secondary schools were established by religious groups.

The private schools, like the public schools, are under the direction of the Ministry of Education and the local city or provincial Board of Education. Both public and private schools charge similar tuition and fees established by the Ministry of Education and receive governmental allocations. The private secondary schools follow the Ministry of Education curricula, and assignment of students and have the same graduation requirements as public schools.

## Non-Formal Education

A number of continuing education programs for adults have been established throughout the Republic of Korea. These are similar to non-formal programs in the United States. Under the supervision of the Ministries of Labor, Agriculture and Fisheries, Education, and Health and Social Affairs, there are many programs offered to the citizens of Korea. Many are privately operated or under the special city or provincial Board of Education.

As in the United States, there are some programs for training in penitentiaries. There is also extensive specialized technical training within the armed forces; and throughout the country there are numerous training programs offered by industry, business and government for various age levels and occupational areas.

信 一 中 高 等 學 校
## Shin-il Junior & Senior High School
193. Mia-Dong, Dobong-Ku
Seoul, Korea
Tel. 989-4151/6

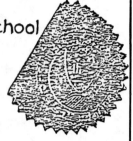

## CERTIFICATION OF GRADUATION

Name:                                                    Date:

Date of Birth:

Permanent Address:

This is to certify that the above mentioned person was graduated from Shin-Il Senior High School, completed the requirements of the whole three year course from                    to

Byung Taik Ahn

Principal (Vice-Principal)

Shin-Il Senior High School

2.3.   Certificate of Graduation, Shin-il Junior and Senior High School

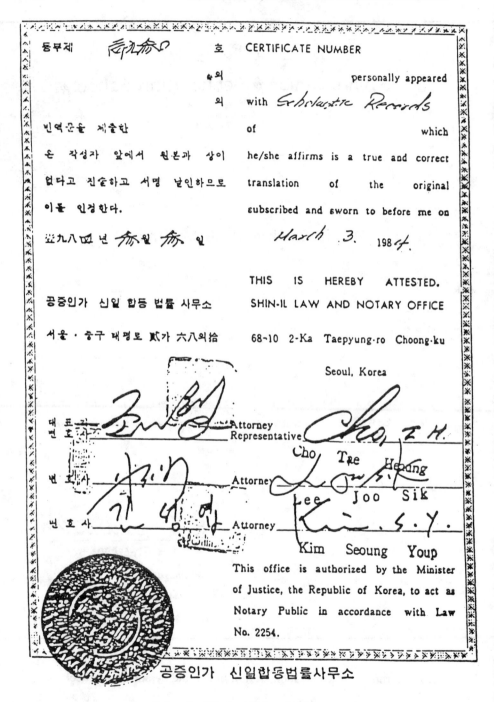

2.4.   Certificate of Notarization

## REQUEST FOR NOTARIZATION

This is to request a notarization of the true and faithful translation

of the    *Transcript of Academic Record*

Requested by:_____

## CERTIFICATE

This is to certify that the translation of the    *Transcript of Academic Record.*

    attached
is true and correct from the Korean original.

Date:  1984 3. 20

등·부 제壽八O壽호
별지 번역된 영문은 별지 국문으로된
원문과 부의 일치하다
이에 이를 증하다
서기 壹九、⑩년 壽월 丼O일
서울특별시중구을지로2가21번지
을지방검찰청소속
·증인

Du Shik Yoo
Notary Public
Seoul District prosecutor's Office
Seoul, Korea

2.5.   Request for and Certificate of Notarization

KYUNGGI HIGH SCHOOL
SEOUL, KOREA                                         ,

DATE: March 7, 1984.

SUBJECT            :    CERTIFICATION OF GRADUATION

NAME IN FULL       :

DATE OF BIRTH      :

PERMANENT DOMICILE:

PRESENT ADDRESS    :

DATE OF ENTRANCE   :  March 2, 1979.

DATE OF GRADUATION :  February 11, 1982.

  This is to certify that the above mentioned person completed the full course of, and duly graduated from, Kyunggi High School, Seoul, Korea.

OFFICIAL SEAL                        SIGNATURE: _____

                                     PRINCIPAL Kim In Suk
                                     KYUNGGI HIGH SCHOOL
                                     SEOUL, KOREA

2.6.   Certificate of Graduation, Kyunggi High School

KYUNGGI HIGH SCHOOL
SEOUL, KOREA

DATE: March 7, 1984.

This is to certify that Mr._____ graduated from Kyunggi High
School On February 11, 1982 and the following is his academic record
throughout his course of the above school.

|  | 1ST YEAR ( 1979 ) | | 2ND YEAR ( 1980 ) | | 3RD YEAR ( 1981 ) | |
|---|---|---|---|---|---|---|
| SUBJECTS | GRADE/HPW | | GRADE/HPW | | GRADE/HPW | |
|  | 1Sem. | 2Sem. | 1Sem. | 2Sem. | 1Sem. | 2Sem. |
| KOREAN LANGUAGE | C/4 | B/4 | C/7 | C/7 | B/6 | B/6 |
| Readers Classics | | | | | | |
| Gram. & Comp. | | | | | | |
| History of Lit. | | | | | | |
| Chinese Classics | C/3 | B/3 | | | A/2 | A/2 |
| SOCIAL STUDIES | A/4 | A/4 | B/4 | B/4 | B/6 | B/6 |
| Moral & Ethics | A/1 | B/1 | A/1 | B/1 | A/1 | B/1 |
| Korean History | B/1 | B/1 | B/1 | C/1 | C/1 | B/1 |
| World History | | | | | | |
| Civics Geography | | | | | | |
| MATHEMATICS | B/3 | B/3 | B/3 | B/3 | A/3 | B/3 |
| Algebra | | | | | | |
| Geometry | | | | | | |
| Calculus | | | | | | |
| SCIENCE | C/2 | C/2 | B/2 | B/2 | B/4 | A/4 |
| Physics/Chemistry | | | | | | |
| Biology/Geology | | | | | | |
| Engineering | | | | | | |
| Technology | | | | | | |
| ENGLISH | C/4 | B/4 | C/3 | B/3 | C/3 | C/3 |
| Readers | | | | | | |
| Side Readers | | | | | | |
| Gram. & Comp. | | | | | | |
| GYMNASTICS | B/3 | C/3 | A/2 | B/2 | A/2 | C/2 |
| MILITARY TRAINING | B/2 | B/2 | C/2 | A/2 | A/2 | B/2 |
| MUSIC | A/2 | A/2 | B/1 | C/1 | ••• | ••• |
| FINE ARTS | C/2 | A/2 | A/1 | B/1 | ••• | ••• |
| BUSINESS | B/2 | B/2 | B/3 | C/3 | A/4 | A/4 |
| GENERAL MANAGEMENT | ••• | ••• | ••• | ••• | ••• | ••• |
| GERMAN | C/2 | B/2 | B/2 | C/2 | A/1 | A/1 |
| FRENCH | ••• | •• | ••• | ••• | ••• | ••• |
| TOTAL | 130. | 140. | 125. | 120 | 149. | 146. |
| AVERAGE | 3.71 | 4.00 | 3.68 | 3.43 | 4.26 | 4.17 |
| CLASS STANDING | 222/920. | 136/920. | 63/259 | 102/258. | 35/255. | 68/253. |

REMARKS: "A" stands for the highest mark obtainable and "D" the lowest
passing mark, respectively. "HPW" stands for Hours taken Per
Week.

OFFICIAL SEAL                    SIGNATURE

PRINCIPAL Kim In Suk
KYUNGGI HIGH SCHOOL
SEOUL, KOREA

2.7.  Transcript of Records, Kyunggi High School

Many educational programs are offered by private organizations, churches, and other groups. While these private institutes or schools may not be accredited by the Ministry of Education, they do carry out an important function in the area of non-formal education for the people of Korea. Table 2.9 lists the numbers of these institutes, their staff, and their enrollment.

**Table 2.9.  Non-Formal Institutes and Courses, 1984**

| Classification | Institutes | Enrollment | Instructors |
|---|---|---|---|
| Arts Courses | 3147 | 107,678 | 4544 |
| Business Courses | 2606 | 226,297 | 4390 |
| Home Economics Courses | 141 | 5227 | 288 |
| Liberal Arts and Science Courses | 456 | 130,830 | 3920 |
| Physical Training Courses | 1170 | 57,230 | 1396 |
| Technical Courses | 1072 | 67,141 | 3734 |
| Other | 249 | 15,506 | 461 |
| Totals | 8841 | 607,909 | 18,713 |

Source: Adapted from Ministry of Education publications, 1983 and 1984.

## Adult Education (*Sung-in Kyoyuk*)

A number of special schools have been established for adults who have left school early for work purposes or other reasons. These schools are sometimes attached to junior high or senior high schools, or to a business or industry, and operate predominantly in the evening. Included in this category are civic schools, higher civic schools, trade schools and senior higher trade schools. Higher civic schools, trade schools and senior higher trade schools have one-, two- and three-year programs with exit points at the first, second or third year of the program.

## Civic Schools (*Kongmin Hakkyo*)

Civic schools offer a range of primary education courses for adults who need literacy training. In past years these schools played a more significant role. Today, because basic literacy is almost universal, the need for these institutions is rapidly diminishing.

Higher civic schools are for graduates of primary schools who did not advance to junior high school. Programs in the higher civic schools span a period of one to three years and cover all requirements of the regular junior high school curriculum. Since the early 1970s, the need for civic schools at the junior high school level has also diminished because of the increasing number of students continuing their education after primary school.

### Trade Schools (*Kisul Hakkyo*)

Trade schools at the junior high school level are occupationally oriented with the duration of courses lasting between one and three years. Admission to a trade school requires graduation from primary school or civic school. The junior high school level trade schools are diminishing in number primarily because a higher number of students are completing regular junior high school and are seeking a general education background at this level. A few of them are under the management of business or industrial establishments, and are used to meet the educational and training needs of their employees, an arrangement that allows for a more flexible approach to training schedules.

### Air and Correspondence Senior High School
(Bangsong Tongshin Kodung Hakkyo)

The education provided by the Air and Correspondence Senior High School is predominantly for working adults. The curriculum requires a student to review text material; listen to subject area radio broadcasts 280 days per year, predominantly in early morning or late evening throughout the week; and attend classes twenty-six days per year at one of the selected senior high schools throughout the country. Upon successful completion of this program, students receive a senior high school diploma (*Kodung Hakkyo Chorupchung*) and may take the SAECE examination to be used for entrance into a college or university.

The curriculum recommended by the Ministry of Education for the senior general high school is the course of study generally followed (see Table 2.7). Testing and evaluation take place throughout the year during the time the student is at the sponsoring senior high school. The work of the student is graded and an official transcript is issued just as though the student had attended a regular senior high school. Competency in subject matter is evaluated in a similar fashion to that of regular day students.

Table 2.10 indicates the size of the enrollment and number of affiliated senior high schools cooperating with the Air and Correspondence Senior High School program.

## Special Education (*Tuksu Hakkyo*)

Special education is offered to students in Korea at the primary, junior high and senior high school levels. These schools are for the physically handicapped

**Table 2.10.  Air and Correspondence Senior High School, 1980–84**

| 1981 | | 1982 | | 1983 | | 1984 | |
|---|---|---|---|---|---|---|---|
| Schools | Enroll. | Schools | Enroll. | Schools | Enroll. | Schools | Enroll. |
| 48 | 32,279 | 49 | 35,096 | 47 | 38,191 | 49 | 42,364 |

Source: Adapted from the Ministry of Education publication, "Education in Korea," 1984.

which includes the blind, deaf and mute. The curriculum parallels that of the primary, junior high and senior high schools. The programs are adapted to help handicapped individuals gain the knowledge necessary to earn a living and make a contribution to their society.

Each of the provinces and special cities provides education for the physically handicapped. In some cases, special education classes are attached to primary and secondary schools.

# Chapter Three

# Tertiary Education

## Overview of Tertiary Education

In the last several years approximately 40% of the senior high school graduating classes entered postsecondary education. Table 3.1 indicates the number of senior high school graduates and the percent entering postsecondary education by type of senior high school attended.

**Table 3.1. High School Graduates Entering Postsecondary Education, 1983**

| Type of High School | Number of High School Graduates | % of Students Entering Higher Education |
|---|---|---|
| General | 299,782 | 56% |
| Vocational | 245,816 | 15% |

Source: Adapted from Ministry of Education publications, 1983 and 1984.

Foreign trade is a major ingredient driving Korea's economic system and making possible an improved standard of living. With the future calling for an increasing commitment to high technology and its trade implications, higher education, particularly in the areas of engineering, science and research, is a major priority for Korea.

During the period from 1974 to 1984, the enrollment in four-year colleges and universities increased from 192,000 to 870,000. The total enrollment in all types of postsecondary educational institutions exceeds one million. This enrollment increase has taxed Korea's ability to meet the needs of students requesting admittance to postsecondary institutions, while attempting to improve the quality of all facets of tertiary education. As has been the case in most developed or developing countries, difficulty has been encountered in determining program emphasis and student program choice and balancing this with employment opportunities and the number of college and university graduates. In most cases allocations for capital outlay, maintenance, equipment and salaries have not been adequate for the standards that Korea would like to achieve.

Campuses of colleges and universities are almost universally located in urban settings with few having dormitory facilities; consequently, the majority of students live in private dwellings, at home, or with an extended family or friends. Campus life is somewhat similar to that found on urban campuses in the United States where the student body is predominantly comprised of commuters.

## Private Education (*Sarip Hakkyo Kyoyuk*)

Historically, in Korea, private education has played a major role in the development of primary, secondary and higher education. Western educational influences were initiated by Protestant and Catholic missionaries in the 1800s. Prior to that time, private education was fostered by influences from Confucian and Buddhist sects and philosophers. The tradition of private education remained as a partial counterbalance to the control of education during the Japanese colonial period.

Many of the four-year colleges and universities and a considerable number of the private junior vocational colleges developed from private senior high schools. In many cases, business, industry, or individuals interested in fostering education developed a foundation in which grants and gifts were used to establish private secondary schools or colleges. Today, most of the private colleges and universities, including junior vocational colleges, predominantly generate their operational and capital funds from tuition with some help from the national government. There is considerable variance in the size of foundations, allocation of their funds to their institutions, and their years of operation.

Over the years this approach to funding has made it very difficult for Korean private higher educational institutions to adequately fund their physical facilities and laboratories. Further, they have fallen behind in compensating faculty competitively with other segments of Korean business, industry or government. With the help of international loans secured through the Korean government, gifts, grants and national allocations, improvement has been steady. A number of the private colleges and universities rank in the top tier academically although many private institutions throughout the country must still contend with inadequate facilities, faculty and staff.

## Public Education (*Kongrip Hakkyo Kyoyuk*)

One of the primary goals of the national government is to help both national and private colleges and universities improve their facilities and make available professional development programs for faculty and staff without increasing the level of tuition too rapidly. The national colleges and universities and the several municipal and provincial institutions have been funded over the years for capital outlay, equipment and staffing on an overall basis at a higher level than that which is available for private colleges and universities. Scholarships, for many years, have been funded at a higher level for the national institutions, which, as a rule, have a higher enrollment per institution and a greater number of programs than their private counterparts. Overall, their laboratories, equipment and faculty are adequate. The range in quality among the national institutions is not as diverse as with private colleges and universities.

## Categories of Korean Tertiary Education

Korean tertiary education divides itself into seven major categories, with the predominant number being private institutions: 1) Junior Vocational Colleges, 2) Miscellaneous Schools, 3) Open Colleges, 4) Korea Correspondence University (Korea Correspondence College), 5) National Teacher's Colleges, 6) Colleges and Universities, and 7) Graduate Schools. For more information on each institution and the programs offered, see Appendix A, "Institutions of Higher Education in Korea."

The Korean military academies for the Army, Navy, and Air Force are not normally listed as a part of tertiary education by the Ministry of Education. There are three four-year military academies, one for each branch of the service, Army, Navy, and Air Force. There are two additional Army military academies that are two years in length. The selection of students for these academies is competitive. The curriculum centers on military science, engineering and technical education.

Table 3.2 indicates the number of tertiary institutions by category and the number of students enrolled.

**Table 3.2. Student Enrollment in Tertiary Institutions in Korea, 1984**

| Classification | | Schools | Students | Teachers |
|---|---|---|---|---|
| Junior Vocational Colleges | National | 17 | 24,967 (14%) | 1,186 (11%) |
| | Private | 105 | 205,315 (38%) | 5,227 (25%) |
| | Total | 122 | 230,282 (35%) | 6,413 (22%) |
| Miscellaneous Schools | Private | 22 | 15,339 (24%) | 391 (18%) |
| Open Colleges | National | 5 | 13,008 (11%) | — |
| Korea Correspondence University (Air and Correspondence College) | National | 1 | 139,100 (37%) | — |
| Primary Teacher's Colleges | National | 11 | 14,282 (84%) | 576 (10%) |

*(continued)*

| | | | | |
|---|---|---|---|---|
| Universities & Colleges | National | 20 | 226,443 (26%) | 7,636 (10%) |
| | Municipal | 1 | 4,713 (12%) | 118 (5%) |
| | Private | 78 | 639,014 (27%) | 16,652 (20%) |
| | Total | 99 | 870,170 (27%) | 24,406 (16%) |
| Graduate Schools | MA | — | 52,801 (18%) | — |
| | Ph.D. | — | 9,442 (13%) | — |

Source: Adapted from Ministry of Education publication "Education in Korea," 1984.
Note: Figures in the parentheses indicate the ratio of female students and teachers.

## Higher Education Governance

All institutions of higher education, public and private, must adhere to educational law and presidential and ministerial decrees. They are under the supervision of the Ministry of Education. The Ministry of Education exercises control over qualifications of teaching staff, curriculum and degree requirements, general education, college and university military training, faculty standards for universities and colleges, establishment of curricula, regulation for the establishment and closure of institutions, fiscal review, inspection of facilities, and the establishment of the overall official quota of students for each higher education institution.

The presidents of public national universities and colleges are appointed by the President of the Republic of Korea on the recommendation of the Ministry of Education. In private universities and in private four-year colleges and junior vocational colleges, the institution's Board of Trustees will nominate the president to the Ministry of Education for approval. After this approval is received, the individual is appointed to the presidency by the Chairman of the Board of Trustees.

As is found in the political structure of the Republic of Korea, the administrative structure of colleges and universities is more formal, and the college and university presidents exercise more power and control than is normally found in their counterparts in colleges and universities in the United States.

## Institutional Evaluation

Over the years the Ministry of Education has established a number of educational committees, councils, and associations. In 1982, the Korean Council for University Education was established and approved by the government. This Council has the following functions:

1. Conducts research on the management and organization of college and university education,

2. Promotes coordination,

3. Advises and makes recommendations to the government on matters relating to higher education and helps in the dissemination of information on new educational policies,
4. Conducts autonomous evaluations of colleges and universities,
5. Advocates and takes positions regarding institutional governance and autonomy, and
6. Conducts studies and projects for member colleges and universities.

## Admission to Tertiary Education (*Taehak Iphak Chedo*)

### Admission Prior to 1980

In the late 1960s and the 1970s, the Ministry of Education administered a Preliminary Examination for College Entrance (PECE). High school graduates wishing to attend a four-year college or university had to first pass this examination. In addition, each institution devised and administered its own entrance examination. The examinations of the most academically prestigious colleges and universities were administered early in the year, and the examinations of the less prestigious colleges and universities followed. Thus, students not accepted by the institution of first choice had to apply for entrance to a less prestigious institution by taking that institution's examination at a later date. Those who passed the Ministry of Education's preliminary entrance examination but failed to pass the examination of the college or university of their choice could choose to prepare themselves for the next year's examination with the help of private tutors.

The passing level for the PECE examination was adjusted each year so that 35% to 45% would receive a passing score, thereby allowing approximately twice as many students to qualify as there were college and university openings.

When senior high schools selected students on the basis of the individual senior high school qualifying examination given to students prior to their admission to senior high school, the students of these top academic senior high schools usually scored highest on the individual college or university test, and consequently filled the quota of the most prestigious colleges and universities. The quality of the college or university the student attended helped to determine the quality of opportunity and level of initial placement in business, industry or government which would take place after graduation.

Senior high school students spent many hours on homework, particularly during their senior year, in anticipation of taking the two tests which determined what college or university they might attend. In addition, many students studied with private tutors. Because of the intensity of study on the part of senior high school students, the expense to their families for tutors, and the adverse impact upon senior high school curriculum and teaching faculty, a law was passed making it illegal for private tutorial services to be offered to students.

### Admission After 1980

In 1981, the entrance examination administered by colleges and universities was discontinued along with the Ministry of Education's qualifying examination. The present Ministry of Education examination may be viewed as a high school academic achievement test. The score is used by the college or university in conjunction with the senior high school record to determine eligibility for admission. This test, the Scholastic Achievement Examination for College Entrance (SAECE), is administered each November or December with test scores being received in December or January.

The period of time allotted for students to select colleges and universities of their choice is divided into three segments. Usually, the first segment is utilized for the most academically prestigious colleges and universities. The second period is used by the second tier of colleges and universities, and the third period is used for the third tier of colleges and universities and junior vocational colleges. These periods fall approximately in mid-January, at the end of January, and in early February. Students with the highest SAECE test scores, in addition to excellent academic records and their senior high school principal's recommendation, are selected by the more academically prestigious institutions in the first period.

This approach to admissions on the part of colleges and universities emphasizes the senior high school record as well as the results of the Scholastic Achievement Examination for College Entrance (SAECE). It is anticipated that this reform of the admissions system to colleges and universities will allow the senior high school to prepare students within institutional purposes and goals rather than having the senior year utilized intensely for preparation for the former examination period.

### Scholastic Achievement Examination for College Entrance (SAECE)
### (*Taehak Iphak Haklyok Kosa*)

The Scholastic Achievement Examination for College Entrance evaluates the student's level of mastery of the senior high school subject matter studied. This SAECE score, in conjunction with the student's senior high school grades, is used by Korean colleges and universities to determine admission eligibility. As a general rule, high school grades are weighted at 30% and the test score at 70% by college and university admissions officers, along with the senior high school principal's recommendation in considering eligibility for admissions.

In 1986, the SAECE will contain an essay component which will be used to analyze the student's writing ability. Postsecondary institutions will be able to evaluate this component with flexibility as it relates to the total SAECE score.

The senior vocational/technical high school curriculum concentration in the area of foreign language, theoretical mathematics, and science is not as intense from the standpoint of time or depth as the curriculum of the senior general high school program. Consequently, SAECE scores in these fields for senior vocational/technical high school students are generally not as high as those of the senior general high school graduate. Table 3.3 lists the subject areas for the SAECE administered at the end of senior high school to all graduating seniors. The chart also differentiates the subject matter emphasis for the two major areas, humanities or science, studied by students in the senior general high schools. There is no special emphasis or division within the test for vocational/technical majors as there is for the humanities or science major. The maximum SAECE test score for 1984 was 340 points with a maximum test score of 320 points projected for 1985.

**Table 3.3. Scholastic Achievement Examination for College Entrance (SAECE)**

| Humanities Major | Science Major | Subjects Tested | 1984 Points | 1985 Points |
|---|---|---|---|---|
| | | Business | 15 | 15 |
| | | Chinese Characters | 5 | 5 |
| | | Ethics | 15 | 15 |
| | | Foreign Language | 50 | 50 |
| | | Korean History | 20 | 20 |
| | | Korean Language I | 45 | 45 |
| | | Mathematics I | 50 | 40 |
| | | Political Science & Economics | 15 | 10 |
| | | Home Economics (F) ⎫ Technology (M) ⎭ | 15 | 15 |
| Required | Not Required | Korean Language II | 15 | 15 |
| Not Required | Required | Mathematics II | 15 | 15 |
| All Required | Select One | Land & Geography | 15 | 10 |
| | | People & Geography | 15 | 10 |
| | | Society & Culture | 15 | 10 |
| | | World History | 15 | 10 |
| Select Two | All Required | Biology | 15 | 15 |
| | | Chemistry | 15 | 15 |
| | | Earth Science | 15 | 15 |
| | | Physics | 15 | 15 |
| | | Subtotal | 320 | 300 |
| | | Athletic Fitness Score | 20 | 20 |
| | | Total | 340 | 320 |

Source: Taken from information provided by private publishers that is used in guiding students.

Student competition for admission into certain programs and departments within a college or university is reflected by the range in SAECE test scores within the same college or university. For example, a student requesting entrance into a law program which may be in high demand will need to have a higher SAECE score than a student requesting entrance into a program which is in less student demand, such as agriculture. It is possible then that a student at an academically prestigious institution might actually have a lower SAECE score than a student entering a slightly less prestigious institution who has selected a program in high student demand. Table 3.4 shows the range of SAECE scores used for selection in four programs in different institutions in 1983.

**Table 3.4.   SAECE Scores in Selected Programs, by Institution**

| College or University | Programs | | | |
|---|---|---|---|---|
| National Universities | Bus. Mgmt. | Electronics | Math | Law |
| Busan National University | 290 | 287 | 252 | 266 |
| Chonnam National University | 268 | 240 | 241 | 255 |
| Chonpook National University | 268 | 256 | 228 | 253 |
| Chungnam National University | 241 | 236 | 217 | 247 |
| Gyeong Sang National University | 234 | – | 214 | 241 |
| Kangweon National University | 224 | 207 | 202 | 223 |
| Kyungpook National University | 272 | 262 | 247 | 276 |
| Seoul National University | 315 | 322 | 302 | 323 |
| Private Universities | | | | |
| Cheongju University | 220 | 200 | – | 218 |
| Chosun University | 231 | 223 | 198 | 221 |
| Chung-ang University | 278 | 265 | 254 | 272 |
| Dankook University | 251 | 244 | – | 260 |
| Dong-A University | 253 | 240 | 220 | 255 |
| Dong-Eui University | 210 | – | 196 | 211 |
| Dong Guk University | 277 | 263 | 226 | 263 |
| Ewha Womans University | 267 | – | 265 | |
| Hanyang University | 280 | 275 | 263 | 279 |
| Hong Ik University | 242 | 238 | – | – |
| Keimyung University | 226 | – | 203 | 240 |
| Kon Kuk University | 281 | 285 | – | 273 |
| Kookmin University | 274 | 258 | – | 276 |
| Korea University | 294 | 287 | 270 | 301 |
| Kyung Hee University | 271 | 235 | 254 | 264 |

| | | | | |
|---|---|---|---|---|
| Kyungnam University | 214 | 202 | 183 | 211 |
| Sogang University | 281 | 279 | 249 | – |
| Sook Myung Women's University | 252 | – | 247 | 245 |
| Sung Kyun Kwan University | 282 | 247 | 239 | 280 |
| Yonsei University | 302 | 283 | 283 | 292 |

Source: Taken from information provided by private publishers that is used in guiding students.

### Graduation Quota System (*Chorup Chongwon Chedo*)

In the early 1980s the graduation quota system was established by the Ministry of Education. Its purpose was to increase the academic rigor for students in each college and university within the country.

Each year the Ministry of Education establishes the enrollment quota of students for each college and university. The number is determined by taking into account availability of institutional facilities, national manpower needs, budget considerations, and political reality. At the time that a class enters a college or university, the quota for graduation is also established. This quota system allows an institution to admit a number of students that exceeds the college or university's graduation quota. This approach intensifies internal academic consideration to reduce the size of the graduating class to fit within the institution's quota.

The quota system approach has met with some difficulties and is under constant evaluation since it requests institutions that have an academically excellent student body and a high retention rate to be more rigorous on their students than an institution with lower academic standards and a high drop-out rate. Thus, a student accepted at an academically prestigious institution in a more rigorous program than that provided by a less prestigious institution might be asked to leave because of academic standards when in fact the student's performance may be well above the average level in another institution.

## Diploma/Degrees Awarded

Students entering tertiary programs in Korea may earn a diploma from a two-, three- or four-year junior vocational college, miscellaneous school, open college or teacher training program. Students who enroll in a four- to six-year degree program (five- to ten-year part-time program at the Korea Correspondence University) at a college or university in Korea may earn a bachelor's degree upon successful completion of 140 or more credit hours of study. Graduate programs in various fields award the master's degree or the doctoral degree. For more details on the diploma or degree awarded by different tertiary institutions in Korea, see the discussion later in Chapter III on different institutions. See also "College Credentials" in Chapter VI.

In 1984, for the first time, some students graduated from colleges and universities without a degree despite the fact that they had successfully fulfilled the required number of credits for graduation. This situation evolved because the Ministry of Education limits the number of bachelor's degrees that may be awarded by each college or university, and allows the recruitment of more students in their freshman year than will be allowed to graduate. At graduation, these students are given a document certifying completion as opposed to diplomas or degrees. In essence, these are no different than a bachelor's degree and should be considered in the United States as equivalent to a bachelor's degree. The only meaningful difference is that the student has a certificate rather than a degree and may be academically in the lower 25% to 30% of the graduating class within a department, rather than in the college or university, as a whole.

## Transfer Within and Between Colleges and Universities

Opportunities for a student to transfer from one department to another within a college or university are difficult inasmuch as it entails a change from one major to another. A vacancy must exist and the student must be qualified to make application. Transfer between different institutions is also difficult because of the enrollment quota system and differences in the academic standards of the institutions. Transfer is further complicated by the limitation of the new graduation quota system established by the Ministry of Education.

In order to transfer between departments within a college or university or to another college or university, a student must have the recommendation of the department or college or university from which the transfer will be made and an acceptance by the department or college or university into which the transfer is requested. There must also be an evaluation of the courses requested for transfer credit. As is the case in the United States, a change in majors many times results in the requirement of additional coursework. Normally, there is only one transfer allowed. The request is usually submitted during the student's third or fourth semester, with implementation generally occurring at the beginning of the student's junior year.

Transfers between departments within a college or university can be a critical issue for Korean students. Many times as they are making their initial selection for admission, in their desire to attend the institution of highest academic prestige for which they are qualified, students may select a major that is their second or third choice. They do this in the hope that they may be allowed to transfer to the field of their first choice at the end of their sophomore year.

## Academic Calendar for Colleges and Universities

The academic college year for Korean colleges and universities follows this general schedule:

1) First semester—March through the latter part of June with summer vacation in the latter part of June to the latter part of August,

2) Second semester—Latter part of August through the latter part of December with winter vacation in the latter part of December through February,

3) Summer semester—Latter part of June to the latter part of August,

4) Commencement—Latter part of February.

The colleges and universities in Korea operate six days per week with a summer and winter vacation and with certain university and national holidays being observed. Class days for an academic college year number 210 or more. The length of study for a bachelor's degree is four years or eight semesters. The maximum time for completion of a degree cannot exceed six years or twelve semesters. Leaves of absence from college can be obtained for illness, military service, or other similar reasons.

## Programs of Study

### Classification of Students

Regular degree students are classified by semester hours of earned credit which places them in the following general categories:

| Classification | Approximate Semester Hours of Earned Credit |
|---|---|
| Freshman | Less than 35 |
| Sophomore | 36 to 70 |
| Junior | 71 to 105 |
| Senior | 106 or more |

### Grading System (*Chaechom Chedo*)

Korean colleges and universities predominantly follow a letter grade system:

| | | | |
|---|---|---|---|
| A | = Excellent or superior | D | = Inferior but passing |
| B | = Good or above average | F | = Failure |
| C | = Fair or average | | |

Other categories used are:

| | | | |
|---|---|---|---|
| FA | = Failure for excessive absences | U | = Unsatisfactory |
| S | = Satisfactory | INC | = Incomplete |
| P | = Passing | W | = Withdrawal |

Variations in marking systems are found such as A +, B +, etc. Normally, one hour of class work per week per semester generates one credit. Two to three hours of laboratory work per week per semester yield one semester of credit.

### Semester Load and Special Requirements

Colleges and universities today are considering a minimum fulltime semester load as twelve credit hours and a maximum load as twenty credit hours. To carry credits beyond this number usually requires approval of college officials. However, an excess of twenty credit hours is still found in most programs, particularly those of a technical or professional emphasis, which are laboratory-intensive. Physical education is a normal requirement for men and women, and general military education is a requirement for men. Documents 3.1, 3.2, and 3.3, Scholastic Records for a Bachelor of Science in Electronic Engineering, a Bachelor of Law, and a Bachelor of Science (Mathematics), respectively, illustrate programs in which the student's credit hours exceeded the maximum 20 credit hour limit.

## Transcripts of Scholastic Records (*Songjokpyo*)

Transcripts sent to postsecondary institutions in the United States from colleges and universities in Korea are normally in English, with grades and the marking system included on the document. To be considered official they should have the chop or seal of the college or university along with the signature of a college official. The transcript, as in the United States, is mailed directly from the sending institution to the requesting admissions officer of the college or university in the United States.

## Institutions

### Junior Vocational Colleges (*Chonmun Taehak*)

There are 122 junior vocational colleges, seventeen national and 105 private junior vocational colleges. Junior vocational college enrollment represents approximately 20% of the students involved in higher education. The fields of major technical interest to the students are the engineering technologies and nursing. See Appendix A, "Institutions of Higher Education in Korea," for a list of junior vocational colleges and the programs they offer.

In the past, the two-year liberal arts junior college was not well received in Korea, and diminished in importance. Many former liberal arts junior colleges were the antecedents for some of the present four-year colleges and universities as well as for a number of junior vocational colleges. The junior college movement developed institutions that became junior teacher's colleges, junior technical colleges, and junior vocational colleges. In 1963, a five-year program was developed for students in junior vocational colleges which included a three-year senior vocational high school program and a two-year junior vocational college program.

The junior vocational colleges of today no longer include a senior high school segment. Their programs are two years in length, with the exception of the

# SCHOLASTIC RECORD

SEOUL NATIONAL UNIVERSITY
SEOUL, KOREA

| No. | | | |
|---|---|---|---|
| Date | | | |
| Name | | | |
| Date of Admission: March 1, 1970 | Date of Birth: | Male / Female | Present Address: February 26, 1974 |

College: College of Law   Degree Received: Bachelor of Law   Dept. of Law   Major: Law   Date of Completion:

## Freshman Course (March 1970–February 1971)

| Subject | 1st Semester Grade | 1st Semester Credits | 2nd Semester Grade | 2nd Semester Credits |
|---|---|---|---|---|
| Korean | B | 3 | C | 3 |
| English | B | 3 | B | 3 |
| Intermediate German | C | 2 | C | 2 |
| Chinese | B | 2 | A | 1 |
| History of Culture | C | 2 | — | — |
| Principles of Economics | C | 2 | B | 2 |
| Biology | B | 2 | — | — |
| Chemistry | A | 2 | A | 2 |
| Physical Education | B | 1 | A | 1 |
| Military Training | B | 1 | A | 1 |
| Introduction to Law | C | 2 | B | 2 |
| Introduction to Political Science | C | 2 | C | 2 |
| Introduction to Philosophy | — | — | C | 2 |
| National Ethics | — | — | A | 2 |
| Introduction to Literature | — | — | C | 2 |
| Mathematics (B) | — | — | B | 2 |
| **Total** | | **21** | | **24** |

## Sophomore Course (March 1971–February 1972)

| Subject | 1st Semester Grade | 1st Semester Credits | 2nd Semester Grade | 2nd Semester Credits |
|---|---|---|---|---|
| Readings in Foreign Book (English) | B | 2 | B | 2 |
| Readings in Foreign Book (German) | B | 2 | A | 2 |
| Constitutional Law | B | 2 | C | 2 |
| Physical Education | B | 1 | A | 1 |
| Military training | B | 1 | B | 1 |
| General Principles of Civil Law | A | 4 | — | — |
| Administrative Law (1) | B | 2 | C | 2 |
| International Law (1) | A | 2 | B | 2 |
| Commercial Law (1) | A | 2 | B | 2 |
| Roman Law | A | 2 | C | 2 |
| Public Administration | B | 2 | B | 2 |
| Law of Property | — | — | A | 4 |
| **Total** | | **22** | | **22** |

## Junior Course (March 1972–February 1973)

| Subject | 1st Semester Grade | 1st Semester Credits | 2nd Semester Grade | 2nd Semester Credits |
|---|---|---|---|---|
| Military Training | A+ | 1 | A- | 1 |
| Law of Contracts & Torts | — | — | A- | 3 |
| Criminal Law (1) | — | — | — | — |
| Commercial Law (2) | Bo | 2 | Bo | 2 |
| Commercial Law (3) | Bo | 2 | B+ | 2 |
| Criminal Procedure (1) | Co | 2 | Co | 2 |
| Civil Procedure (1) | C+ | 2 | Ao | 2 |
| Administrative Law (2) | B+ | 2 | A- | 2 |
| Anglo-American Law | B+ | 2 | B+ | 2 |
| Economic Law | A- | 2 | B+ | 2 |
| Criminal Law (2) | — | — | C+ | 4 |
| **Total** | | **15** | | **22** |

## Senior Course (March 1973–February 1974)

| Subject | 1st Semester Grade | 1st Semester Credits | 2nd Semester Grade | 2nd Semester Credits |
|---|---|---|---|---|
| Family & Succession Law | Do | 3 | — | — |
| Legal Philosophy | B- | 3 | — | — |
| Insurance & Marine Law | B+ | 3 | — | — |
| Seminar in Civil Law | A- | 2 | — | — |
| Seminar in Constitutional Law | B- | 2 | — | — |
| German | A- | 2 | — | — |
| Introduction to Law of Contracts & Torts | C- | 3 | — | — |
| Criminal Law | C+ | 3 | — | — |
| Labor law | — | — | B+ | 3 |
| Korean Legal History | — | — | B+ | 3 |
| Criminology | — | — | B+ | 3 |
| Seminar in Administrative law | — | — | Bo | 2 |
| Special Seminar | — | — | Bo | 2 |
| **Total** | | **21** | | **13** |

Total Number of Credits: 160
Grade Point: 473.5
Grade Point Average: 2.9

### Remarks:

1. Hours Per Week.
   One hour class work per week for 1 semester makes 1 credit.
   Two or more hours of laboratory work per week for 1 semester makes 1 credit.
2. Weeks Per Year:
   15 weeks make 1 semester and 2 semesters one academic year.
3. Grades:
   "A" for 100-90, "B" for 89-80, "C" for 79-70, and "D" for 69-60.
4. Required Credits:
   Minimum 160 credits (on- Bachelor of Law, Doctor)
5. Following subclassified grade point system is in effect since the 1972 academic year.

   A+ 4.3   B+ 3.3   C+ 2.3   D+ 1.3   F Failure
   A 4.0   B 3.0   C 2.0   D 1.0   I Incomplete
   A- 3.7   B- 2.7   C- 1.7   D- 0.7
6. Lowest passing grade point average for graduation is 2.0.
   Grades A, B, C and D assigned before 1972 are deemed A, B, C and D respectively.

Office of the Academic Dean
Seoul National University

Seal

3.1. Scholastic Record for a Bachelor of Science in Engineering, Seoul National University

## SCHOLASTIC RECORD

**SEOUL NATIONAL UNIVERSITY**
SEOUL, KOREA

| No. | | |
|---|---|---|
| Date | | |

| Name | Male / Female | Date of Birth: | Present Address: |
|---|---|---|---|

**College:** College of Engineering    **Dept. of Electronic Engineering**    **Major:** Electronic Engineering

| Date of Admission: March 1, 1977 | | |
|---|---|---|
| Date of Degree Received: | Present Address: February 26, 1981 | Date of Completion: |

**Degree Received:** Bachelor of Science in Engineering

Columns: **1st Semester — Grade / Credits** | **2nd Semester — Grade / Credits**

### Freshman Course (March 1977–February 1978)

| Subject | 1st Sem Grade | Credits | 2nd Sem Grade | Credits |
|---|---|---|---|---|
| National Ethics | C+ | 3 | | |
| Composition | Ao | 1 | | |
| English (1) | Ao | 3 | | |
| Mathematics (1) | A+ | 4 | | |
| Physics & Lab. (1) | Ao | 4 | | |
| Chemistry & Lab. (1) | B- | 4 | | |
| Physical Education | Co | 1 | | |
| Military Training | C+ | 1 | | |
| Korean | | | Co | |
| English (2) | | | D+ | |
| Physics & Lab. (2) | | | A- | |
| Chemistry & Lab. (2) | | | B+ | |
| Mathematics (2) | | | A+ | |
| **Total** | | 20 | | 18 |

### Sophomore Course (March 1978–February 1979)

| Subject | 1st Sem Grade | Credits | 2nd Sem Grade | Credits |
|---|---|---|---|---|
| Introduction to economics | B- | 3 | | |
| Applied Mathematics (1) | B+ | 3 | | |
| Electromagnetism (1) | A+ | 3 | | |
| Electric Circuits | Bo | 3 | | |
| Electricity Magnetism Lab. | B- | 2 | | |
| Physical Education | A- | 1 | | |
| Korean History | Bo | 1 | | |
| Introduction to Philosophy | | | A- | 2 |
| Applied Mathematics (2) | | | Bo | 3 |
| Introduction to Modern Physics | | | Bo | 3 |
| Physical Electronics (1) | | | A- | 3 |
| Circuit Theory | | | Bo | 3 |
| Electricity Magnetism Lab. | | | C+ | 2 |
| Physical Education | | | A- | 1 |
| Military Training | | | Ao | |
| **Total** | | 16 | | 21 |

### Junior Course (March 1979–February 1980)

| Subject | 1st Sem Grade | Credits | 2nd Sem Grade | Credits |
|---|---|---|---|---|
| FORTRAN & Practice | Co | 3 | | |
| Electronic Circuits (1) | C+ | 3 | | |
| Electronic Engineering Lab. (1) | Bo | 2 | | |
| Network Theory | A+ | 3 | | |
| Semiconductor Devices (1) | C- | 3 | | |
| Digital Computer Organization (1) | B+ | 3 | | |
| Military Training | B- | 1 | | |
| Electromagnetism (2) | | | Ao | 3 |
| Electronic Circuits (2) | | | Ao | 3 |
| Electronic engineering Lab. (2) | | | Ao | 2 |
| Semiconductor Devices (2) | | | Co | 3 |
| Basic Communication Theory (1) | | | B+ | 3 |
| Military Training | | | Ao | 1 |
| **Total** | | 18 | | 15 |

### Senior Course (March 1980–February 1981)

| Subject | 1st Sem Grade | Credits | 2nd Sem Grade | Credits |
|---|---|---|---|---|
| Electronic Engineering Lab. (3) | Co | 3 | | |
| Design of Analog Electronic Circuits | Ao | 3 | | |
| Pulse and Switching Circuits | C+ | 3 | | |
| Microwaves (1) | Ao | 3 | | |
| Basic Communication Theory (2) | B+ | 3 | | |
| Communication Systems (1) | Ao | 3 | | |
| Semiconductor Electronics | | | Ao | 3 |
| Electric Power Engineering | | | C+ | 3 |
| Electronic Engineering Lab. (4) | | | B- | 2 |
| Communication systems (2) | | | A- | 3 |
| Design of Digital Electronic Circuits | | | Bo | 3 |
| Control Engineering (1) | | | B+ | 3 |
| Graduation Thesis | | | S | |
| **Total** | | 20 | | 14 |

**Total Number of Credits:** 142
**Grade Point:** 453.10
**Grade Point Average:** 3.19

**Remarks:**

1. Hours-Per-Week:
   One hour class work per week for 1 semester makes 1 credit.
   Two or more hours of laboratory work per week for 1 semester makes 1 credit.

2. Weeks-Per-Year:
   15 weeks make 1 semester and 2 semesters one academic year.

3. Grades:
   "A" for 100-90, "B" for 89-80, "C" for 79-70, and "D" for 69-60.

4. Required Credits:
   Minimum 140 credits for ~~Doctor~~ Bachelor of Science in Engineering.

5. Following subclassified grade-point system is in effect since the 1972 academic year.

   A+ 4.3   B+ 3.3   C+ 2.3   D+ 1.3   F Failure
   A° 4.0   B° 3.0   C° 2.0   D° 1.0   I Incomplete
   A- 3.7   B- 2.7   C- 1.7   D- 0.7
   Lowest passing grade point average for graduation is 2.0.

6. Grades A, B, C and D assigned before 1972 are deemed A°, B°, C° and D° respectively.

Office of the Academic Dean
Seoul National University

Seal

3.2.   Scholastic Record for a Bachelor of Law, Seoul National University

**EWHA WOMANS UNIVERSITY**
SEOUL, KOREA

Office of the
Academic Administration

**TRANSCRIPT OF ACADEMIC RECORD**

Name:
Date of Birth:
Address:

Major: Mathematics
Date of Admission: March 2, 1977
Date of Withdraw:
Date of Graduation: February 23, 1981
Degree Received: Bachelor or Science

| Year | Subject | 1st Semester Credit | 1st Semester Grade | 2nd Semester Credit | 2nd Semester Grade |
|---|---|---|---|---|---|
| 1977 Fresh. | Korean | 2 | C | 2 | A |
| | Eng. | 2 | C | 2 | C |
| | German | 2 | D | 2 | D |
| | National Ethics | 1 | P | | |
| | Physical Education | 1 | A | 1 | B |
| | General Biology | 2 | A | 2 | C |
| | General Chemistry | 2 | B | 2 | B |
| | General Biology: Lab. Work | 1.5 | B | | |
| | General Chemistry: Lab. Work | 1.5 | B | 1.5 | A |
| | Economics | 3 | C | 3 | B |
| | Intro. to Christianity | | | 2 | A |
| | Principles of Education | | | 2 | C |
| | Typing | | | 1 | A |
| | Subtotal | 16.5 | | 18.5 | |
| | Year Total | | | 35 | |
| 1978 Sopho. | Korean | 2 | B | 2 | D |
| | Eng. | 2 | C | 2 | B |
| | Philosophy | 3 | B | | |
| | Plant Taxonomy | 2 | B | 2 | A |
| | Plant Taxonomy: Lab. Work | 1 | A | | |
| | Set Theory | 3 | B | 3 | A |
| | Calculus | 3 | A | | |
| | School & Community | 2 | B | | |
| | Intro. to Christianity | | | 2 | B |
| | Cultural History of Korea | | | 3 | B |
| | Differential Equations | | | 3 | A |
| | Health Education & Recreation | | | 2 | A |
| | National Ethics | | | 2 | C |
| | Subtotal | 18 | | 20 | |
| | Year Total | | | 38 | |

Note: Interpretation of our grading system is as follows:

90 to 100-A    60 to 69-D
80 " 89-B    0 " 59-F (Failure)
70 " 79-C

**EWHA WOMANS UNIVERSITY**
SEOUL, KOREA

Name:

| Year | Subject | 1st Semester Credit | 1st Semester Grade | 2nd Semester Credit | 2nd Semester Grade |
|---|---|---|---|---|---|
| 1979 Junior | Algebra & Geometry | 3 | B | | |
| | Linear Algebra | 3 | B | | |
| | Statistics | 3 | A | | |
| | Computer Science | 3 | A | | |
| | Logic | 3 | A | | |
| | Ancient Western Philosophy | | | 3 | A |
| | Advanced Calculus | | | 3 | A |
| | Abstract Algebra | | | 3 | B |
| | Modern Geometry | | | 3 | B |
| | Complex Analysis | | | 3 | C |
| | Symbolic Logic | | | 3 | A |
| | Modern Western Philosophy | | | 3 | B |
| | Curriculum Construction | | | 3 | A |
| | Sub-total | 18 | | 21 | |
| | Year Total | | | 39 | |
| 1980 Senior | Real Analysis | 3 | A | 3 | B |
| | Topology | 3 | A | | |
| | Complex Analysis | 3 | B | | |
| | History of Chinese Philosophy | 3 | B | | |
| | History of Korean Philosophy | 3 | C | | |
| | Method of Teaching | 3 | A | | |
| | Abstractive Algebra | | | 3 | A |
| | Ethics | | | 3 | B |
| | History of Education | | | 3 | A |
| | Human Growth & Development | | | 2 | A |
| | Student Teaching | | | 2 | A |
| | Sub-total | 18 | | 17 | |
| | Year Total | | | 35 | |

Date:

Bong Jo Rho
Hwon Jo Rho
Director   Assistant Dean
Academic Administration

3.3. Transcript of Academic Record for a Bachelor of Science (Mathematics), Ewha Womans University

fisheries/marine colleges which offer an additional six-month course for naviga-
tion practice, and the nursing program which is three years in length.

Junior vocational colleges offer programs in these general categories: 1) commerce, 2) kindergarten education, 3) engineering/technical, e.g., agriculture,
fisheries, civil and electronics technology, 4) liberal arts, 5) nursing and health
careers, and 6) textiles and design. The majority of programs for training kindergarten teachers is offered by junior vocational colleges although the program is
also offered by a few four-year colleges and universities.

Table 3.5 presents a curriculum for the two-year civil technology program at a
junior vocational college. This table indicates the required courses and credits
but omits the elective subjects.

**Table 3.5.   Two-Year Diploma in Civil Technology at a Junior Vocational College**

| Required Courses | Credits | Required Courses | Credits |
|---|---|---|---|
| | | Freshman Year | |
| English | 3 | Military Drill (Men) | 3 |
| Korean | 3 | National Ethics | 3 |
| Korean History | 2 | Physical Education | 2 |
| Mathematics | 3 | Related Elective in Major | 6 |
| Methodology of Practical | | Theory of Education | 2 |
| Education | 2 | | |
| | | Sophomore Year | |
| Architectural Code I | 2 | Architectural Design | |
| Architectural Construction I | 3 | and Drawing II | 3 |
| Architectural Design I | 3 | Architectural Materials | 3 |
| Architectural Design II | 3 | Architectural Structure I | 2 |
| Architectural Design | | Ferrous Concrete Structure | 4 |
| and Drawing I | 3 | Planning of Architecture I | 2 |
| | | Structural Mechanics I | 2 |
| Required courses total credits | 59 | | |
| Total credits required to graduate | 80 or more | | |

Source: Adapted from Ministry of Education publications, 1983 and 1984.

During the last two decades, senior vocational/technical high schools have
been the primary producer of skilled workers for Korea. With the increased
sophistication of technology during the 1970s, which is anticipated to continue,
Korea is placing a major emphasis on the junior vocational college to produce
and retrain its manpower for the future. As a result, institutions are being
upgraded through professional development for faculty, upgrading and additions to laboratories, and budget allocation for overall improvement of quality.
Junior vocational colleges are distributed across the country. Many of them are

developing or incorporating curricula that relate directly to industry or business in their service area.

The distribution of subjects in the curricula in these institutions is approximately 40% general academic and 60% technical courses. The technical subject matter is divided approximately equally between theory and applied study. A number of these colleges in the last several years have developed evening programs and have initiated work study programs in cooperation with industry.

## Miscellaneous Schools (*Kakchong Hakkyo*)

The category Miscellaneous Schools was established by the Ministry of Education to indicate institutions which are highly specialized and are not broadly diversified in their academic program. Another interpretation for miscellaneous schools could be various schools that do not fall into a college or university category. As a rule, these schools lack a sufficient liberal arts core or basic general education programs which meet the standards for an accepted undergraduate college or university program in Korea. They are predominantly theological or single purpose institutions and are not authorized by the Ministry of Education to grant a bachelor's degree.

Miscellaneous schools that have received Ministry of Education approval are four years in length and the students receive a diploma at the completion of their program. Theological schools in this category that have Ministry of Education approval are four years in length and may submit their four-year program to colleges or universities for acceptance at the graduate level in theology.

The three other schools on the institutional list in this specialized category are: 1) Chugye School of Arts, 2) Gannam Social Welfare College, and 3) the Korea Judo College. These three schools have received approval from the Ministry of Education, are four years in length, and the students receive a diploma at the completion of their studies. Students who receive a diploma from the Korea Judo College are certified to teach judo. Students from these institutions may be accepted for graduate level work by colleges or universities if their specialized training is adequate preparation for graduate level studies in Korea.

Some schools that have not received approval from the Ministry of Education have exit points of less than four years. In most instances, the coursework of these schools is not recognized for transfer purposes by the Ministry of Education or other colleges and universities. Entrance requirements vary, and many times acceptance is based on the specialized skill of the student; however, students have taken the SAECE test and have completed high school.

There are a number of these schools in Korea that are not listed by the Ministry of Education or in this book because, in many instances, they have been recently established or there is very little information regarding their operation. As is also found in the United States, many of these institutions that are not listed may be on the very edge of acceptability.

The military operates a number of highly specialized schools that are established to serve their specific needs for training of paraprofessionals. In most cases, the program is not transferable; however, if the student passes the country's test for certification for licensing, the individual may be allowed to practice in the field of specialty.

## Open Colleges (*Kaebang Taehak*)

In 1981, the first open college was established. Gyeong-gi Vocational Technical College, now renamed Kyeong-gi Technical Open College, was selected to establish this new approach. Four other vocational/technical institutions have also been selected as national open colleges. An open college for commerce has been established in Kwangju. The two- and four-year programs of these colleges concentrate on technical or commerce programs that lead to a diploma for the two-year programs or a bachelor's degree for four-year programs.

Educational planning for the future dictates that Korea should have more flexible two- and four-year technical programs. Open colleges are the institutions that are being developed to fit this need. These institutions will have the responsibility to develop model programs with industry and business such as short-cycle training programs, cooperative work study programs, and in-plant educational programs. They also will have the responsibility to develop programs that are sensitive to the needs of part-time students who work during the day or evening.

Special features of the open college are the junior and senior years that lead to a bachelor's degree in technology or commerce. These junior/senior programs in technology and commerce are for a select group of open college students and are open to similarly qualified students from two-year junior vocational colleges. The four-year technical programs that have been developed are more theoretical than the two-year technician program but less theoretical than four-year engineering college or university programs. They provide an opportunity for more applied and industrial experience. These institutions will allow qualified junior vocational students to continue their work for a bachelor's degree in a technical or commercial field rather than transfer to a four-year engineering college with the possibility of appreciable loss of credit. It is possible that these institutions will also serve as sources for in-service training for faculty of junior vocational colleges and for vocational teachers in secondary schools.

Table 3.6 indicates the required courses and credits for a four-year bachelor's degree in electronics from Kyeong-gi Technical Open College.

## Korea Correspondence University (Bangsong Tongshin Taehak)

The Korea Correspondence University, formerly known as the Korea Correspondence College or the Korea Air and Correspondence College or University, started as a junior college within Seoul National University in 1972. In 1982, when it separated from Seoul National University, it developed a five-year university program offering a bachelor's degree. Now the Korea Corre-

**Table 3.6. Bachelor of Science in Electronics, Kyeong-gi Technical Open College**

| Required Courses | Credits | Required Courses | Credits |
|---|---|---|---|
| | | Freshman Year | |
| Foreign Languages | 6 | Military Drill (Men) | 2 |
| Korean Composition | 1 | National Ethics | 2 |
| Korean History | 2 | Natural Science | 2 |
| Korean Language | 2 | Physical Education | 2 |
| | | Sophomore Year | |
| Advanced Calculus I | 3 | Human Science | 3 |
| Advanced Calculus II | 3 | Introduction to Computer System | |
| Circuit Theory | 3 | and Programming | 3 |
| Electric Circuits | 3 | Introduction to Electro-Magnetic | |
| Electrical and Electronic | | Fields | 3 |
| Engineering Lab I | 2 | Military Drill (Men) | 2 |
| Electrical and Electronic | | National Ethics | 2 |
| Engineering Lab II | 2 | Physical Electronics | 3 |
| | | Junior Year | |
| Electrical Engineering | 3 | Junior Electronic | |
| Electronic Circuits I | 3 | Engineering Lab II | 3 |
| Electronic Circuits II | 3 | Modern Physics I | 3 |
| Junior Electronic | | Modern Physics II | 3 |
| Engineering Lab I | 3 | Network Analysis and | |
| | | Synthesis I | 3 |
| | | Senior Year | |
| Electronic Engineering | | Electronic Engineering | |
| Lab I | 2 | Lab II | 2 |
| Required courses total credits | 79 | | |
| Total credits required to graduate | 140 or more | | |

Source: Kyeong-gi Technical Open College catalog, 1983.

spondence University operates as an independent institution. During the past ten years, over 530,000 men and women have applied for admission, of whom only 153,000 were admitted. In the last ten years this institution has graduated a number of students who have completed the first two years of college; 33,800 have successfully completed the prescribed courses of study. In 1985, the University will graduate its first class with a bachelor's degree.

Attendance at one of the over forty participating institutions located throughout the country is required for each student. The normal time generally required for class participation and examination is during the regular college and university vacation periods.

Students requesting admission to the Korea Correspondence University are initially screened by computer lottery from among graduates or prospective graduates of senior high schools, or those with an earned senior high school equivalency. The lottery system takes into consideration departmental and area quotas throughout the country, the senior high school academic standings, and the results of the Scholastic Achievement Examination for College Entrance (SAECE). Because of this approach and limited acceptance, only academically above-average senior high school graduates have been admitted to the program. Tuition is considerably less than for the public or private colleges or universities.

Bachelor's degrees are offered in nine fields: 1) administrative science, 2) agriculture science, 3) business administration, 4) economics, 5) education (early childhood education at the diploma level and elementary education at the bachelor's degree level), 6) engineering (computer science and statistics), 7) home economics, 8) jurisprudence, and 9) library science. Each program includes general education (English is offered as a course and future plans call for the addition of French and Chinese) and specialized areas divided into elective and required courses. A candidate for a Bachelor of Arts degree is required to take forty-one or more credits in general education, seventy-two or more credits in required specialized courses, and twenty-seven or more credits in a cognate area. The minimum requirement for graduation is 140 credit hours. A student admitted to the four-year program to study for a bachelor's degree is expected to finish in five to ten years. This time period is more extensive than that allocated for fulltime students in college or university programs.

Table 3.7 indicates the required courses for a bachelor's degree in business administration from the Korea Correspondence University.

**Table 3.7.  Bachelor's Degree Program in Business Administration, Korea Correspondence University**

| Required Courses | Credits | Required Courses | Credits |
|---|---|---|---|
| | | Freshman Year | |
| English I | 2 | Introduction to Sociology | 3 |
| English II | 2 | Korean Language | 3 |
| Introduction to Natural Science | 3 | Physical Education and Health | 1 |
| Introduction to Philosophy | 3 | World Civilization | 3 |
| Introduction to Psychology | 3 | | |
| | | Sophomore Year | |
| Financial Accounting | 3 | National Ethics | 2 |
| Introduction to Computer | 3 | Principles of Accounting | 3 |
| Korean History | 3 | Quality Control Management | 3 |
| Management Statistics | 3 | | |
| Methodology of Social Science | 3 | | |

### Junior Year

| Cost Accounting | 3 | Management Policy | 3 |
|---|---|---|---|
| Financial Management | 3 | Marketing Management | 3 |
| Investment Theory | 3 | Personnel Management | 3 |
| Labor Relations | 3 | Principles of Marketing | 3 |
| Management of Organizational | | Production Management | 3 |
| Behavior | 3 | | |

### Senior Year

| Management Accounting | 3 |
|---|---|

| Required courses total credits | 79 |
|---|---|
| Total credits required to graduate | 140 or more |

Source: Ministry of Education publication, 1983.

The only diploma-level program offered is in early childhood education and requires eighty credits with twenty-six credits in general education, and fifty-four or more credits in the major field for completion of the program. It takes two to five years for a student to complete the diploma program.

The Korea Correspondence University program requires the following for an admitted student: 1) the use and completion of self-study materials and text-books developed by the Korea Correspondence University, 2) listening to radio lectures of thirty minutes each day, and 3) sixteen days of attendance at lecture sessions per year related to coursework. In addition, students complete three to five term papers per semester, use lectures printed in the Correspondence University newspaper along with study guides, and attend supplemental lectures in their local area. It is anticipated that within a few years the Korea Correspondence University programs will be supplemented with the addition of educational television.

The drop-out rate for the Korea Correspondence University is approximately 70%. The program is rigorous. The students who complete the program demonstrate a high standard of academic excellence as compared with junior vocational college students taking the national qualifying examination for third year enrollment at four-year colleges or universities. The success rate for the Korea Correspondence University graduates in 1980 was 25%.

The Korea Correspondence University program is facing some of the same difficulties that other segments of Korean postsecondary education are realizing. These include the rapid increase in admissions and its impact on the quality of the program, and the intense utilization of the central staff and faculty, and the staff and facilities of the participating colleges and universities.

## National Teacher's Colleges (*Kyoyuk Taehak*)

There are eleven four-year primary teacher colleges in Korea. They are national in their funding and control. These colleges are distributed across Korea by province and by major cities. An elementary school is a part of each of the eleven national teacher's colleges, making available practice teaching opportunities for student teachers. For more information on primary teacher colleges, see "Preschool and Primary Teacher Training Programs" in Chapter IV.

## Colleges and Universities (*Taehak* and *Taehakkyo*)

There are twenty national, one municipal and seventy-eight private colleges and universities, for a total of ninety-nine four-year colleges and universities. A minimum of 140 credits is required to receive a bachelor's degree. Bachelor's degrees in four-year colleges and universities are offered in the following twenty-five programs:

| | |
|---|---|
| 1) Administrative Science | 13) Jurisprudence |
| 2) Agriculture Science | 14) Library Science |
| 3) Business Administration | 15) Literature |
| 4) Commerce | 16) Medicine |
| 5) Dentistry | 17) Music |
| 6) Economics | 18) Nursing |
| 7) Education | 19) Oriental Medicine |
| 8) Engineering | 20) Pharmacology |
| 9) Fine Arts | 21) Physical Science |
| 10) Fisheries | 22) Political Science |
| 11) Gymnastics | 23) Sanitation |
| 12) Home Economics | 24) Theology |
| | 25) Veterinary Medicine |

In order to be classified as a university by the Ministry of Education, the institution must have three or more colleges, and at least one graduate school which offers in-depth, research-oriented classes for graduate students. Four-year colleges and universities provide programs leading to the bachelor's degree. A number of colleges and universities have programs in medicine, oriental medicine, and dentistry. These programs are six years in length. (For a discussion of programs in medicine, oriental medicine, and dentistry, see Chapter V, "Other Professional Preparation.") A four-year college may also have a graduate program.

In general, one credit requires one hour of class attendance per week per semester. Laboratory work requires two hours per unit per week. Generally, the maximum amount that can be earned per semester is twenty-four credits with a minimum 140 credits required for graduation. All programs are divided into required courses and electives. The required courses emphasize general education such as cultural history, ethics, Korean language, natural science, philosophy, and physical education. The curriculum for the bachelor's degree is divided approximately into three areas: one-third for required core courses, one-third for electives and one-third in the field of specialty. In most instances a senior thesis or its equivalent is required.

Documents 3.1, 3.2, and 3.3—transcripts for programs in electronic engineering, law, and mathematics, respectively—illustrate sample programs in these different fields. See Document 6.1 in Chapter VI for a business administration program from Yonsei University, and Document 6.2 for a bachelor's of science program from the Korea Military Academy.

## Graduate Education (*Taehagwon Kyoyuk*)

Korean graduate education, like other segments of postsecondary education, is going through extensive change and growth. The range in the quality of equipment, facilities, and staffing between and among colleges and universities appears to be quite wide. In addition to these difficulties, many new programs are being established. Most institutions do not have a fulltime graduate faculty. Few colleges or universities have an adequate number of fulltime staff assigned to this level of education. However, the rate of improvement in staffing and in the sophistication of staff members has been rapid in the last several years, particularly in the fields of social science and humanities.

Because of the rapid increase in student enrollment in the areas of science and engineering, there is a scarcity of faculty members with experience and advanced degrees in engineering and science. One of Korea's primary goals for the future is to develop the areas of basic research and high technology. This, coupled with the increased needs of industry for talented scientists and engineers, has placed the postsecondary educational system in these fields under severe stress to recruit experienced faculty. As in the United States, well-trained and experienced faculty are scarce, enrollments are increasing, and the quality of physical facilities is adequate at best. This segment of higher education is receiving high priority for study and improvement by the Ministry of Education and the colleges and universities.

Each year many faculty and staff at colleges and universities are sent to other countries to update their education and skills or to pursue advanced degrees. This commitment to professional development has made it possible for Korea to establish a potential for excellence at the graduate level. Within the next ten years a number of Korean programs at the graduate level should rank high in their productivity and their quality of research in comparison with other institutions worldwide.

The following tables, 3.8 and 3.9, respectively, indicate the rate of growth of graduate school enrollment and the status of graduate schools in 1984.

## Master's Degree Program (*Suksa Kwajong*)

Students requesting admission to a graduate school must have a bachelor's degree or its equivalent from a college or university with approved standing. The student normally has adequate undergraduate preparation for the requested graduate specialized field. If there are deficiencies in this area, background

58

**Table 3.8. Graduate School Enrollment by Year**

Number of students

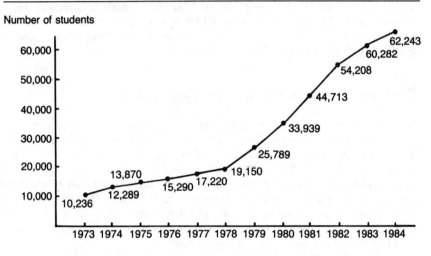

**Table 3.9. The Status of Graduate Schools, 1983**

| Academic Field | Total Schools | Total Students | M.A. Courses Schools | M.A. Courses Students | Ph.D. Courses Schools | Ph.D. Courses Students |
|---|---|---|---|---|---|---|
| Graduate Schools (General) | 68 | 39,044 | 68 | 28,707 | 49 | 10,377 |
| Graduate Schools (Professional) | 102 | 25,530 | 102 | 25,530 | – | – |
| Bus. Mgmt. | 18 | 5030 | 18 | 5030 | – | – |
| Bus. Mgmt. & Gov't Admin. | 2 | 470 | 2 | 470 | – | – |
| Correspondence | 1 | 200 | 1 | 200 | – | – |
| Education | 32 | 9920 | 32 | 9920 | – | – |
| Environmental Protection | 4 | 620 | 4 | 620 | – | – |
| Food Development | 1 | 200 | 1 | 200 | – | – |
| Gov't Admin. | 12 | 2850 | 12 | 2850 | – | – |
| Health & Sanitation | 3 | 400 | 3 | 400 | – | – |
| Indust. Arts | 2 | 500 | 2 | 500 | – | – |

| | | | | | | |
|---|---|---|---|---|---|---|
| Indust. Relns. | 8 | 2090 | 8 | 2090 | – | – |
| Int'l Bus. Mgmt. | 1 | 250 | 1 | 250 | – | – |
| Int'l Trade | 4 | 1000 | 4 | 1000 | – | – |
| Lang. Interpretation | 1 | 140 | 1 | 140 | – | – |
| Regional Development | 3 | 480 | 3 | 480 | – | – |
| Theology (Ecumenical) | 9 | 1260 | 9 | 1260 | – | – |
| Urban Admin. | 1 | 120 | 1 | 120 | – | – |
| Total | 170 | 64,574 | 170 | 54,237 | 49 | 10,337 |

Source: Adapted from information provided by Ministry of Education publications, 1983 and 1984.

courses will be requested. The master's degree is four semesters or two years of fulltime study and should not take more than three years to complete. Advisory committees are appointed by the college or university accepting the student at the master's level.

Usually a minimum of twenty-four credit hours is required, and prior to receiving a master's degree, the candidate must pass a comprehensive examination and an examination in one foreign language as well as complete a master's thesis. A student is required to have a 3.0 or B average or better to successfully complete the master's program. Documents 3.4 and 3.5 provide sample course listings for master's programs in mathematics and education respectively. The Transcript of Academic Record from Ewha Womans University (Document 3.4) gives the grade point average and foreign language requirement and shows courses taken for no credit for which a pass mark was awarded. Grading scales and time requirements are noted on the Scholastic Record from Seoul National University (Document 3.5).

## Doctoral Degree Program (*Paksa Kwajong*)

Students requesting admission to a doctoral program must have a master's degree or its equivalent, a scholarly background in the field of specialty with some demonstrated research experience, and recommendations from individuals in the master's degree field of speciality. Generally, a doctoral program requires a minimum of sixty credits taken over three or more years. Students must pass a foreign language test to demonstrate an ability to comprehend and write in two foreign languages. They must also pass a comprehensive examination, complete the coursework with a 3.0 or B average or better, submit a dissertation and have it accepted, and pass an oral examination. An advisory committee is appointed at the doctoral level.

The Graduate School                                   No. _____
Ewha Womans University
Seoul, Korea            ACADEMIC TRANSCRIPT

Date: Jan. 4, 1984

Name:
Date of Birth:
Address:
Department: Mathematics          ( Master's program)
Date Entered: Mar. 6, 1981
Degree Conferred:  Master of Science
                         (Date:  Feb. 28, 1983

| Year | 1st Semester | | | 2nd Semester | | |
|------|--------------------|----------------|-------|--------------------|----------------|-------|
|      | Course Description | Credit Hours | Grade | Course Description | Credit Hours | Grade |
| 1981 | Algebra | 3 | A | Advanced Numerical Analysis | 3 | B |
|      | Analysis | 3 | A | Topics in Algebra | 3 | A |
|      |  |  |  | Topics in Analysis | 3 | B |
|      | Topology | 3 | A | Algebraic Topology | 3 | A |
| 1982 | Theory of Numbers | 3 | A | Algebraic Topology | 3 | A |
|      | Topics in Topology | 3 | A |  |  |  |
|      | Thesis Style & Format | NC | P | Seminar in Thesis | 3 | A |
|      | Comprehensive Examination | NC | P | Thesis | NC | P |

Total Credit Hours:    33        Average Grade Point: 3.8/4.0

Foreign Language Requirements (      English      ) completed.

Remarks:    NC:Non-Credit
            P:Passed

Note: Grade below "C" is a failure in the course          CHANG Sang
concerned, "U.g." stands for an undergraduate          Director of Academic Affairs
course.                                                The Graduate School

(This transcript is not official, if not sealed.)

3.4.  Academic Transcript for a Master of Science in Mathematics, Ewha Womans University

**SCHOLASTIC RECORD**

**SEOUL NATIONAL UNIVERSITY**
SEOUL, KOREA

Dept. of Education
Major: Education
Date of Completion:

No. _____
Date. _____

| Name | | | Date of Birth | Present Address: | |
|---|---|---|---|---|---|
| | Male | | | Graduate School | |
| | Female | | | | |

Date of Admission: March 1, 1978    Date of Degree Received: February 26, 1981    Degree Received: Master of Education

| Subject: | 1st Semester Grade | 1st Semester Credits | 2nd Semester Grade | 2nd Semester Credits |
|---|---|---|---|---|
| **First Year Course(March 1978-February 1979 )** | | | | |
| Research Methods in Education | A+ | 3 | - | - |
| Human Characteristics and Education | A+ | 3 | - | - |
| Statistical Methods in Education | - | - | Ao | 3 |
| Theories of Personality and Education | - | - | A- | 3 |
| Total | | 6 | | 6 |
| **Second Year Course(March 1979-February 1980 )** | | | | |
| Process of Educational Innovation | A+ | 3 | - | - |
| Theories of School Counseling (1) | - | - | A- | 3 |
| Cognitive Development and Education | - | - | A- | 3 |
| Studies in Thesis | - | - | S | (2) |
| Total | | 3 | | 6 |
| **Third Year Course(March 1980-February 1981 )** | | | | |
| Educational and Psychological Measurement | A+ | 3 | - | - |
| Socialization Process and Education | A- | 3 | - | - |
| Studies in Thesis | S | (2) | - | - |
| Master's Thesis | - | - | A | - |
| Total | | 6 | | |

Remarks:
1. Hours-Per-Week:
120(90) minutes class work per week for 1 semester makes 2 credits.
(120(90) minutes of laboratory work per week for 1 semester makes 1 credit).
2. Weeks-Per-Year:
15 weeks make 1 semester and 2 semesters a year.
3. Grade:
"A" for 100-90, "B" for 89-80, "C" for 79-70.
The lowest passing grade is C(70).

Seal

Office of the Academic Dean
Seoul National University

3.5.   Scholastic Record for a Master of Education, Seoul National University

# Chapter Four

# Teacher Education

## Teacher Training Institutions (*Kyosa Taehak Kyoyuk Kikwan*)

Teacher education (*Kyosa Taehak Kyoyuk*) is conducted at the postsecondary level in Korea by junior vocational colleges; national teacher's colleges; colleges of education; colleges and universities with educational departments; the Korea Correspondence University; graduate schools of education; and other in-service organizations, national, provincial, municipal and private.

## Teacher Training Programs

### Preschool and Primary Teacher Training Programs

- Kindergarten teachers' training programs are delivered primarily in the junior vocational colleges, the eleven national teacher's colleges for primary level education, and in some private and public four-year colleges and universities;
- Primary school teacher training is conducted in the eleven national teacher's colleges and in some public and private four-year colleges and universities.

Prior to 1961 there were three types of institutions that trained primary teachers, i.e., normal schools, junior teacher's colleges, and four-year colleges. Normal schools trained high school graduates to be primary school teachers; junior teacher's colleges trained teachers for junior high school; and the four-year colleges trained teachers for senior high schools. In 1961, the normal schools were discontinued, and junior colleges were given the responsibility for training teachers for primary education. Junior high and senior high school teachers were trained in four-year colleges or universities.

The national junior teacher's colleges that were established to train primary teachers for certification required two years of college with specialized training before the student was awarded a diploma and certified as a primary teacher. A bachelor's degree was required for junior high and senior high school teacher certification. Admission to the national junior teacher's colleges was competitive. These institutions had dormitories. Tuition, room and board were made available to students who qualified, with the understanding that after graduation they would teach for at least two years in the elementary schools to which they were assigned.

In 1981, these eleven national junior teacher's colleges were changed to four-year institutions. Graduates from these colleges receive a bachelor's degree and certification to teach in primary education. Students who are provided tuition, room and board to attend national four-year teacher's colleges are obligated, upon graduation, to teach for at least four years in the primary schools to which they are assigned. Table 4.1 indicates the enrollment for primary teacher training programs in the various institutions and the type of license awarded. Table 4.2 indicates the curiculum for the four-year teacher education program at Seoul National Teacher's College.

**Table 4.1. Primary Teacher Training Institutions, 1981–82**

| Type of Institution | Number of Colleges | Enrollment | Type of License |
|---|---|---|---|
| Teacher's Colleges (National) | 11 | 4,560 | 2nd class |
| Ewha Womans University | 1 | 65 | 2nd class |
| Korea Correspondence University | 1 | | Associate |

Source: Adapted from Ministry of Education publications, 1983 and 1984.

**Table 4.2. Bachelor of Arts Curriculum in Primary Education, Seoul National Teacher's College, 1983**

| Required Courses | Credits | Required Courses | Credits |
|---|---|---|---|
| | | Freshman Year | |
| English | 4 | Natural Science II | 1 |
| History of Education | 3 | Physical Education | 2 |
| Korean History | 3 | Principles of Education | 3 |
| Korean Language | 4 | Studies of Art Text | 0.5 |
| Mathematics I | 1 | Studies of Music Text | 0.5 |
| Mathematics II | 2 | Teaching Methods of Art | 0.5 |
| Military Drill (Men) | 2 | Teaching Methods of Music | 0.5 |
| National Ethics | 4 | World Civilization | 3 |
| Natural Science I | 2 | | |
| | | Sophomore Year | |
| Art Studies | 2 | Music Studies | 2 |
| Childhood Education (Women) | 4 | Physical Education | 2 |
| Curriculum and | | Physical Education Studies | 1 |
| Educational Evaluation | 3 | Practical Teaching Methods | 1 |
| Educational Psychology | 3 | Practicum | 1 |
| Introduction to Korean Language | 3 | Second Foreign Language | 2 |
| Introduction to Korean Literature | 3 | Teaching Methods of Art | 1 |
| Military Drill (Men) | 2 | Teaching Methods of Music | 1 |
| | | Teaching Methods of Physical Education | 1 |

*(continued)*

#### Junior Year

| | | | |
|---|---|---|---|
| Arithmetic Studies | 1 | Social Studies | 2 |
| Educational Sociology | 3 | Teaching Methods of Arithmetic | 2 |
| History of Korean Literature | 3 | Teaching Methods of Korean | |
| Korean Grammar Studies | 3 | Language | 2 |
| Korean Language Studies | 2 | Teaching Methods of Natural | |
| Natural Science Studies | 2 | Science | 2 |
| Participant Observation | | Teaching Methods of Physical | |
| Practice (2 weeks) | 1 | Education | 1 |
| Physical Education Studies | 2 | Teaching Methods Practicum | 1 |
| Practicum | 1 | Teaching Methods of Social | |
| School and Classroom | | Studies | 2 |
| Administration | 3 | | |

#### Senior Year

| | | | |
|---|---|---|---|
| Korean Education | 3 | Seminar in Korean Language | |
| Korean Semantics | 3 | and Korean Literature | 3 |
| Classroom Management | 3 | Studies of Ethics Text | 2 |
| Practice Teaching | 2 | Teaching Methods of Ethics | 2 |
| Pre-service Training | | Theory of Teaching | 2 |
| (2 weeks) | 1 | Topics in Korean Literature | 3 |

Required courses total credits      125
Total credits required to graduate   140 or more

Source: Seoul National Teacher's College catalog, 1983.

## Secondary Teacher Training Programs

- Secondary school teachers must have a bachelor's degree with a major in a discipline and professional coursework in teacher education. Programs for secondary teachers are found in colleges of education, universities, and four-year colleges.
- Secondary vocational teachers are trained in four-year colleges and universities or may be graduates of junior vocational colleges; however, both must be certified as technicians in accordance with the National Technical Qualification Law.
- The majority of nurses assigned to primary and secondary schools are trained in a three-year program in junior vocational colleges. There are a number of four-year Bachelor of Science programs in nursing at colleges and universities. A school nurse must have graduated from a three- or four-year nursing program and be certified.
- Librarians must be graduates of a Library Science Department in a four-year college or university. (This group is also required to complete a course of study for the teaching profession.)
- A new national university called the National Teacher's University has been approved. Its primary purpose will be to offer undergraduate and graduate degrees for primary and secondary teachers.

Table 4.3 indicates types of teacher training institutions for secondary school teachers.

**Table 4.3.  Secondary Teacher Training Institutions, 1981–82**

| Institution | Control | Colleges | Enrollment | License |
|---|---|---|---|---|
| Colleges of | National | 10 | 6,270 | 2nd class |
| Education | Private | 23 | 6,825 | 2nd class |
| Departments of | National | 7 | 1,050 | 2nd class |
| Education | Private | 35 | 3,420 | — |
| Courses for | National | 18 | 22,695 | 2nd class |
| Teacher Educ. | Private | 62 | 69,905 | — |
| Graduate Schools | National | 10 | 928 | 2nd class |
| of Educ. | Private | 22 | 2,029 | — |

Source: Adapted from Ministry of Education publications, 1983 and 1984.

# Teacher and Administrative Certification
## (*Kyosa Chakyokchung*)

Teacher certification for primary and secondary education is categorized into one of the five following types: 1) associate, 2) second class, 3) first class, 4) vice-principal, and 5) principal.

The associate teacher's license is a special license for teachers who have not completed the professional education segment of a two-year or bachelor's degree program; however, they are qualified in an academic area. For example, an individual with a Bachelor of Science in Engineering teaching in a vocational high school would not normally have taken preprofessional academic degree courses. A provincial or special city board may administer a competency examination based upon professional educational criteria; those who pass may receive an associate's license.

Teachers who have completed a program for teaching primary school or the program for junior high school or senior high teaching from a college or university receive a second class teacher certificate at graduation in addition to a bachelor's degree. The first class certificate is granted after three years of experience plus 240 hours of professional in-service training.

A vice-principal's (*Kyogam*) certificate may be awarded to a teacher who has earned a first class certificate and has been recommended for in-service training by the superintendent. Selection for in-service training is competitive. For appointment to the position of vice-principal, which is a non-teaching position, the candidate must have a vice-principal's certificate in addition to specific in-service training. Selection for the position is based upon the recommendation of the superintendent and the concurrence of the provincial or special city Board of Education.

A principal's (*Kyojang*) certificate is awarded to the vice-principal's certificate holder who is selected as a principal. In public institutions, selection of the principal is based upon the recommendation of the superintendent and the concurrence of the provincial or special city Board of Education. This recommendation is then forwarded to the Ministry of Education. The President of Korea then confirms the appointment. In private primary and secondary schools, selection of the principal is made upon the recommendation of the chairman of the Board of Trustees of the private school to the superintendent of the Board of Education of the special city or province.

## Master Teachers

In primary, junior high and senior high schools with a large enrollment a master teacher who is a well qualified senior professional may be appointed in both administrative and academic areas such as science, discipline, moral education and physical education. The number of master teachers is limited according to the size of enrollment. Large schools have up to eight master teachers while smaller schools may have at least three master teachers, one each in the areas of discipline, academics, and counseling.

# Chapter Five

# Other Professional Preparation

## Health Education (*Bogun Kyoyuk*)

With the exception of theology, professional and paraprofessional specializations in Korea are primarily related to the health education field. Table 5.1 is a flowchart which diagrams the pathway for education and licensure followed by individuals selecting a health career program. The flowchart developed on health education in Korea provides an overview and will be helpful to review prior to reading about each professional or paraprofessional program.

### Paraprofessional Education

Training for 1) dental technologists, 2) dieticians, 3) laboratory technologists, 4) physical therapists, 5) radiology technologists, and 6) sanitarians is predominantly found in two-year vocational colleges affiliated with a medical hospital or university. Other programs which offer this type of training are junior vocational colleges or programs that are hospital-centered for senior high school graduates. The programs vary in length from one year to three years. Upon completion of the program and passing a national licensure examination, the graduate is allowed to practice in the specialty.

Table 5.2 describes a two-year program for a radiology technologist at the Junior College of Public Health and Medical Technology, Korea University.

### Nursing Education (*Kanho Kyoyuk*)

The majority of nursing education programs are offered in junior vocational colleges. Several of these colleges are totally dedicated to nursing programs. Programs are three years in length and lead to a diploma. Other institutions offering nurse's training are four-year colleges or universities with programs leading to a Bachelor of Science degree. To be certified as a registered nurse all graduates of nursing programs must complete a designated period of service in a hospital and pass a national nursing licensure examination.

Entrance into nursing programs is competitive and students must take the Scholastic Achievement Examination for College Entrance (SAECE). Students who have completed their three-year program and who wish to continue their education to receive a bachelor's degree at a four-year college or university may

**Table 5.1.  Health Education in Korea, 1985**

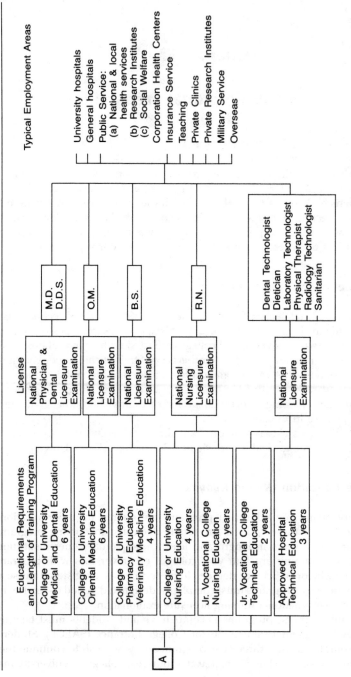

Source: Adapted from catalog of Junior College of Public Health and Medical Technology, Korea University.

 Admission determined by the Scholastic Achievement Examination for College Entrance (SAECE) and the senior high school academic record.

**Table 5.2. Diploma Program in Radiology Technology, Junior College of Public Health and Medical Technology, Korea University**

| Required Courses | Credits | Required Courses | Credits |
|---|---|---|---|
| | | Year One | |
| Biology | 2 | Pathology | 3 |
| Culture/History | 2 | Physical Education | 2 |
| Electronics | 2 | Physics | 1 |
| English | 4 | Physiology | 2 |
| Home Economics (Women) | 2 | Practice of Radiographic | |
| Human Anatomy | 3 | Positioning | 3 |
| Inorganic Chemistry | 1 | Public Health | 2 |
| Introduction to Medicine | 1 | Radiation Physics | 4 |
| Introduction to | | Radiographic Positioning (I) | 2 |
| Radiotechnology | 2 | Radiologic Technology | 4 |
| Korean Language | 1 | Social Science | 2 |
| Military Training (Men) | 2 | X-Ray Equipment | 4 |
| Organic Chemistry | 1 | | 52 |
| | | Year Two | |
| Biochemistry | 1 | Practice of X-Ray Equipment | 2 |
| Military Training (Men) | 2 | Public Health Law | 1 |
| Nuclear Medicine | 2 | Radiation Control | 1 |
| Physical Education (Women) | 2 | Radiation Dosimetry | 4 |
| Practice of Nuclear | | Radiation Therapy | 4 |
| Medicine | 2 | Radiobiology | 4 |
| Practice of Radiation | | Radiographic Anatomy | 2 |
| Dosimetry | 2 | Radiographic | |
| Practice of Radiation | | Positioning (II) | 4 |
| Therapy | 2 | Special Radiography | 2 |
| Practice of Radiographic | | X-Ray Equipment | 2 |
| Positioning | 10 | | 50 |
| Practice of Radiology | | | |
| Technology | 1 | | |

Source: Junior College of Public Health and Medical Technology, Korea University catalog, 1983.
Note: These programs are highly intensive and the credit hours leave very little time for electives.

do so. As is the case in the United States, this is a difficult procedure. Usually there is loss of credit and additional general education courses are required. Students in the four-year program meet the requirements for a bachelor's degree in general education. In the three-year program, approximately 80% of the curriculum deals with professional nursing courses. Table 5.3 indicates the curriculum for a three-year nursing program from a junior vocational college.

**Table 5.3. Sample Three-Year Nursing Curriculum in a Junior Vocational College**

| Required Courses | Credits | Required Courses | Credits |
|---|---|---|---|
| | | Freshman Year | |
| Anatomy I | 3 | Microbiology & Lab II | 1 |
| Anatomy II | 2 | Military Drill (Men) | 2 |
| Chemistry | 2 | National Ethics I | 2 |
| English I | 2 | National Ethics II | 1 |
| English II | 1 | Nursing Philosophy & | |
| Introductory Psychology | 2 |   Ethics | 2 |
| Introductory Sociology | 2 | Physical Education | 2 |
| Introduction to | | Physiology & Lab I | 2 |
|   Pathological Nursing | 3 | Physiology & Lab II | 1 |
| Microbiology & Lab I | 2 | Religion | 2 |
| | | Sophomore Year | |
| Adult Nursing Practice I | 3 | Maternity Nursing & | |
| Adult Nursing Practice II | 3 |   Practice | 4 |
| Clinical Pharmacy | 3 | Military Drill (Men) | 1 |
| Introduction to Nursing | | Nursing Philosophy & | |
|   & Lab I | 3 |   Ethics | 1 |
| Introduction to Nursing | | Pediatric Nursing & | |
|   & Lab II | 3 |   Practice | 4 |
| Introduction to Nursing | | Principles & Practice in | |
|   & Lab III | 3 |   Community Health Nursing I | 2 |
| Korean | 2 | Principles & Practice in | |
| | |   Community Health Nursing II | 3 |
| | | Junior Year | |
| Adult Nursing Practice III | 3 | Pediatric Nursing & | |
| Adult Nursing Practice IV | 3 |   Practice | 1 |
| Health Assessment & | | Principles & Practice in | |
|   Practice I | 1 |   Community Health Nursing III | 2 |
| Health Assessment & | | Principles & Practice in | |
|   Practice II | 2 |   Community Health Nursing IV | 1 |
| Korean | 1 | Psychiatric Nursing & | |
| Korean History | 2 |   Practice | 1 |
| Maternity Nursing & | | | |
|   Practice | 1 | | |

Total credits generally required to graduate   125 or more

Source: Information taken from several college catalogs, 1983.

## Pharmacy and Veterinary Medicine (*Yakhak* and *Su-uikwa*)

Pharmacy and veterinary medicine are offered as programs within a university or college, and are four years in length compared with five- to seven-year programs in the United States. They are similar in their professional coursework to U.S. programs. However, students trained in institutions in the United States have more background in basic sciences and more breadth and depth in their professional field both from the standpoint of laboratory experience and time spent on courses in the professional field. Upon successful completion of the program, students receive a Bachelor of Science degree and, after passing the licensure examination, are qualified to practice.

## Medicine and Dentistry (*Uikwa* and *Chikwa*)

Students in medicine, oriental medicine, and dentistry are trained in six-year programs which are a part of a college or university. The first two years involve undergraduate preparation and the last four years are professional in orientation. Upon completing their academic work, students receive a Bachelor of Science in Medicine, Oriental Medicine, or Dentistry.

Internships and residencies follow the same procedure as in the United States; internships are one year in length, and residencies are two to three years in length following an internship.

Table 5.4 indicates the curriculum for a Bachelor of Science in Medicine at Seoul National University.

**Table 5.4. Bachelor of Science in Medicine, Seoul National University**

| Required Courses | Credits | Required Courses | Credits |
|---|---|---|---|
| First Year | | | |
| Foreign Language | 6 | National Ethics | 4 |
| Korean Composition | 1 | Physical Education | 2 |
| Korean Language | 2 | Science | 22 |
| Military Drill | 2 | | |
| Second Year | | | |
| Foreign Language | 3 | Military Drill | 2 |
| Human Science | 3 | Science | 19 |
| Korean History | 2 | Social Science | 3 |

*(continued)*

### Third Year

| | | | |
|---|---|---|---|
| Anatomy | 7 | Neuroanatomy | 2 |
| Behavioral Science | 1 | Parasitology | 1 |
| Biochemistry | 2 | Pathology | 3 |
| Embryology | 1 | Pharmacology | 2 |
| Histology | 4 | Physiology | 6 |
| Medical History | 1 | Preventive Medicine | 2 |
| Microbiology | 2 | Reproductive Medicine | 1 |

### Fourth Year

| | | | |
|---|---|---|---|
| Cardiology | 2.5 | Neuropsychiatry | 2 |
| Endocrinology | 2.5 | Oncology | 1 |
| Gastroenterology | 2.5 | Parasitology | 2 |
| Hematology | 2.5 | Pathology | 6 |
| Immunology | 2 | Pharmacology | 4 |
| Introduction to Clinics | 1 | Preventive Medicine | 4 |
| Microbiology | 4 | Respiratory System | 2.5 |
| Nephrology | 2.5 | | |

### Fifth Year

| | | | |
|---|---|---|---|
| Anesthesiology | 1 | Oncology | 1.5 |
| Community Medicine | 1 | Orthopedic Surgery | 3 |
| General Surgery | 4 | Pediatrics | 6 |
| Immunology | 1.5 | Psychiatry | 4 |
| Internal Medicine | 12 | Radiology | 2 |
| Neurosurgery | 3 | Thoracic Surgery | 3 |
| Obstetrics and Gynecology | 6 | | |

### Sixth Year

| | | | |
|---|---|---|---|
| Anesthesiology | 1 | Neurology | 1 |
| Dermatology | 3 | Nutrition | 1 |
| Emergency Medicine | 1 | Obstetrics and Gynecology | 2 |
| Forensic Medicine and | | Ophthalmology | 3 |
|    Medical Legislation | 1 | Orthopedic Surgery | 1 |
| General Surgery | 1 | Otolaryngology | 3 |
| General Surgery (Plastic | | Pediatrics | 2 |
|    Surgery): (6 hours one time) | 1 | Psychiatry | 2 |
| Internal Medicine | 1 | Radiology | 1 |
| | | Urology | 3 |

Total credits required to graduate   230 or more

Source: Seoul National University catalog, 1983.

## Theological Education (*Shinhak Kyoyuk*)

Theological education is offered in a number of colleges and universities through the Department of Theology or in Miscellaneous Schools. Theological programs offered in a college or university lead to a bachelor's degree in theology. In those institutions with master's and doctoral programs, the degrees conferred are a Master's of Theology and a Doctor of Philosophy. Generally, Miscellaneous Schools offer a diploma.

A bachelor's degree in theology in Korea may follow two different pathways:

1. A four-year bachelor's degree program offered by a four-year college or university in the College or Department of Theology;
2. A three-year bachelor's degree program in a Department or College of Theology at a four-year college or university following successful completion of a four-year liberal arts program. This program is for students who are changing their major, and who, upon completion of the theology program, will have a bachelor's degree in their original major and a bachelor's degree in theology.

Theological Miscellaneous Schools offer four-year diploma programs in theology. In most cases the work is not transferable nor is the degree accepted by other institutions. If the institution has Ministry of Education approval, students may be accepted for graduate work in theology.

The master's and Ph.D. degrees in the field of theology have the same requirements as those in other disciplines. Table 5.5 presents the Bachelor of Arts in Theology program from the Methodist Theological Seminary. Table 5.6 lists the required courses and programs for a Master's Degree in Theology and a Doctor of Philosophy in Theology at the Methodist Theological Seminary in Seoul.

**Table 5.5  Bachelor of Arts in Theology from the Methodist Theological Seminary in Seoul**

| Required Courses | Credits | Required Courses | Credits |
|---|---|---|---|
| | | Freshman Year | |
| Chorus (Women) | 1 | Korean History | 3 |
| Church Practice | 1 | Military Drill (Men) | 2 |
| Study of Communism | 2 | National Ethics | 2 |
| English | 4 | Philosophy | 3 |
| German | 4 | Physical Education | 2 |
| Korean | 3 | Piano (Women) | 1 |

*(continued)*

### Sophomore Year

| | | | |
|---|---|---|---|
| Chorus (Women) | 1 | Introduction to Old Testament | 3 |
| Church History I | 3 | Introduction to Theology | 3 |
| Church & the Rapidly | | Latin or Hebrew | 3 |
|   Changing Society | 3 | Military Drill (Men) | 2 |
| Church Practice | 1 | Physical Education | 2 |
| English | 4 | Piano (Women) | 1 |
| Greek | 3 | Theories of Church Education | 3 |
| Intro. to New Testament | 3 | | |

### Junior Year

| | | | |
|---|---|---|---|
| Christian's Ethical | | Counseling | 3 |
|   Resolution | 3 | History of Christian Doctrine I | 3 |
| Church History II | 3 | Modern Theology | 3 |
| Church Practice | 1 | Systematic Theology | 3 |
| Interpretation of Old Testament | 3 | Theology of the Gospel | 3 |

### Senior Year

| | | | |
|---|---|---|---|
| Church Practice | 1 | Methodist Theology | 3 |
| Discipline of the | | Old Testament Theology | 3 |
|   Korean Methodist Church | 3 | Paul's Theology | 3 |
| Homiletics | 3 | Theories of Instruction | 3 |

Total credits required to graduate   140 or more

Source: Methodist Theological Seminary catalog, 1983.

**Table 5.6.   Master's Degree and Ph.D. Degree Programs in Theology, Methodist Theological Seminary in Seoul**

| Required Courses | Credits | Required Courses | Credits |
|---|---|---|---|
| **Master's Degree in Theology** | | | |
| Year One | | | |
| Modern Christian Theory | 3 | Natural Law and Human Rights | 3 |
| Modern View of God | 3 | Seminar of Early | |
| | |   Christian Theory | 3 |
| Year Two | | | |
| Korean Church and Society | 3 | Sociology of Religion | 3 |
| Seminar of Social Thoughts | 3 | | |

Required courses total credits   21
Total credits generally required to graduate   33 or more and a Master's Thesis

## Ph.D. Degree in Theology

### Year One

| | | | |
|---|---|---|---|
| Korean Church & Society | 3 | Seminar of Early | 3 |
| Modern Christian Theory | 3 | Christian Theory | |
| Modern View of God | 3 | Seminar of Social Thoughts | 3 |
| Natural Law and Human Rights | 3 | | |

Required courses total credits     18
Total credits generally required to graduate   57 or more and a Doctoral Dissertation

Source: Methodist Theological Seminary catalog, 1983.

# Chapter Six

# Guidelines and Placement Recommendations

## Overview

At the end of the Korean War in 1953, Korea's educational system was in total disarray. Over the last thirty years, however, the quality of Korean education has undergone extensive change. Significant improvements are taking place each year in the quality of instruction, staff and facilities. Although the structure of the educational system is very similar to that found in the United States, it is important to be aware of some of the differences.

Primary education in Korea today is compulsory and free, and by the early 1990s, the junior high school may also achieve this goal. The birth rate is decreasing and the number of primary students has stabilized. With primary education beginning to stabilize, a clear effort is being made to evaluate and improve the quality of this segment of education. Secondary high school enrollment is increasing rapidly, possibly because of affluence, availability, and the desire on the part of families to have their children complete their senior high school education. The number and size of higher education institutions has grown significantly in the last ten years, and the number of qualified senior high school students has increased even more rapidly as the Korean population in the last twenty-five years has almost doubled in size. Emigration is permitted and encouraged by the Korean government.

Two major streams of activities have been apparent in the last few years. First, increased training has resulted in more sophisticated faculties and the educational facilities have been upgraded. In the years prior to 1960, opportunities for students to attend secondary schools were limited. Consequently, the student selection process at these schools was academically rigorous. Today, primary and secondary education is widely available. This expansion of opportunity has been enthusiastically accepted by the people of Korea; however, meeting this demand has placed tremendous stress on developing adequate staff and facilities.

Second, the demographics of the students requesting admission to institutions in the United States has changed, particularly in the last several years. More students qualify for college than there are openings in Korean colleges and universities. Also, students face limitations in the programs of their choice when they apply for admission to a Korean college or university. Until the mid-1970s, only students who were graduates of a college or university or who had finished at least two years in a college or university in a science field were allowed to study abroad after they had taken the Overseas Study Qualification Examination. This test primarily examined the student's ability to translate and write in the language of the country in which the student wished to study. The pass rate on this examination was not high. Therefore, students applying to institutions in the United States were a very select group. Since 1984, it has been the policy of the Korean government to be more liberal regarding student requests for study abroad than has previously been the case. This may change in the future. However, at the present time there are many students who have completed senior high school, without prior college work, who are applying for admission to colleges and universities in the United States.

In the last several years admissions officers in the United States have received many more requests for admission than previously from students whose level of competency covers a broader range. Today's Korean student wishing to study abroad is no longer required to take the Overseas Study Qualification Examination. Colleges and universities in the United States are receiving applications from both men and women who have just graduated or are anticipating graduation from senior high school. Since graduation from Korean senior high schools takes place in February, there is a limited time for applications to be processed prior to the fall term at colleges and universities in the United States.

The desire on the part of Korea for students to receive training in science, engineering and technology is being fostered. Because of competition for entrance into colleges and universities, and the limited opportunities for the student to study in the institution of choice in Korea, many students are requesting the opportunity for overseas study. The increase in the college-age population, the increasing affluence of the Korean society, and the lessening of restrictions by the Korean government have all helped to rapidly increase the number of students requesting admission to every type of college and university in the United States.

The Korean student requesting admission to a college or university in the United States may then request a passport. Even though all events proceed on schedule, this request may take considerable time both for in-country clearance and visa approval by the United States Embassy.

Exchange of mail is usually a twenty-five to thirty day process. Adding to that the time required for the translation of credentials from Korean to English, the admissions officer must anticipate the greater time lapse required for the receipt of the transcript from Korea than would normally be allowed for this type of request from a senior high school or college or university in the United States.

## Evaluation of Credentials

Many Koreans who have studied in or visited the United States on a professional basis are now teaching in Korean senior high schools, colleges and universities. Their understanding of the educational system of the United States gives them the background to translate transcripts from Korean senior high schools and postsecondary institutions into English and to indicate the patterns of grading for that institution. The transcripts will have the official chop or identification mark of the institution, and will be signed by an official of the senior high school or a college. Letters of recommendation will be in English, and if requested, a Certificate of Notarization regarding the accuracy of the translation will accompany the material. This document is completed for a fee which the student must pay. Also, if requested, certificates of graduation will be attached to the transcripts. Document 6.1, a transcript for a bachelor's program in business administration, indicates the courses taken, credits awarded, and grading system used, and is verified by the institutional chop and signature of the college official. Document 6.2 is a transcript for a bachelor's of science program from the Korea Military Academy. Document 6.3 is a Certificate of Graduation from Seoul National University.

The score of the Scholastic Achievement Examination for College Entrance (SAECE), administered since 1981, is not normally a part of the high school or college transcript. It may be requested by the admissions officer, and this request may be honored by the high school principal. By studying the SAECE test scores and the student's previous admissions record and academic success, admissions officers at Korean colleges and universities have a predictability rate of student success within their institution of over 80%. U.S. admissions officers, in the evaluation of credentials and the use of the SAECE score, should remember that this test is not a qualifying examination for colleges or universities; consequently, there is no cut-off score used.

### Senior High School Credentials (*Kodung Hakkyo Songjuk*)

There are two major types of Korean senior high schools: vocational and general. The senior high school may be named for a city, an individual or for something else, as is the case in the United States. The transcript usually will designate the type of curriculum the student is following. Credentials and diplomas are similar in format to credentials in educational institutions in the United States. In reviewing senior high school credentials it is important for the admissions officer to keep in mind that Korean students in senior high schools have few electives, and that mathematics, science and English are required. The level of abstraction taught in these courses will vary in the following three major streams for general senior high school: 1) vocational/technical, 2) humanities, and 3) science. In the case of the vocational/technical and commercial student in the technical high schools, courses in science and mathematics are more applied than theoretical, and less time is spent on a foreign language than is the case with the student in the senior general high school.

### College Credentials (*Taehak Songjuk*)

A student in Korea today selects fewer credit hours than in the past. In the last several years the emphasis, particularly in the more academically rigorous in-

# YONSEI UNIVERSITY
## SEOUL, KOREA

Permanent Academic Record
File No. ___

| | |
|---|---|
| Name in Full: | Sex: |
| Date of Birth: | Address: |
| Date of Entrance: | Department: Business Administration |
| Status: | Major: |
| | Minor: |
| | Degree: Bachelor of Business Administration |

| Title of Course | 1st Sem Credit | Grade | 2nd Sem Credit | Grade |
|---|---|---|---|---|
| **1980–81 (Freshman year)** | | | | |
| Korean Language | 3 | B | – | – |
| English I | 3 | A | 3 | C |
| Korean History | 3 | D | 3 | W |
| Principles of Economics I | 3 | D | – | – |
| Military Science I | 1 | C | – | – |
| Elementary English Conversation | 3 | C | – | – |
| Mathematics for Economics I | 3 | A | – | – |
| English II | – | – | 3 | B |
| Introduction to the Christian Thought | – | – | 3 | C |
| National Ethics | – | – | 3 | B |
| Military Science | – | – | 1 | B |
| Principles of Economics | – | – | 3 | C |
| Mathematics for Economics | – | – | 3 | C |
| Cultural Anthropology | – | – | 3 | C |
| **1981–82 (Sophomore year)** | | | | |
| Military Science | 1 | A | – | – |
| Principles of Business Administration | 3 | A | – | – |
| Principles of Accounting | 3 | C | – | – |
| Business Forecasting | 3 | B | – | – |
| Statistics | 1 | B | – | – |
| Soccer | 1 | B | – | – |
| Military Science | 3 | D | – | – |
| Natural Science | – | – | 3 | A |
| English III | – | – | 3 | A |
| Military Science | – | – | 3 | B |
| Sociology | – | – | 3 | B |
| Principles of Accounting II | – | – | 3 | B |
| Organization Behavior | – | – | 3 | B |
| Statistics II | – | – | 3 | B |
| Principles of Management | – | – | 3 | A |
| **1982–83 (Junior year)** | | | | |
| Intermediate Accounting | 3 | A | – | – |
| Personnel Management | 3 | A | – | – |
| Business Finance | 3 | B | – | – |
| Environment of International Business | 3 | A | – | – |
| | 140 | | | |

| Title of Course | 1st Sem Credit | Grade | 2nd Sem Credit | Grade |
|---|---|---|---|---|
| General Principles of Commercial Law | 3 | C | – | – |
| Marketing | 3 | B | 3 | C |
| Financial Management | – | – | 3 | W |
| Marketing Strategy | – | – | 3 | B |
| Introduction to Political Science | – | – | 3 | B |
| Consumer Behavior | – | – | 3 | B |
| Production Management | – | – | 3 | B |
| Quantitative Business | – | – | 3 | B |
| **1983–34 (Senior year)** | | | | |
| Models for Production Management | 3 | A | – | – |
| Industrial Relations | 3 | A | – | – |
| International Business Policy | 3 | B | – | – |
| Business Policy | 3 | A | – | – |
| Insurance Theory | 3 | B | – | – |
| Investment Theory | 3 | C | – | – |
| Health Education I | 1 | C | 1 | A |
| Thesis | – | – | 3 | B |
| Principles of Public Relations | – | – | 3 | B |
| Human Relations | – | – | 3 | B |
| Psychology | – | – | 3 | A |
| Marketing Strategy | – | – | 3 | C |
| Fundamental of Film Making | – | – | | |

* Remark:

Grading System:
A: Excellent, carries 4 grade points
B: Good, carries 3 grade points
C: Satisfactory, carries 2 grade points
D: Passing, carries 1 grade point
F: Failed, carries no point(0)
Minimum Credits required for a degree.

Note: Copies of this document are not official unless Dean's signature and seal are affixed.

Date: January 24, 1984

Signed: Young-Hoon Kim

Dean of the University

6.1. Permanent Academic Record for a Bachelor of Business Administration, Yonsei University

## KOREA MILITARY ACADEMY
### Wharang Dae Seoul Korea
### SCHOLASTIC RECORD

No. :  
Date DEC 12 1983

Name in Full:  
Date of Birth:  
Present Address:

Date of Admission : Mar. 2, 1968  
Date of Graduation : Mar. 30, 1972  
Degree Received : Bachelor of Science

| Year | Subject | 1st Semester Grade | 1st Semester Credit | 2nd Semester Grade | 2nd Semester Credit |
|---|---|---|---|---|---|
| Freshman | Differential & Integral Calculus | B | 3 | B | 6 |
| | Algebra | A | 3 | C | 3 |
| | Natural Science | B | 1.5 | B | 1.5 |
| | Graphics | D | 2 | D | 2 |
| | Korean Literature | B | 4 | B | 4 |
| | English | B | 3 | C | 3 |
| | Cultural History | B | 0.5 | A | 0.5 |
| | Tactics | B | 1 | B | 1 |
| | Physical Education | | | | |
| | Total | | 21 | | 21 |
| Sophomore | Integral Equation | B | 4.5 | | - |
| | Advanced Calculus | | - | B | 1.5 |
| | Statistics | | - | B | 3 |
| | Physics | B | 3.5 | C | 3.5 |
| | Physics Lab | B | 0.5 | C | 0.5 |
| | Chemistry | B | 3.5 | B | 3.5 |
| | Chemistry Lab | B | 0.5 | B | 0.5 |
| | Korean Literature | C | 2 | C | 2 |
| | Philosophy | B | 2 | D | 2 |
| | English | B | 2 | B | 2 |
| | 2nd Foreign Language | A | 2 | A | 2 |
| | Surveying | C | 1.5 | C | 1.5 |
| | Physical Education | C | 1 | B | 1 |
| | Tactics | B | 0.5 | B | 0.5 |
| | Total | | 23.5 | | 23.5 |
| Junior | Psychology | B | 2 | B | 2 |
| | Economics | C | 2 | C | 2 |
| | Law | B | 2 | C | 2 |
| | Politics | B | 2 | B | 2 |
| | English | C | 2 | B | 2 |
| | Physics | B | 2 | C | 2 |
| | Thermodynamics | B | 2.5 | | |
| | Thermodynamics Lab | B | 0.5 | | |
| | Fluid Mechanics | | - | C | 2.5 |
| | Fluid Mechanics Lab | | - | B | 0.5 |
| | Strength of Materials | | - | B | 2.5 |
| | Strength of Materials Lab | | - | B | 0.5 |

| Year | Subject | 1st Semester Grade | 1st Semester Credit | 2nd Semester Grade | 2nd Semester Credit |
|---|---|---|---|---|---|
| | Engineering Mechanics | C | 3 | | - |
| | Network Analysis | B | 4 | | 4 |
| | Electronics | B | 1 | B | 1 |
| | Physical Education | A | 2 | C | 1 |
| | Tactics | A | 2 | A | 2 |
| | Total | | 25 | | 25 |
| Senior | Military Arts | C | 4 | D | 4 |
| | Comparative Sociology | C | 2 | C | 2 |
| | Environmental Science | B | 1 | B | 2 |
| | Ordnance Engineering | | - | C | 4 |
| | Civil Engineering | B | 3 | B | 3 |
| | National Security Seminar | D | 1 | C | 2 |
| | Physical Education | D | 1 | A | 1 |
| | Tactics | B | 2 | C | 2 |
| | French Conversation | B | 2 | A | 1 |
| | History of French Literature | B | 2 | | |
| | Intermediate French | A | 4 | | |
| | French Reading | | | A | 2 |
| | Total | | 24 | | 23 |
| | Grand Total 186 | | | | |

Cho, Hyo Nam  
Lt. Colonel  
Korean Military Academy

Remarks:  
1. Hours Per Week: One hour class work per week for 1 semester make 1 credit. Two or more hours of laboratory work per week for 1 semester makes 1 credit.
2. Week Per Year: 15 weeks make 1 semester and 2 semesters one academic year.
3. Grades: "A" for 100-90, "B" for 89-80, "C" for 79-70, and "D" for 69-60.
4. Required Credits: Minimum 150 credits for Bachelor of Science.
5. Lowest-passing grade point average for graduation is 67 %.

6.2.    Scholastic Record for a Bachelor of Science, Korea Military Academy

# Seoul National University

### Seoul, Korea

Office of
The Academic Dean

No. _____

Date __July 18, 1983__

Subject                     Graduation Certificate

To

Name in Full

Date of Birth

Present Address

Permanent Address          Seoul, Korea.

Date of Admission          March 1, 1977

Date of Graduation         February 26, 1981

Major                      Electronic Engineering

Degree Received            Bachelor of Science in Engineering

This is to certify that the above mentioned graduated from the Department of _____

__Electronic Engineering__ , College of ____ __Engineering__ ____ , Seoul National University.

Office of the Academic Dean

Seoul National University

6.3.   Graduation Certificate, Seoul National University

stitutions, has been on fewer credit hours and more on outside preparation and the use of research and the library approach to learning. Credentials of students graduating prior to the early 1980s will have an average semester credit load of twenty to twenty-four credits. Programs in technical, engineering and nursing fields normally have a higher semester credit hour load as is true for these programs in the United States.

In reviewing credentials from Miscellaneous Schools, it is appropriate to take into consideration that these schools are specialized in their makeup and are predominantly theological. However, they may have an adequate academic program. A careful review of the program in conjunction with academic testing is recommended. This approach is taken by some colleges and universities in Korea.

The admissions officer should keep in mind when reviewing credentials of students in programs such as pharmacy and veterinary medicine that the program in Korea is four years in length. These programs are predominantly specific in their vocational training with more limited general education in comparison to similar programs in the United States.

Students with a master's or doctoral degree should have their programs reviewed very carefully. Many of the graduate programs in Korea have been initiated in the last several years. The graduate faculties may not be in sufficient number or have time allocations to carry out their graduate responsibilities. However, there are a number of programs in the more academically prestigious institutions that are on a par with similar programs in the United States.

It is important for admissions officers to evaluate the credentials of students very carefully inasmuch as 1) there is as great a difference in the quality of colleges and universities in Korea as there is in the United States; 2) these differences in quality range from department to department within a college as well as between and among colleges and universities; 3) it is possible that a student from a top college or university in Korea may be in the bottom 30% of the class and thus may not receive a degree while at the same time a student from a less academically rigorous institution may receive a degree; 4) for employment purposes in Korea, there is not a great deal of difference between the degree and the document certifying completion. The academic reputation of the college or university is more significant in relationship to employment.

## Transfer of Credit

In past years in Korea, transfer from one institution to another, and even within the same institution, has been difficult. Transferring is still a difficult process within Korea. In recent years many applications have been made to institutions in the United States by Korean students requesting what appears to be a significant change in their academic program. In many instances this has taken place because students were unable to gain entrance to the Korean program of first choice, and now wish to change to an area reflecting their initial or new academic interest when they apply to a college or university in the United States.

Consequently, at the end of the sophomore year in recognition of the present transfer difficulties in the Korean system, many students now look for opportunities to transfer to U.S. institutions and choose a program that was their first choice at the start of their Korean college or university program.

## English Proficiency

Korean students study English from junior high through senior high school and many continue their studies at the postsecondary level. The teaching of English has improved significantly in Korea over the last several years. However, students with good academic standings in Korea and with excellent marks in English usually have skills primarily in the areas of reading, writing and grammar. Conversational skills, speaking, inferential reading, and listening skills are generally not as high. These areas of competency should be carefully reviewed with the student because of their importance for classroom and laboratory work in colleges and universities in the United States. The student should be required to take the Test of English as a Foreign Language examination (TOEFL). A review of the senior high school grades, the examination score of the SAECE and the TOEFL, may allow the admissions officer to make an adequate evaluation of the student's potential for academic success in an institution in the United States. However, further testing in the areas of speaking, listening, and inferential reading of the English language is highly recommended. U.S. institutions that have programs in English as a Second Language should review the student's competency in English and possibly offer or require additional English language study during the first academic year of study.

## Academic Standards

It is very difficult to accurately compare and evaluate educational standards in Korea and the United States inasmuch as both systems are going through dramatic changes regarding the competencies of their students at the primary and secondary levels, and the quality of their programs at the postsecondary level. Overall, the Korean primary and secondary system is more rigorous, has a longer academic year, and requires greater time spent on homework than is the case in the United States. However, in Korea, the science laboratories and other facilities are often not as adequate, and the classrooms are crowded.

In Korea, at the primary and secondary levels the difference between the quality of the private and public institutions is rather narrow. At the postsecondary level, however, the difference will be more extensive, and the quality of the faculties and facilities may vary. It may be helpful for the admissions officer to obtain, if possible, the SAECE score for the student seeking admission. This information, however, may be difficult for the student to obtain. A serious student in the United States in an academically rigorous senior high school would have, as a rule, higher overall scholastic competency. Korean senior high school graduates, because of time on subject, on the average will be more competent in theoretical mathematics and science. It is important, however, that these factors be considered only as a general guideline.

The senior high school principal's recommendation and the senior high school academic transcript will usually give background information regarding the student's citizenship.

## Inter-Institutional Cooperation, Networking and Sister College Affiliations

During the last several years a number of relationships between colleges and universities have developed within Korea and with institutions in other countries. For example, intra-country cooperation at the graduate level has taken place between Sogang, Yonsei and Ewha Womans Universities, making possible a cross-registration program. Students in Korea who have participated in programs of this type will have academic records showing that they have taken coursework at another university, but within an integrated program.

A number of U.S. colleges and universities as well as colleges and universities in other countries have established sister college affiliations with colleges and universities in Korea. Examples of this are the sister affiliations of Korea University with Waseda University and Tokyo University in Japan and the University of Washington in the United States. These affiliations involve faculty and student exchanges.

In recent years, a number of U.S. colleges and universities have established joint programs in affiliation with colleges and universities in Korea. There are also private secondary schools established for foreign residents as well as English language centers and programs sponsored by U.S. institutions for U.S. military personnel in Korea. From time to time, Korean students are involved in these programs and their credentials may indicate this attendance. For example, Korea, Ewha Womans, and Yonsei Universities have developed agreements with foreign universities to recognize credits foreign students earn for summer courses offered in Korea. These three institutions have agreements regarding credit acknowledgement with a number of foreign universities primarily in the United States, Japan and Australia and are looking forward to developing agreements with institutions in Europe. These summer courses are approximately six weeks in length with six to nine credit hours, and they are primarily concerned with the Korean language, history and political science. A number of Korean university students will be spending time during the summer vacation period at foreign institutions where their credits earned will be recognized by the Korean institutions.

# Placement Recommendations

## The Role of the National Council on the Evaluation of Foreign Educational Credentials

The placement recommendations that follow have been approved by the National Council on the Evaluation of Foreign Educational Credentials. In order that these recommendations may be of maximum use to admissions officers, the following information on the development of the terminology used in stating the recommendations, along with instructions for their use, is offered by the Council and the World Education Series Committee.

The recommendations deal with all levels of formal education in roughly chronological order up through the highest degree conferred. Recommendations, as developed through discussion and consensus in the Council, are not directives. Rather, they are general guidelines to help admissions officers determine the admissibility and appropriate level of placement of students from the country under study.

The recommendations should be applied flexibly rather than literally. Before applying the recommendations, admissions officers should read the supporting pages in the text and take into account their own institutional policies and practices. For example, a recommendation may be stated as follows: ". . . may be considered for up to 30 semester hours of transfer credit. . ." The implication is that the U.S. institution may consider giving less than or as much as one year of transfer credit, the decision to be based on various factors—the currentness of the applicant's transfer study, applicability of the study to the U.S. curriculum, quality of grades, and the receiving institution's own policies regarding transfer credit. Similarly, the recommendation ". . . may be considered for freshman admission" indicates possible eligibility only; it is not a recommendation that the candidate be admitted. Although consideration for admission at the same level may be recommended for holders of two different kinds of diplomas, use of identical phrasing in the recommendation does not mean that the two diplomas are identical in nature, quality, or in the quantity of education they represent.

In most cases, the Council will not have attempted to make judgments about the quality of individual schools or types of educational programs within the system under study. Quality clues are provided by the author and must be inferred from a careful reading of the text.

Certain phrases used repeatedly in the recommendations have acquired, within Council usage, specific meanings. For example, "through a course-by-course analysis" means that in dealing with transfer credit, each course taken at the foreign institution is to be judged on an individual basis for its transferability to the receiving institution. Another phrase "where technical training is considered appropriate preparation" suggests that the curriculum followed by the candidate is specialized, and this wording is often a hint that within the foreign system the candidate's educational placement options are limited to certain curriculums. However, while the Council is aware of the educational policies of the country under study, the Council's policies are not necessarily set in conformity with that country's policies. Rather, the recommendations reflect U.S. philosophy and structure of education.

In voting on individual recommendations, Council decisions are made by simple majority. Although consistency among volumes is sought, some differences in philosophy and practice may occur from volume to volume.

# Placement Recommendations

## A. Secondary Credentials

| Credential | Entrance Requirements | Length of Study | Gives Access in Country to | Placement Recommendation |
|---|---|---|---|---|
| 1. Junior High School Diploma (*Chung Hakkyo Chorupchung*, p. 15) | Completion of Primary School (*Kukmin Hakkyo Chorupchung*) | 3 years | General or vocational high school, dependent upon passing of qualifying examination. | May be considered for placement in grade 10. |
| 2. General High School Diploma, Air and Correspondence Senior High School Diploma (*Kodung Hakkyo Chorupchung*, pp. 23, 33) | Junior High School Diploma (*Chung Hakkyo Chorupchung*) | 3 years | Tertiary education, dependent upon SAECE test score and academic record. | May be considered for freshman admission. |
| 3. Vocational High School Diploma (*Kodung Hakkyo Chorupchung*, p. 23) | Junior High School Diploma (*Chung Hakkyo Chorupchung*) | 3 years | Tertiary education, dependent upon SAECE test score and academic record. | May be considered for freshman admission where the secondary program followed is considered appropriate preparation. |
| 4. Diploma from Normal School* (p. 62) | Junior High School Diploma (*Chung Hakkyo Chorupchung*) | 3 years | Tertiary education | May be considered for freshman admission. |

* Program discontinued in 1961

## B. Undergraduate Credentials

Students who have completed some coursework for any of the programs listed below may be considered for undergraduate admission with up to a maximum of 30 semester hours of transfer credit for each year, determined through a course-by-course analysis. When length of the program is cited, it refers to the standard length of the program when pursued fulltime. Actual period of attendance may vary.

| | | | | |
|---|---|---|---|---|
| 1. Junior Vocational College Diploma, Open College Diploma, or two year teacher training program (*Chonmun Taehak Choruphchung*, pp. 43, 51, 52, 55, 62, 67) | High School Diploma (general, vocational, or air and correspondence), based on SAECE score and quality of high school academic record. | 2 years | Junior year of Open College determined by individual Open College examination. Sophomore year level up to junior year level determined through a course-by-course analysis at a four-year college or university. | May be considered for undergraduate admission with up to two years of transfer credit determined through a course-by-course analysis. |
| 2. Nursing Diploma from a Junior Vocational College (*Sillop Chongmun Taehak Kanhokwa Choruphchung*, pp. 50, 67, 69–70) | High School Diploma (general, vocational, or air and correspondence), based on SAECE score and quality of high school academic record. | 3 years | May receive up to three years' college credit towards bachelor's degree in nursing at a four-year college or university. | May be considered for undergraduate admission and placement as are graduates of U.S. hospital schools of nursing. |
| 3. Diplomas from Miscellaneous Schools (*Kakchong Hakkyo Choruphchung*, pp. 43, 51, 73, 82) | | | | |
| a. Fine Arts | High School Diploma (general, vocational, or air and correspondence), based on SAECE score and quality of high school academic record. | 4 years | employment | May be considered for admission and placement as are students from fine arts schools in the U.S. |

| Credential | Entrance Requirements | Length of Study | Gives Access in Country to | Placement Recommendation |
|---|---|---|---|---|
| b. Physical Education (Judo)† | High School Diploma (general, vocational or air and correspondence), based on SAECE score and quality of high school academic record. | 4 years | employment | Primarily a professional qualification. |
| c. Social Welfare | High School Diploma (general, vocational or air and correspondence), based on SAECE score and quality of high school academic record. | 4 years | employment | May be considered for graduate admission where the program followed is considered appropriate preparation. |
| d. Theology | High School Diploma (general, vocational or air and correspondence), based on SAECE score and quality of high school academic record. | 4 years | employment | May be considered for graduate admission where the program followed is considered appropriate preparation. |
| 4. Bachelor's Degree (*Haksa*, pp. 43, 46, 52, 73–74) | High School Diploma (general, vocational, or air and correspondence), based on score of SAECE and quality of high school academic record. | 4 years | Graduate study | May be considered for graduate admission. |

† Information on the content of this program was not available to the Council so no more detailed recommendation could be made.

| | | | | |
|---|---|---|---|---|
| 5. | Bachelor's Degree from a teacher's college (*Haksa*, pp. 63–64, 65) | High School Diploma (general, vocational, or air and correspondence), based on score of SAECE and quality of high school academic record. | 4 years | Graduate study | May be considered for graduate admission. |
| 6. | Bachelor's Degree from Korea Correspondence University (Air and Radio Correspondence College) (*Haksa*, p. 43, 46, 52–55) | High School Diploma (general, vocational, or air and correspondence), based on score of SAECE and quality of high school academic record. | 5 years to 10 years of part-time study | Graduate study | May be considered for graduate admission. |
| 7. | Bachelor's Degree from a college or university (*Haksa*, pp. 43, 46, 56, 67, 69, 71, 73) | High School Diploma (general, vocational, or air and correspondence), based on score of SAECE and quality of high school academic record. | 4 years | Graduate study | May be considered for graduate admission. |
| 8. | Document signifying completion (p. 44) | High School Diploma (general, vocational, or air and correspondence), based on score of SAECE and quality of high school academic record. | 4 years | Graduate study | May be considered comparable to a bachelor's degree. |
| 9. | B.S. in Medicine (*Haksa*), B.S. in Dentistry (*Haksa*, pp. 56, 71–72) | High School Diploma (general, vocational, or air and correspondence), based on score of SAECE and quality of high school academic record. | 6 years | Graduate study | May be considered to have a first professional degree in the field in the U.S.; may be considered for graduate admission. |

| Credential | Entrance Requirements | Length of Study | Gives Access in Country to | Placement Recommendation |
|---|---|---|---|---|
| 10. B.S. in Oriental Medicine (*Haksa*, pp. 56, 71) | High School Diploma (general, vocational, or air and correspondence), based on score of SAECE and quality of high school academic record. | 6 years | Graduate study | May be considered for graduate admission. |

### C. Graduate Credentials

| Credential | Entrance Requirements | Length of Study | Gives Access in Country to | Placement Recommendation |
|---|---|---|---|---|
| 1. Master's Degree (*Suksa*, pp. 57, 59, 73–74) | Bachelor's Degree | 2 years | Further graduate studies | May be considered to have a degree comparable to a master's degree in the U.S. |
| 2. Ph.D. Degree (*Paksa*, pp. 59, 73, 74–75) | Master's Degree | 3 or more years | Post-doctoral studies | May be considered to have a degree comparable to an earned doctorate in the U.S. |

# Appendix A

## Institutions of Higher Education in Korea

The following alphabetical listing of institutions of higher education in Korea includes, where available, the name of the institution in Korean (given in parentheses following the Anglicized name of the institution), the address, date of establishment, source of control, and 1983-84 enrollment figures. Information on four-year open colleges and four-year colleges and universities includes departments that offer bachelor's degrees. In addition, the list of four-year colleges and universities includes information on colleges within the universities, and graduate schools and the degrees they offer. Programs of study are provided for public and private junior vocational colleges, for national four-year teacher's colleges, and for public and private miscellaneous tertiary schools, with the latter also listing the diploma awarded.

The following symbols precede each institutional listing and will be used to identify the type of institution:

- (A) Public and Private Four-Year Colleges and Universities;
- (B) National Four-Year Teacher's Colleges;
- (C) Public and Private Junior Vocational Colleges;
- (D) Public and Private Miscellaneous Tertiary Schools;
- (E) Four–Year Open Colleges;
- (F) Korea Correspondence University.

The list was developed from primary data gathered from across Korea. Military academies are not included in this list.

1. **(C) Agricultural Co-operative Federation, Junior College** (Nongop Hyopdong Chohap Chonmun Dae Hak), San 38-27, Wondang-ri, Wondang-up, Koyang-gun, Kyonggi-do, 100-63, Korea. *Established:* 1979. *Institution:* Junior Vocational. *Control:* Private. *Enrollment:* 200.

   Program: Agriculture.

2. **(A) A-Jou University** (A-Ju Dae Hak Kyo), San 5, Wonchon-dong, Suwon, Kyonggi-do, 170, Korea. *Established:* 1973. *Institution:* 4-year. *Control:* Private. *Enrollment:* Undergraduate, 6136; Graduate, 80.

   Colleges within the University: 1) Business Administration, 2) Engineering, 3) Liberal Arts and Social Science.

   Departments that offer bachelor's degrees: 1) Business Administration, 2) Economics, 3) Engineering, 4) Literature.

   | Graduate Schools: | Master's | Doctor's |
   |---|---|---|
   | General | x | x |

3. **(A) Andong National College** (Andong Dae Hak), Myongryun-dong, Andong, Kyongbuk, 660, Korea. *Established:* 1979. Institution: 4-year. *Control:* National. *Enrollment:* Undergraduate, 3068.

Departments that offer bachelor's degrees: 1) Administrative Science, 2) Business Administration, 3) Commerce, 4) Education, 5) Fine Arts, 6) Gymnastics, 7) Home Economics, 8) Literature, 9) Music.

4. **(C) Andong Nursing Junior College** (Andong Kanho Chonmun Dae Hak), 470, Pukmun-dong, Andong-shi, Kyong Puk, 660, Korea. *Established:* 1979. *Institution:* Junior Vocational. *Control:* Private. *Enrollment:* 360.

   Programs: 1) Health Careers, 2) Nursing.

5. **(C) Anseong Agricultural Junior College** (Ansong Nongop Chonmun Dae Hak), 67, Sokjong-dong, Ansong-up, Ansong-gun, Kyonggi-do, 180-20, Korea. *Previous Name:* Anseong Agricultural School (Ansong Nongop Hak Kyo). *Established:* 1930. *Institution:* Junior Vocational. *Control:* National. *Enrollment:* 960.

   Programs: 1) Commerce, 2) Communication, 3) Engineering/Technical, 4) Home Economics.

6. **(C) Anseong Agricultural School** (*See also* No. 5, Anseong Agricultural Junior College). *Institution:* Junior Vocational.

7. **(C) Anyang Technical Junior College** (Anyang Kongop Chonmun Dae Hak), 39-1, Anyang-dong, Anyang-shi, Kyonggi-do, 171, Korea. *Previous Name:* Yeongseong School (Yongsong Hak Won). *Established:* 1976. *Institution:* Junior Vocational. *Control:* Private. *Enrollment:* 2640.

   Programs: 1) Commerce, 2) Education, 3) Engineering/Technical.

8. **(A) Asian Union Theological College** (Asia Yonhab Shinhak), San 15-1, Ashinri, Okchunmyun, Yangpyongkun, Kyonggi-do, 130-84, Korea. *Established:* 1974. *Institution:* 4-year. *Control:* Private. *Enrollment:* Undergraduate, 364.

   Department that offers a bachelor's degree: Theology.

9. **(C) Baesong Vocational Junior College** (*See also* No. 293, Seo-il Technical Junior College). *Institution:* Junior Vocational.

10. **(C) Baewha Women's Junior College** (Paewha Yoja Chonmun Dae Hak), 12, Pilun-dong, Chongno-gu, Seoul, 110, Korea. *Established:* 1977. *Institution:* Junior Vocational. *Control:* Private. *Enrollment:* 960.

    Programs: 1) Commerce, 2) Education, 3) Engineering/Technical, 4) Liberal Arts, 5) Textiles.

11. **(D) Boy's Orphanage Technical School** (Sonyonui Chip Kongop Chonmun Hak Kyo), 7-62, Amnam-dong, So-gu, Pusan, 600, Korea. *Established:* N.A. *Institution:* Specialized Institution—Miscellaneous. *Control:* Private. *Enrollment:* N.A. Discontinued operation in 1983.

    Diploma Program: Technical.

12. **(C) Bucheon Technical Junior College** (Puchon Kongop Chonmun Dae Hak), 424, Shimgok-dong, Puchon-shi, Kyonggi-do, 150-71, Korea. *Established:* 1978. *Institution:* Junior Vocational. *Control:* Private. *Enrollment:* 2560.

    Programs: 1) Commerce, 2) Design, 3) Education, 4) Engineering/Technical, 5) Home Economics.

13. **(C) Busan Junior College of Business and Commerce** (Pusan Kyong-Sang Chonmun Dae Hak), 73-2, Yonsan 8-dong, Tonglae-gu, Pusan, 607, Korea. *Established:* 1979. *Institution:* Junior Vocational. *Control:* Private. *Enrollment:* 2000.

    Programs: 1) Commerce, 2) Design, 3) Education, 4) Fine Arts.

14. **(B)** **Busan National Teacher's College** (Pusan Kyoyuk Dae Hak), 263, Koje-dong, Tongnae-gu, Pusan, 607, Korea. *Previous Name:* Busan National Teacher's School (Busan Sabum Hak Kyo). *Established:* 1946. *Institution:* 4-year teacher's college. *Control:* National. *Enrollment:* 1701.

Program: Primary Education.

15. **(B)** **Busan National Teacher's School** (*See also* No. 14, Busan National Teacher's College). *Institution:* 4-year teacher's college.

16. **(A)** **Busan National University** (Pusan Dae Hak Kyo), 30, Changjon-dong, Tongnae-gu, Pusan, 607, Korea. *Established:* 1946. *Institution:* 4-year. *Control:* National. *Enrollment:* Undergraduate, 22,324; Graduate, 2335.

Colleges within the University: 1) Business Administration, 2) Education, 3) Engineering, 4) Home Economics, 5) Law and Political Science, 6) Liberal Arts and Science, 7) Medicine, 8) Pharmacy.

Departments that offer bachelor's degrees: 1) Administrative Science, 2) Business Administration, 3) Commerce, 4) Dentistry, 5) Economics, 6) Education, 7) Engineering, 8) Fine Arts, 9) Home Economics, 10) Jurisprudence, 11) Literature, 12) Medicine, 13) Music, 14) Nursing, 15) Pharmacology, 16) Physical Science, 17) Political Science.

| Graduate Schools: | Master's | Doctor's |
| --- | --- | --- |
| General | x | x |
| Business Administration | x | – |
| Education | x | – |
| Industry | x | – |
| Public Administration | x | – |

17. **(C)** **Busan Technical Junior College** (Pusan Kongop Chonmun Dae Hak), 1268, Taeyon-dong, Nam-gu, Pusan, 608, Korea. *Established:* 1973. *Institution:* Junior Vocational. *Control:* National. *Enrollment:* 2640.

Programs: 1) Design, 2) Engineering/Technical.

18. **(E)** **Busan Technical Open College** (Pusan Kongop Kaebang Dae Hak), 1268 Taeyon-dong, Nam-gu, Pusan, 608, Korea. *Established:* 1983. *Institution:* 4-year Open. *Control:* National. *Enrollment:* N.A.

Department that offers a bachelor's degree: Engineering.

19. **(D)** **Busan Theological Seminary** (Pusan Shin Hak Kyo), 316-3, Taeyon-dong, Nam-gu, Pusan, 601-01, Korea. *Established:* 1967. *Institution:* Specialized Institution— Miscellaneous. *Control:* Private. *Enrollment:* Day—330; Evening—240.

Diploma Program: Theology.

20. **(C)** **Busan Women's Junior College** (Pusan Yoja Chonmun Dae Hak), 74, Yangjong-dong, Pusanjin-gu, Pusan, 607, Korea. *Established:* 1969. *Institution:* Junior Vocational. *Control:* Private. *Enrollment:* 2720.

Programs: 1) Commerce, 2) Education, 3) Fine Arts, 4) Engineering/Technical, 5) Home Economics, 6) Liberal Arts, 7) Music, 8) Tourism.

21. **(A)** **Catholic College** (Catholic Dae Hak), 505, Panpo-dong, Kangnam-gu, Seoul, 135, Korea. *Established:* 1947. *Institution:* 4-year. *Control:* Private. *Enrollment:* Undergraduate, 1088; Graduate, 299.

Departments that offer bachelor's degrees: 1) Dentistry, 2) Medicine, 3) Nursing, 4) Theology.

Graduate Schools:                                      Master's Doctor's

General                                                        x        x

22. **(D) Central Theological Seminary** (Chunbu Shin Hak Kyo), Kumsan, Shungnam, Korea. *Established:* 1984. *Institution:* Specialized Institution—Miscellaneous. *Control:* Private. *Enrollment:* 40 est.

Diploma Program: Theology.

23. **(C) Changweon Junior College of Business and Commerce** (Changwon Kyong-Sang Chonmun Dae Hak), 196, Tudae-dong, Changwon-shi, Kyong Nam, 615, Korea. *Established:* 1978. *Institution:* Junior Vocational. *Control:* Private. *Enrollment:* 2560.

Programs: 1) Commerce, 2) Education, 3) Engineering/Technical, 4) Home Economics, 5) Liberal Arts, 6) Textiles, 7) Tourism.

24. **(A) Changwon College** (*See also* No. 245, Kyungnam University). *Institution:* 4-year.

25. **(A) Changwon National University** (Changwon Dae Hak), Toechon-dong, Changwon, Kyongnam, 615, Korea. *Established:* 1969. *Institution:* 4-year. *Control:* National. *Enrollment:* Undergraduate: 3484.

Departments that offer bachelor's degrees: 1) Administrative Science, 2) Business Administration, 3) Commerce, 4) Economics, 5) Engineering, 6) Fine Arts, 7) Home Economics, 8) Jurisprudence, 9) Literature, 10) Music.

26. **(B) Cheju National Teacher's School** (*See also* No. 171, Jeju National Teacher's College). *Established:* 4-year teacher's college.

27. **(C) Cheonan Technical Junior College** (Chonan Kongop Chonmun Dae Hak), 275-1, Pudae-dong, Chonan-shi, Chung Nam, 330, Korea. *Established:* 1972. *Institution:* Junior Vocational. *Control:* National. *Enrollment:* 600.

Program: Engineering/Technical.

28. **(C) Cheongju Junior College** (Chongju Chonmun Dae Hak), San 53, Sanchang-dong, Chongju-shi, Chung Puk, 310, Korea. *Established:* 1982. *Institution:* Junior Vocational. *Control:* National. *Enrollment:* 440.

Programs: 1) Education, 2) Nursing.

29. **(B) Cheong Ju National Teacher's College** (Chong Ju Kyoyuk Dae Hak), 135, Sugok-dong, Chongju-shi, Chung Puk, 310, Korea. *Established:* 1962. *Institution:* 4-year teacher's college. *Control:* National. *Enrollment:* 804.

Department that offers a bachelor's degree: Primary Education.

30. **(A) Cheong Ju University of Education** (Chongju Sabom Dae Hak), Mochung-dong, Chongju, Chungbuk 310, Korea. *Previous Names:* Cheongju Women's College (Chongju Yoja Dae Hak), Cheongju Teacher's College (Chongju Sabom Dae Hak). *Established:* 1969. *Institution:* 4-year. *Control:* Private. *Enrollment:* Undergraduate, 4212.

Department that offers a bachelor's degree: Education.

31. **(A) Cheongju University** (Chongju Dae Hak Kyo), 36, Naedok-dong, Chongju-shi, Chungbuk, 310, Korea. *Established:* 1946. *Institution:* 4-year. *Control:* Private. *Enrollment:* Undergraduate, 10,504; Graduate, 335.

Colleges within the University: 1) Arts, 2) Economics and Commerce, 3) Education, 4) Law and Political Science, 5) Liberal Arts, 6) Science and Engineering.

Departments that offer bachelor's degrees: 1) Administrative Science, 2) Business Administration, 3) Commerce, 4) Economics, 5) Education, 6) Engineering, 7) Fine Arts, 8) Jurisprudence, 9) Library Science, 10) Literature, 11) Political Science.

| Graduate Schools: | Master's | Doctor's |
|---|---|---|
| General | x | x |
| Industry | x | – |

32. (A) **Cheongju Women's College** (*See also* No. 30, Cheong Ju University of Education). *Institution:* 4-year.

33. (A) **Cheonwon Technical Junior College** (*See also* No. 152, Hoseo College). *Institution:* 4-year.

34. (A) **Chinju Junior Agricultural College** (*See also* No. 128, Gyeong Sang National University). *Institution:* 4-year.

35. (B) **Chinju National Teacher's School** (*See also* No. 183, Jinju National Teacher's College). *Institution:* 4-year teacher's college.

36. (B) **Chon Ju National Teacher's College** (Chon Ju Kyoyuk Dae Hak), 128, Tongsohak-dong, Chonju-shi, Chon Puk, 520, Korea. *Previous Name:* Chonju Teacher's School (Chonju Sabum Hak Kyo). *Established:* 1936. *Institution:* 4-year teacher's college. *Control:* National. *Enrollment:* 800.

Department that offers a bachelor's degree: Primary Education.

37. (C) **Chonbuk University, Nursing Junior College** (Chonbuk Dae Hak Kyo Uikwa Dae Hak Pusok Kanho Chonmun Dae Hak), 13, Kyongwon-dong, 3 ka, Chonju-shi, Chon Puk, 520, Korea. Attached to College of Medicine. *Established:* 1974. *Institution:* Junior Vocational. *Control:* National. *Enrollment:* 240.

Program: Nursing.

38. (A) **Chong Shin College and Seminary** (Chong Shin Dae Hak), San 31-3, Sadang-dong, Tongjak-gu, Seoul, 151, Korea. *Established:* 1955. *Institution:* 4-year. *Control:* Private. *Religious Affiliation:* Presbyterian. *Enrollment:* Undergraduate, 1456; Graduate, 220.

Departments that offer bachelor's degrees: 1) Education, 2) Music, 3) Theology.

| Graduate Schools: | Master's | Doctor's |
|---|---|---|
| General | x | x |
| Theology | x | – |

39. (A) **Chongju Junior Agricultural College** (*See also* No. 61, Chung Buk National University). *Institution:* 4-year.

40. (D) **Chonju Hanil Women's Theological Seminary** (Chonju Hanil Yoja Shin Hak Kyo), 155, 1-ka, Chunghwasan-dong, Chonju-shi, Chon Puk, 520, Korea. *Established:* 1973. *Institution:* Specialized Institution—Miscellaneous. *Control:* Private. *Enrollment:* Day—520.

Diploma Program: Theology.

41. (B) **Chonju National Teacher's School** (*See also* No. 36, Chon Ju National Teacher's College). *Institution:* 4-year teacher's college.

42. (A) **Chonnam National University** (Chonnam Dae Hak Kyo), Puk-gu, Kwangju, Chonnam, 500-05, Korea. *Established:* 1952. *Institution:* 4-year. *Control:* National. *Enrollment:* Undergraduate, 22,372; Graduate, 2125.

Colleges within the University: 1) Agriculture, 2) Arts, 3) Business Administration, 4) Dentistry, 5) Education, 6) Engineering, 7) Law and Public Administration, 8) Liberal Arts and Social Science, 9) Medicine, 10) Pharmacy, 11) Science.

Departments that offer bachelor's degrees: 1) Administrative Science, 2) Agriculture Science, 3) Business Administration, 4) Commerce, 5) Dentistry, 6) Economics, 7) Education, 8) Engineering, 9) Fine Arts, 10) Home Economics, 11) Jurisprudence, 12) Library Science, 13) Literature, 14) Medicine, 15) Pharmacology, 16) Physical Science, 17) Political Science, 18) Veterinary Medicine.

| Graduate Schools: | Master's | Doctor's |
| --- | --- | --- |
| General | x | x |
| Business Administration | x | – |
| Education | x | – |
| Public Administration | x | – |

43. (C) **Chonnam University, Nursing Junior College** (Chonnam Dae Hak Uikwa Dae Hak Pusok Kanho Chonmun Dae Hak), 8, Hak 1-dong, Tong-gu, Kwangju-shi, Chon Nam, 500, Korea. Attached to College of Medicine. *Established:* 1962. *Institution:* Junior Vocational. *Control:* National. *Enrollment:* 160.

Program: Nursing.

44. (A) **Chonpook National University** (Chonbuk Dae Hak Kyo), 2-ka, Tokjin-dong, Chonju, Chonbuk, 520, Korea. *Established:* 1951. *Institution:* 4-year. *Control:* National. *Enrollment:* Undergraduate, 18,664; Graduate, 1565.

Colleges within the University: 1) Agriculture, 2) Business Administration, 3) Dentistry, 4) Education, 5) Engineering, 6) Law and Political Science, 7) Liberal Arts, 8) Medicine, 9) Science.

Departments that offer bachelor's degrees: 1) Administrative Science, 2) Agriculture Science, 3) Business Administration, 4) Commerce, 5) Dentistry, 6) Economics, 7) Education, 8) Engineering, 9) Jurisprudence, 10) Library Science, 11) Literature, 12) Medicine, 13) Political Science, 14) Veterinary Medicine.

| Graduate Schools: | Master's | Doctor's |
| --- | --- | --- |
| General | x | x |
| Business Administration | x | – |
| Education | x | – |
| Public Administration | x | – |

45. (C) **Choongju Technical Junior College** (Chungju Kongop Chonmun Dae Hak), 123, Komdan-ri, Iryu-myon, Chungwon-gun, Chung Puk, 320-79, Korea. *Previous Name:* Choongju Technical Junior (Chungju Kongop Chogup Dae Hak). *Established:* 1962. *Institution:* Junior Vocational. *Control:* National. *Enrollment:* 2000.

Programs: 1) Commerce, 2) Engineering/Technical.

46. (A) **Cho-sun Political Science School** (*See also* No. 197, Kon Kuk University). *Institution:* 4-year.

47. (A) **Chosun Theological Seminary** (*See also* No. 140, Han Shin University). *Institution:* 4-year.

48. (A) **Chosun University** (Choson Dae Hak Kyo), Sosok-dong, Tong-gu, Kwangju, Chonnam, 500, Korea. *Established:* 1946. *Institution:* 4-year. *Control:* Private. *Enrollment:* Undergraduate, 20,800; Graduate, 1270.

Colleges within the University: 1) Arts, 2) Dentistry, 3) Economics and Commerce, 4) Education, 5) Engineering, 6) Foreign Languages, 7) Gymnastics, 8) Law and Political Science, 9) Liberal Arts and Science, 10) Medicine, 11) Pharmacology, 12) Women's Industry.

Departments that offer bachelor's degrees: 1) Administrative Science, 2) Business Administration, 3) Commerce, 4) Dentistry, 5) Economics, 6) Education, 7) Engineering, 8) Fine Arts, 9) Gymnastics, 10) Jurisprudence, 11) Literature, 12) Medicine, 13) Nursing, 14) Pharmacology, 15) Physical Science, 16) Political Science.

| Graduate Schools: | Master's | Doctor's |
|---|---|---|
| General | x | x |
| Education | x | – |
| Industry | x | – |

49. (C) **Chosun University, Nursing Junior College** (Choson Dae Hak Kyo Pyongsol Kanho Chonmun Dae Hak), 280, Sosok-dong, Kwangju-shi, Chon Nam, 400, Korea. *Established:* 1971. *Institution:* Junior Vocational. *Control:* Private. *Enrollment:* 480.

Program: Nursing.

50. (C) **Chosun University, Technical Junior College** (Choson Dae Hak Kyo Pyongsol Kongop Chonmun Dae Hak), 306, Sosok 2-dong, Kwangju-shi, Chon Nam, 500, Korea. *Established:* 1965. *Institution:* Junior Vocational. *Control:* Private. *Enrollment:* 4000.

Programs: 1) Commerce, 2) Engineering/Technical.

51. (A) **Choyang Teacher Training College** (*See also* No. 239, Kyonggi University). *Institution:* 4-year.

52. (C) **Christian Hospital, Nursing Junior College** (Kidok Pyongwon Kanho Chonmun Dae Hak), 67, Yanglim-dong, So-gu, Kwangju-shi, Chon Nam, 500, Korea. *Previous Name:* Supia Nursing Junior College (Supia Kanho Chonmun Hak Kyo). *Established:* 1973. *Institution:* Junior Vocational. *Control:* Private. *Enrollment:* 280.

Program: Nursing.

53. (D) **Chugye School of Arts** (Chugye Yesul Hak Kyo), 190-1, Puk-a-hyon-dong, Sodaemun-gu, Seoul, 120, Korea. *Established:* 1973. *Institution:* Specialized Institution—Miscellaneous. *Control:* Private. *Enrollment:* Day—1050. *Ministry of Education:* Accreditation.

Diploma Program: Fine Arts.

54. (B) **Chuncheon National Teacher's College** (Chunchon Kyoyuk Dae Hak), 339, Soksa-dong, Chunchon-shi, Kang-won-do, 200, Korea. *Previous Name:* Chuncheon Teacher's School (Chunchon Sabum Hak Kyo). *Established:* 1939. *Institution:* 4-year teacher's college. *Control:* National. *Enrollment:* 807.

Department that offers a bachelor's degree: Primary Education.

55. (B) **Chuncheon National Teacher's School** (*See also* No. 54, Chuncheon National Teacher's College). *Institution:* 4-year teacher's college.

56. (A) **Chunchon Agricultural College** (*See also* No. 189, Kangweon National University). *Institution:* 4-year.

57. (C) **Chunchon Nursing Junior College** (Chunchon Kanho Chonmun Dae Hak), 34, Chung-ang-ro, 3 ka, Chunchon-shi, Kangwon-do, 200, Korea. *Established:* 1939. *Institution:* Junior Vocational. *Control:* Private. *Enrollment:* 440.

Programs: 1) Engineering/Technical, 2) Health Careers, 3) Nursing.

**58. (A) Chung Ang Buddhist College** (*See also* No. 97, Dong Guk University). *Institution:* 4-year.

**59. (A) Chung-ang University** (Chung-ang Dae Hak Kyo), Huksok-dong, Kwan-ak-gu, Seoul, 151, Korea. *Previous Name:* Chung-ang Women's Junior College (Chung-ang Yoja Chonmun Hak Kyo). *Established:* 1918. *Institution:* 4-year. *Control:* Private. *Enrollment:* Undergraduate, 19,724; Graduate, 2290.

> Colleges within the University: 1) Agriculture, 2) Arts, 3) Business Administration, 4) Economics and Political Science, 5) Education, 6) Engineering, 7) Foreign Languages, 8) Home Economics, 9) Law, 10) Liberal Arts, 11) Medicine, 12) Pharmacology, 13) Social Science.

> Departments that offer bachelor's degrees: 1) Administrative Science, 2) Agriculture Science, 3) Business Administration, 4) Commerce, 5) Economics, 6) Education, 7) Engineering, 8) Fine Arts, 9) Home Economics, 10) Jurisprudence, 11) Library Science, 12) Literature, 13) Medicine, 14) Pharmacology, 15) Physical Science, 16) Political Science.

| Graduate Schools | Master's | Doctor's |
| --- | --- | --- |
| General | x | x |
| Education | x | – |
| International Business Administration | x | – |
| Mass Communication | x | – |
| Social Development | x | – |

**60. (A) Chung-ang Women's Junior College** (*See also* No. 59, Chung-ang University). *Institution:* 4-year.

**61. (A) Chung Buk National University** (Chungbuk Dae Hak Kyo), 48 Kaeshin-dong, Chongju, Chungbuk, 310, Korea. *Previous Name:* Chongju Junior Agricultural College (Chongju Cho gup Dae Hak). *Established:* 1951. *Institution:* 4-year. *Control:* National. *Enrollment:* Undergraduate, 14,040; Graduate, 593.

> Colleges within the University: 1) Agriculture, 2) Education, 3) Engineering, 4) Liberal Arts, 5) Pharmacy, 6) Science, 7) Social Science.

> Departments that offer bachelor's degrees: 1) Administrative Science, 2) Agriculture Science, 3) Business Administration, 4) Commerce, 5) Economics, 6) Education, 7) Engineering, 8) Jurisprudence, 9) Literature, 10) Pharmacology, 11) Physical Science, 12) Political Science.

| Graduate Schools | Master's | Doctor's |
| --- | --- | --- |
| General | x | x |
| Education | x | – |

**62. (C) Chungcheong Vocational Junior College** (Chungchong Shilop Chonmun Dae Hak), 330, Wolgok-ri, Kangnae-myon, Chongwon-gun, Chung Puk, 320-23, Korea. *Established:* N.A. *Institution:* Junior Vocational. *Control:* Private. *Enrollment:* 1040.

> Programs: 1) Commerce, 2) Education, 3) Design, 4) Engineering/Technical, 5) Liberal Arts, 6) Textiles, 7) Tourism.

**63. (C) Chung-Nam Junior College of Business and Commerce** (Chung-Nam Kyong-Sang Chonmun Dae Hak), San 11, Chang-dong, Chung-gu, Taejon-shi, Chung Nam, 300-31, Korea. *Established:* 1980. *Institution:* Junior Vocational. *Control:* Private. *Enrollment:* 1200.

> Programs: 1) Commerce, 2) Education.

**64. (A) Chungnam National University** (Chungnam Dae Hak Kyo), Munwha-dong, Chung-gu, Taejon-shi, Chungnam 300-01, Korea. *Established:* 1952. *Institution:* 4-year. *Control:* National. *Enrollment:* Undergraduate, 19,216; Graduate, 1,688.

Colleges within the University: 1) Economics and Commerce, 2) Law, 3) Liberal Arts, 4) Medicine, 5) Science, 6) Technical Education.

Departments that offer bachelor's degrees: 1) Administrative Science, 2) Agriculture Science, 3) Business Administration, 4) Commerce, 5) Economics, 6) Education, 7) Engineering, 8) Fine Arts, 9) Jurisprudence, 10) Library Science, 11) Literature, 12) Medicine, 13) Music, 14) Nursing, 15) Pharmacology, 16) Physical Science, 17) Veterinary Medicine.

| Graduate Schools | Master's | Doctor's |
| --- | --- | --- |
| General | x | x |
| Business Administration | x | – |
| Education | x | – |
| Public Adminstration | x | – |

**65. (C) Chun Hae Nursing Junior College** (Chun Hae Kanho Chonmun Dae Hak), 504-417, Chonpo-dong, Pusanjin-gu, Pusan, 601, Korea. *Established:* 1968. *Institution:* Junior Vocational. *Control:* Private. *Enrollment:* 360.

Program: Nursing.

**66. (D) Chushin Theological Seminary** (Chunshin Shin Hak Kyo), Kuro, Seoul, Korea. *Established:* 1981. *Institution:* Specialized Institution—Miscellaneous. *Control:* Private. *Enrollment:* 120 est.

Diploma Program: Theology.

**67. (C) Daedong Nursing Junior College** (Taedong Kanho Chonmun Dae Hak), San 60-1, Pugok-dong, Tonglae-gu, Pusan, 07, Korea. *Established:* 1971. *Institution:* Junior Vocational. *Control:* Private. *Enrollment:* 369.

Program: Nursing.

**68. (A) Daegu Catholic College** (Daegu Catholic Dae Hak), Pongduk-dong, Nam-gu, Taegu, 634, Korea. *Established:* 1982. *Institution:* 4-year. *Control:* Private. *Enrollment:* Undergraduate, 208.

Department that offers a bachelor's degree: Theology.

**69. (C) Daegu Health College** (Taegu Pokon Chonmun Dae Hak), San 7, Chilgok 1-dong, Puk-gu, Taegu, 630-51, Korea. *Established:* 1971. *Institution:* Junior Vocational. *Control:* Private. *Enrollment:* 2880.

Program: Health Careers.

**70. (B) Daegu National Teacher's College** (Taegu Kyoyuk Dae Hak), 1797-6, Taemyong-dong, Nam-gu, Taegu, 634, Korea. *Established:* 1962. *Institution:* 4-year teacher's college. *Control:* National. *Enrollment:* 1131.

Department that offers a bachelor's degree: Primary Education.

**71. (A) Daegu Oriental Medical College** (Daegu Han-ui-kwa Dae Hak), Ap-ryang-myon, Kyongsan-gun, Kyongbuk, 632-14, Korea. *Established:* 1980. *Institution:* 4-year. *Control:* Private. *Enrollment:* Undergraduate, 1488.

Departments that offer bachelor's degrees: 1) Economics, 2) Literature, 3) Oriental Medicine.

**72. (C) Daegu Technical Junior College** (Taegu Kongop Chonmun Dae Hak), 813, Ponri-dong, So-gu, Taegu, 630-11, Korea. *Previous Name:* Daegu Vocational Junior College (Taegu Shilop Chonmun Dae Hak). *Established:* 1976. *Institution:* Junior Vocational. *Control:* Private. *Enrollment:* 2560.

Programs: 1) Commerce, 2) Design, 3) Education, 4) Engineering/Technical, 5) Home Economics.

73. **(D) Daegu Theological Seminary** (Taegu Shin Hak Kyo), San 332, Taemyong-dong, Nam-gu, Taegu, 634, Korea. *Established:* 1952. *Institution:* Specialized Institution—Miscellaneous. *Control:* Private. *Enrollment:* Day—320.

Diploma Program: Theology.

74. **(A) Daegu University** (Daegu Dae Hak Kyo), Taemyong-dong, Nam-gu, Taegu, 634, Korea. *Previous Names:* Korea Social Work College (Hankuk Sahoe Sa-op Dae Hak), Hansa University (Hansa Dae Hak). *Established:* 1956. *Institution:* 4-year. *Control:* Private. *Enrollment:* Undergraduate, 15,128; Graduate, 378.

Colleges within the University: 1) Arts, 2) Economics and Commerce, 3) Education, 4) Home Economics, 5) Law and Public Administration, 6) Liberal Arts, 7) Science and Engineering, 8) Social Science.

Departments that offer bachelor's degrees: 1) Administrative Science, 2) Business Administration, 3) Commerce, 4) Economics, 5) Education, 6) Engineering, 7) Fine Arts, 8) Home Economics, 9) Jurisprudence, 10) Library Science, 11) Literature, 12) Physical Science.

| Graduate Schools: | Master's | Doctor's |
|---|---|---|
| General | x | x |
| Social Development | x | – |

75. **(C) Daegu Vocational Junior College** (See also No. 72, Daegu Technical Junior College). *Institution:* Junior Vocational.

76. **(C) Daegu Women's Junior College** (Taegu Yoja Chonmun Dae Hak), San 9, Taejon-dong, Puk-gu, Taegu, 630-51, Korea. *Established:* 1978. *Institution:* Junior Vocational. *Control:* Private. *Enrollment:* 1480.

Programs: 1) Education, 2) Engineering/Technical, 3) Fine Arts, 4) Health Careers, 5) Liberal Arts, 6) Nursing, 7) Tourism.

77. **(A) Daegun Theological College** (Daegon Shinhak Dae Hak). So-gu, Kwangju-shi, Chonnam, 505, Korea. *Established:* 1965. *Institution:* 4-year. *Control:* Private. *Religious Affiliation:* Catholic. *Enrollment:* Undergraduate, 364; Graduate, 20.

Department that offers a bachelor's degree: Theology.

| Graduate Schools: | Master's | Doctor's |
|---|---|---|
| General | x | x |

78. **(D) Dae Han Theological Seminary** (Tae Han Shin Hak Kyo), 33-2, Sogye-dong, Yongsan-gu, Seoul, 140, Korea. *Established:* 1952. *Institution:* Specialized Institution—Miscellaneous. *Control:* Private. *Enrollment:* 1920. *Ministry of Education:* Accreditation.

Diploma Program: Theology.

79. **(C) Daeheon Electronic Technical Junior College** (*See also* No. 80, Daeheon Technical Junior College). *Institution:* Junior Vocational.

80. **(C) Daeheon Technical Junior College** (Taehon Kongop Chonmun Dae Hak), 8, Songlim-dong, Tong-gu, Inchon, 160, Korea. *Previous Name:* Daeheon Electronic

Technical Junior College (Taehon Chonja Kongop Chonmun Hak Kyo). *Established:* 1970. *Institution:* Junior Vocational. *Control:* Private. *Enrollment:* 1200.

Programs: 1) Commerce, 2) Communication Electrical Transmission, 3) Education, 4) Engineering/Technical.

81. (C) **Daeil Vocational Junior College** (*See also* No. 229, Kyeongbuk Vocational Junior College). *Institution:* Junior Vocational.

82. (A) **Dae Jeon College** (Dae Jon Dae Hak), Yongun-dong, Tong-gu, Taejon-shi, Chungnam, 300, Korea. *Established:* 1980. *Institution:* 4-year. *Control:* Private. *Enrollment:* Undergraduate, 2600.

Departments that offer bachelor's degrees: 1) Administrative Science, 2) Business Administration, 3) Commerce, 4) Economics, 5) Literature, 6) Oriental Medicine.

83. (C) **Daejon Health Junior College** (Taejon Pokon Chonmun Dae Hak), 77-3, Kayang 2-dong, Tong-gu, Taejon-shi, Chung Nam, 300, Korea. *Established:* 1977. *Institution:* Junior Vocational. *Control:* Private. *Enrollment:* 1320.

Programs: 1) Education, 2) Engineering/Technical, 3) Health Careers, 4) Liberal Arts.

84. (C) **Daejeon Nursing Junior College** (Taejon Kanho Chonmun Dae Hak), 22, Taehung-dong, Chung-gu, Taejon-shi, Chung Nam, 300, Korea. *Established:* 1973. *Institution:* Junior Vocational. *Control:* Private. *Enrollment:* 480.

Program: Nursing.

85. (C) **Daejeon Technical Junior College** (Taejon Kongop Chonmun Dae Hak), 305-3, Samsong 2-dong, Tong-gu, Taejon-shi, Chung Nam, 300, Korea. *Established:* 1979. *Institution:* Junior Vocational. *Control:* National. *Enrollment:* 1000.

Programs: 1) Commerce, 2) Engineering/Technical.

86. (E) **Daejeon Technical Open College** (Taejon Kongop Kaebang Dae Hak), 305-3, Samsong 2-dong, Tong-gu, Taejon-shi, Chung Nam, 300, Korea. *Established:* 1983. *Institution:* 4-year Open. *Control:* National. *Enrollment:* N.A.

Department that offers a bachelor's degree: Engineering/Technical.

87. (C) **Daejeon Vocational Junior College** (Taejon Shilop Chonmun Dae Hak), 226-2, Chayang-dong, Tong-gu, Taejon-shi, Chung Nam, 300, Korea. *Established:* 1963. *Institution:* Junior Vocational. *Control:* Private. *Enrollment:* 3320.

Programs: 1) Commerce, 2) Design, 3) Education, 4) Engineering/Technical, 5) Textiles.

88. (C) **Daelim Industrial Junior College** (Taelim Kongop Chonmun Dae Hak), 526-7, Pisan-dong, Anyang-shi, Kyonggi-do, 171, Korea. *Established:* 1978. *Institution:* Junior Vocational. *Control:* Private. *Enrollment:* 2560.

Programs: 1) Commerce, 2) Liberal Arts.

89. (C) **Dae-yu Industrial Junior College** (Tae-yu Kongop Chonmun Dae Hak), 425, Pokjong-dong, Songnam-shi, Kyonggi-do, 130-14, Korea. *Established:* 1978. *Institution:* Junior Vocational. *Control:* Private. *Enrollment:* 2880.

Programs: 1) Design, 2) Engineering/Technical.

90. (A) **Dankook University** (Dankuk Dae Hak Kyo), San 8-2, Hannam-dong, Yongsan-gu, Seoul, 140, Korea. *Established:* 1947. *Institution:* 4-year. *Control:* Private. *Enrollment:* Undergraduate, 18,668; Graduate, 1610.

Colleges within the University: 1) Economics and Commerce, 2) Education, 3) Engineering, 4) Law and Political Science, 5) Liberal Arts and Science.

Departments that offer bachelor's degrees: 1) Administrative Science, 2) Business Administration, 3) Commerce, 4) Economics, 5) Education, 6) Engineering, 7) Fine Arts, 8) Home Economics, 9) Jurisprudence, 10) Literature, 11) Political Science.

| Graduate Schools: | Master's | Doctor's |
|---|---|---|
| General | x | x |
| Business Administration | x | – |
| Education | x | – |
| Public Administration | x | – |

91. (C) **Dejeon Vocational Junior College** (*See also* No. 187, Junggyeong Technical Junior College). *Institution:* Junior Vocational.

92. (A) **Dong-A University** (Dong-A Dae Hak Kyo), 3-ka, Tongdaeshin-dong, So-gu, Pusan, 600, Korea. *Established:* 1947. *Institution:* 4-year. *Control:* Private. *Enrollment:* Undergraduate, 20,384; Graduate, 1100.

Colleges within the University: 1) Agriculture, 2) Economics and Commerce, 3) Engineering, 4) Gymnastics, 5) Law and Political Science, 6) Liberal Arts, 7) Science.

Departments that offer bachelor's degrees: 1) Administrative Science, 2) Agriculture Science, 3) Business Administration 4) Commerce, 5) Economics, 6) Engineering, 7) Gymnastics, 8) Jurisprudence, 9) Literature, 10) Physical Science, 11) Political Science.

| Graduate Schools: | Master's | Doctor's |
|---|---|---|
| General | x | x |
| Business Administration | x | – |
| Education | x | – |

93. (A) **Dong-Ah Polytechnic Academy** (*See also* No. 145, Hanyang University). *Institution:* 4-year.

94. (A) **Dongduck Women's University** (Dongdok Yoja Dae Hak), Hawolgok-dong, Songbuk-gu, Seoul, 132, Korea. *Established:* 1950. *Institution:* 4-year. *Control:* Private. *Enrollment:* Undergraduate, 3692; Graduate, 40.

Departments that offer bachelor's degrees: 1) Business Administration, 2) Commerce, 3) Education, 4) Fine Arts, 5) Home Economics, 6) Library Science, 7) Literature, 8) Pharmacology.

| Graduate Schools: | Master's | Doctor's |
|---|---|---|
| General | x | – |

95. (C) **Dong-eui Technical Junior College** (Tong-ui Kongop Chonmun Dae Hak), San 72, Yangjong-dong, Pusanjin-gu, Pusan, 601, Korea. *Established:* 1972. *Institution:* Junior Vocational. *Control:* Private. *Enrollment:* 3600.

Programs: 1) Design, 2) Engineering/Technical.

96. (A) **Dong-Eui University** (Dong-Ui Dae Hak Kyo), San 24, Kaya-dong, Pusanjin-gu, Pusan, 601, Korea. *Established:* 1978. *Institution:* 4-year. *Control:* Private. *Enrollment:* Undergraduate, 8008.

Colleges within the University: 1) Arts, 2) Economics and Commerce, 3) Engineering, 4) Law and Public Administration, 5) Liberal Arts, 6) Science.

Departments that offer bachelor's degrees: 1) Administrative Science, 2) Business Administration, 3) Commerce, 4) Economics, 5) Engineering, 6) Fine Arts, 7) Gymnastics, 8) Home Economics, 9) Jurisprudence, 10) Library Science, 11) Literature, 12) Music, 13) Physical Science.

97. (A) **Dong Guk University** (Dong Kuk Dae Hak Kyo), 3-ka, Pil-dong, Chung-gu, Seoul, 100, Korea. *Previous Names:* Myong Jin School (Myong Jin Hak Kyo), Chung Ang Buddhist College (Chung Ang Pulkyo Hak Kyo), Hye Wha College (Hye Wha Chonmun Hak Kyo). *Established:* 1906. *Institution:* 4-year. *Control:* Private. *Religious Affiliation:* Buddhist. *Enrollment:* Undergraduate, 17,492; Graduate, 1501.

Colleges within the University: 1) Agriculture, 2) *Bulkyo* (Buddhistics), 3) Economics and Commerce, 4) Education, 5) Engineering, 6) Law and Political Science, 7) Liberal Arts and Science.

Departments that offer bachelor's degrees: 1) Administrative Science, 2) Agriculture Science, 3) Business Administration, 4) Commerce, 5) Economics, 6) Education, 7) Engineering, 8) Fine Arts, 9) Jurisprudence, 10) Literature, 11) Oriental Medicine, 12) Physical Science, 13) Political Science.

| Graduate Schools: | Master's | Doctor's |
|---|---|---|
| General | x | x |
| Business Administration | x | – |
| Education | x | – |
| Public Administration | x | – |

98. (C) **Dongju Women's Junior College** (Tongju Yoja Chonmun Dae Hak), San 15-l, Koejong-dong, So-gu, Pusan, 600-02, Korea. *Established:* 1976. *Institution:* Junior Vocational. *Control:* Private. *Enrollment:* 2360.

Programs: 1) Commerce, 2) Design, 3) Education, 4) Engineering/Technical, 5) Fine Arts, 6) Liberal Arts, 7) Textiles.

99. (A) **Dongkuk Junior College** (*See also* No. 226, Kwangwoun University). *Institution:* 4-year.

100. (C) **Donglae Women's Junior College** (Tonglae Yoja Chonmun Dae Hak), 640, Pan-song-dong, Haeundae-gu, Pusan, 607-05, Korea. *Established:* 1978. *Institution:* Junior Vocational. *Control:* Private. *Enrollment:* 2320.

Programs: 1) Commerce, 2) Education, 3) Engineering/Technical, 4) Home Economics, 5) Liberal Arts, 6) Textiles, 7) Tourism.

101. (C) **Dongmyeong Culture School** (*See also* No. 105, Dongwon Technical Junior College). *Institution:* Junior Vocational.

102. (C) **Dongnam Health Junior College** (Tongnam Pokon Chonmun Dae Hak), 695-1, Chongja-dong, Suwon-shi, Kyonggi-do, 170, Korea. *Established:* 1973. *Institution:* Junior Vocational. *Control:* Private. *Enrollment:* 1840.

Programs: 1) Education, 2) Engineering/Technical, 3) Health Careers, 4) Nursing.

103. (C) **Dongsan Nursing Junior College** (Tongsan Kanho Chonmun Dae Hak), 194, Tongsan-dong, Chung-gu, Taegu, 630, Korea. *Established:* 1974. *Institution:* Junior Vocational. *Control:* Private. *Religious Affiliation:* Presbyterian. *Enrollment:* 120.

Program: Nursing.

104.(C) **Dongsin Vocational Junior College** (Tongshin Shilop Chonmun Dae Hak), 771, Tuam-dong, Puk-gu, Kwangju-shi, Chon Nam, 500, Korea. *Established:* 1975. *Institution:* Junior Vocational. *Control:* Private. *Enrollment:* 2720.

Programs: 1) Commerce, 2) Design, 3) Education, 4) Engineering/Technical, 5) Health Careers, 6) Tourism.

105.(C) **Dongwon Technical Junior College** (Tongwon Kongop Chonmun Dae Hak), 505, Yongdang-dong, Nam-gu, Pusan, 608, Korea. *Previous Name:* Dongmyeong Culture School (Tongmyong Munwha Hak Kyo). *Established:* 1977. *Institution:* Junior Vocational. *Control:* Private. *Enrollment:* 2640.

Programs: 1) Commerce, 2) Design, 3) Engineering/Technical, 4) Liberal Arts.

106. (C) **Dongwoo Junior College** (Tong-U Chonmun Dae Hak), 244, Nohaksan-dong, Sokcho-shi, Kangwon-do, 210-20, Korea. *Previous Name:* Sokcho Junior College (Sokcho Chonmun Dae Hak). *Established:* 1981. *Institution:* Junior Vocational. *Control:* Private. *Enrollment:* 2490.

Programs: 1) Commerce, 2) Education, 3) Engineering/Technical, 4) Liberal Arts, 5) Nursing, 6) Textiles.

107. (C) **Dongyang Technical Junior College** (Tongyang Kongop Chonmun Dae Hak), 62-160, Kochok-dong, Kuro-gu, Seoul, 150-04, Korea. *Established:* 1965. *Institution:* Junior Vocational. *Control:* Private. *Enrollment:* 2880.

Program: 1) Communication, 2) Engineering/Technical.

108. (A) **Donus Paster Theological College** (*See also* No. 68, Daegu Catholic College). Note: Donus Paster Theological College has been renamed Daegu Catholic College as of March 1985. *Institution:* 4-year.

109. (A) **Duksung Women's College** (Doksong Yoja Dae Hak), Ssangmon-dong, Dobong-gu, Seoul, 110, Korea. *Established:* 1950. *Institution:* 4-year. *Control:* Private. *Enrollment:* Undergraduate, 5200; Graduate, 50.

Departments that offer bachelor's degrees: 1) Business Administration, 2) Commerce, 3) Education, 4) Fine Arts, 5) Home Economics, 6) Library Science, 7) Literature, 8) Pharmacology.

Graduate Schools:                                        Master's Doctor's

General                                                         x        —

110. (A) **Ewha School** (*See also* No. 111, Ewha Womans University). *Institution:* 4-year.

111. (A) **Ewha Womans University** (Ewha Yoja Dae Hak Kyo), 11-1, Tahyon-dong, Sodaemun-gu, Seoul, 120, Korea. *Previous Name:* Ewha School (Ewha Hak Tang). *Established:* 1886. *Institution:* 4-year. *Control:* Private. *Religious Affiliation:* Methodist. *Enrollment:* Undergraduate, 17,540; Graduate, 2160.

Colleges within the University: 1) Education, 2) Fine Arts, 3) Gymnastics, 4) Home Economics, 5) Liberal Arts, 6) Medicine, 7) Music, 8) Nursing, 9) Pharmacology, 10) Science.

Departments that offer bachelor's degrees: 1) Administrative Science, 2) Business Administration, 3) Economics, 4) Education, 5) Fine Arts, 6) Gymnastics, 7) Home Economics, 8) Jurisprudence, 9) Library Science, 10) Literature, 11) Medicine, 12) Music, 13) Nursing, 14) Pharmacology, 15) Physical Science, 16) Political Science.

| Graduate Schools: | Master's | Doctor's |
|---|---|---|
| General | x | x |
| Education | x | – |
| Industrial Arts | x | – |

112. **(D) Full Gospel Theological Seminary** (Sunbokum Shin Hak Kyo), 604-5, Tangjong-ri, Kunpo-up, Shihung-gun, Kyonggi-do, 171, Korea. *Established:* 1982. *Institution:* Specialized Institution—Miscellaneous. *Control:* Private. *Enrollment:* Day—150.

Diploma Program: Theology.

113. **(C) Gaejeong Nursing Junior College** (Kaejong Kanho Chonmun Dae Hak), 413, Kaejong-dong, Kunsan-shi, Chon Puk, 511, Korea. *Established:* 1974. *Institution:* Junior Vocational. *Control:* Private. *Enrollment:* 360.

Program: Nursing.

114. **(D) Gangnam Social Welfare College** (Kangnam Sahoe Pogji Hak Kyo), San 6-2, Kugal-ri, Kihyung-myon, Yongin-gun, Kyonggi-do, 170-73, Korea. *Established:* 1981. *Institution:* Specialized Institution—Miscellaneous. *Control:* Private. *Enrollment:* Day—2770; Evening—570. *Ministry of Education:* Accreditation.

Program: Social Welfare.

115. **(C) Gangreung Nursing Junior College** (Kangrung Kanho Chonmun Dae Hak), 164-1, Nammun-dong, Kangrung-shi, Kangwon-do, 210, Korea. *Established:* 1972. *Institution:* Junior Vocational. *Control:* Private. *Enrollment:* 240.

Program: Nursing.

116. **(C) Gijeon Women's Junior College** (Kijon Yoja Chonmun Dae Hak), 177-1, 1 ka, Chungwasan-dong, Chonju-shi, Chon Puk, 520, Korea. *Established:* 1973. *Institution:* Junior Vocational. *Control:* Private. *Religious Affiliation:* Christian—Non-denominational. *Enrollment:* 1640.

Programs: 1) Commerce, 2) Education, 3) Engineering, 4) Home Economics, 5) Liberal Arts, 6) Textiles.

117. **(C) Gimcheon Health Junior College** (Kimchon Pokon Chonmun Dae Hak), 754, Kumsan-dong, Kimchon-shi, Kyong Puk, 660, Korea. *Established:* 1979. *Institution:* Junior Vocational. *Control:* Private. *Enrollment:* 1440.

Program: Health Careers.

118. **(C) Gimcheon Nursing Junior College** (Kimchon Kanho Chonmun Dae Hak), 480, Samhak-dong, Kimchon-shi, Kyong Puk, 640, Korea. *Established:* 1972. *Institution:* Junior Vocational. *Control:* Private. *Enrollment:* 360.

Program: Nursing.

119. **(D) God's Church Theological Seminary** (Hananimui Kyohoe Shin Hak Kyo), Kuro, Seoul, Korea. *Established:* 1981. *Institution:* Specialized Institution—Miscellaneous. *Control:* Private. *Enrollment:* 120 est.

Diploma Program: Theology.

120. **(C) Golromban Nursing Junior College** (*See also* No. 296, Seongshin Nursing Junior College). *Institution:* Junior Vocational.

121. **(C) Gonju Nursing Junior College** (Kongju Kanho Chonmun Dae Hak), 326, Ok-ryong-dong, Kongju-up, Kongju-gun, Chun Nam, 301, Korea. *Established:* 1963. *Institution:* Junior Vocational. *Control:* National. *Enrollment:* 520.

Programs: 1) Education, 2) Nursing.

122. **(C) Gospel Nursing Junior College** (Pokum Kanho Chonmun Dae Hak), 34, Amnam-dong, So-gu, Pusan, 600, Korea. *Established:* 1968. *Institution:* Junior Vocational. *Control:* Private. *Religious Affiliation:* Gospel. *Enrollment:* 320.

Program: Nursing.

123. **(C) Gunsan Fisheries Junior College** (Kunsan Susan Chonmun Dae Hak), 1044-2, Soryong-dong, Kunsan-shi, Chon Puk, 511, Korea. *Established:* 1962. *Institution:* Junior Vocational. *Control:* National. *Enrollment:* 1120.

Programs: 1) Commerce, 2) Communication, 3) Engineering/Technical.

124. **(A) Gun San National University** (Kun San Dae Hak), 420-1 Tonghungnam-dong, Kunsan, Chonbuk, 511, Korea. *Established:* 1979. *Institution:* 4-year. *Control:* National. *Enrollment:* Undergraduate, 3276.

Departments that offer bachelor's degrees: 1) Administrative Science, 2) Business Administration, 3) Commerce, 4) Fine Arts, 5) Gymnastics, 6) Literature, 7) Music, 8) Physical Science.

125. **(C) Gunsan Technical Junior College** (*See also* No. 292, Seohae Technical Junior College). *Institution:* Junior Vocational.

126. **(C) Gunsan Vocational Junior College** (Kunsan Shilop Chonmun Dae Hak), 832-1, O-ryong-dong, Kunsan-shi, Chon Puk, 511, Korea. *Established:* 1973. *Institution:* Junior Vocational. *Control:* Private. *Enrollment:* 2080.

Programs: 1) Commerce, 2) Education, 3) Engineering/Technical, 4) Health Careers, 5) Tourism.

127. **(A) Gwan Dong College** (Kwan Dong Dae Hak), Naegok-dong, Kangnung-shi, Kangwon-do, 210, Korea. *Established:* 1955. *Institution:* 4-year. *Control:* Private. *Enrollment:* Undergraduate, 5928.

Departments that offer bachelor's degrees: 1) Administrative Science, 2) Business Administration, 3) Commerce, 4) Economics, 5) Education, 6) Engineering, 7) Jurisprudence, 8) Literature.

128. **(A) Gyeong Sang National University** (Kyong Sang Dae Hak Kyo), Chilam-dong, Chinju 620, Kyongnam, Korea. *Previous Name:* Chinju Junior Agricultural College (Chinju Chonmun Nongkwa Dae Hak). *Established:* 1948. *Institution:* 4-year. *Control:* National. *Enrollment:* Undergraduate, 12,800; Graduate, 546.

Colleges within the University: 1) Agriculture, 2) Education, 3) Engineering, 4) Law and Economics, 5) Liberal Arts and Science, 6) Medicine.

Departments that offer bachelor's degrees: 1) Administrative Science, 2) Agriculture Science, 3) Business Administration, 4) Commerce, 5) Economics, 6) Education, 7) Engineering, 8) Jurisprudence, 9) Physical Science, 10) Veterinary Medicine.

| Graduate Schools: | Master's | Doctor's |
|---|---|---|
| General | x | x |
| Education | x | — |

**129. (A) Hae In College** (*See also* No. 245, Kyungnam University). *Institution:* 4-year.

**130. (A) Hallym College** (Hallim Dae Hak), Okchon-dong, Chunchon, Kangwon-do, Korea. *Established:* 1982. *Institution:* 4-year. *Control:* Private. *Enrollment:* Undergraduate, 2168.

Departments that offer bachelor's degrees: 1) Engineering, 2) Literature, 3) Medicine.

**131. (A) Hankook Aviation College** (Hankuk Hanggong Dae Hak), 200-l, Hwajon-ri, Shindo-myon, Koyang-gun, Kyonggi-do, Korea. *Established:* 1952. *Institution:* 4-year. *Control:* Private. *Enrollment:* Undergraduate, 1512; Graduate, 40.

Departments that offer bachelor's degrees: 1) Business Administration, 2) Engineering.

| Graduate Schools: | Master's | Doctor's |
|---|---|---|
| General | x | – |

**132. (A) Hankook University of Foreign Studies** (Hankuk Oe-Kuk-O Dae Hak Kyo), Imun-dong, Tongdaemun-gu, Seoul, 131, Korea. *Established:* 1954. *Institution:* 4-year. *Control:* Private. *Enrollment:* Undergraduate, 14,560; Graduate, 1250.

Colleges within the University: 1) Economics and Commerce, 2) Education, 3) Law and Political Science, 4) Liberal Arts and Science, 5) Oriental Languages, 6) Western Languages.

Departments that offer bachelor's degrees: 1) Administrative Science, 2) Business Administration, 3) Commerce, 4) Economics, 5) Education, 6) Jurisprudence, 7) Literature, 8) Physical Science, 9) Political Science.

| Graduate Schools: | Master's | Doctor's |
|---|---|---|
| General | x | x |
| Education | x | – |
| International Trade | x | – |
| Translation | x | – |

**133. (A) Hankuk Night College** (*See also* No. 168, International University). *Institution:* 4-year.

**134. (A) Hankuk Theological Seminary** (*See also* No. 140, Han Shin University). *Institution:* 4-year.

**135. (A) Han Nam College** (Han Nam Dae Hak), Ojong-dong, Tong-gu, Taejon, Chungnam, 300, Korea. *Previous Names:* Soong Jun University at Taejon (Sungjun Dae Hak Kyo Taejon Campus). *Established:* 1959. *Institution:* 4-year. *Control:* Private. *Enrollment:* Undergraduate, 8892; Graduate: N.A.

Departments that offer bachelor's degrees: 1) Administrative Science, 2) Business Administration, 3) Commerce, 4) Economics, 5) Engineering, 6) Fine Arts, 7) Home Economics, 8) Jurisprudence, 9) Library Science, 10) Literature, 11) Physical Science.

| Graduate Schools: | Master's | Doctor's |
|---|---|---|
| General | x | – |
| Regional Science | x | – |

**136. (A) Han Nam College** (*See also* No. 135, Han Nam College). *Institution:* 4-year.

**137. (A) Hansa University** (*See also* No. 74, Daegu University). *Institution:* 4-year.

**138. (D) Hanseong Theological Seminary** (Hansong Shin Hak Kyo), 21-2, Kajang-dong, Chung-gu, Taejon-shi, Chung Nam, 300-01, Korea. *Established:* 1981. *Institution:* Specialized Institution—Miscellaneous. *Control:* Private. *Enrollment:* Day—250; Evening—60. *Ministry of Education:* Accreditation.

Diploma Program: Theology.

**139. (A) Han Shin College** (*See also* No. 140, Han Shin University). *Institution:* 4-year.

**140. (A) Han Shin University** (Han Shin Dae Hak), San 129, Suyu-dong, Tobong-gu, Seoul, 132, Korea. *Previous Names:* Choson Theological Seminary (Choson Shin-hak Dae Hak), Hankuk Theological Seminary (Hankuk Shinhak Dae Hak), Han Shin College (Han Shin Dae Hak). *Established:* 1947. *Institution:* 4-year. *Control:* Private. *Enrollment:* Undergraduate, 2340; Graduate, 226.

Departments that offer bachelor's degrees: 1) Business Administration, 2) Commerce, 3) Economics, 4) Literature, 5) Theology.

| Graduate Schools: | Master's | Doctor's |
|---|---|---|
| General | x | x |
| Theology | x | – |

**141. (A) Hansung College** (*See also* No. 142, Hansung University). *Institution:* 4-year.

**142. (A) Hansung University** (Hansung Dae Hak), 2-ka, Samson-dong, Songbuk-gu, Seoul, 132, Korea. *Previous Names:* Hansung Women's College (Hansong Yoja Dae Hak), Hansung College (Hansong Dae Hak). *Established:* 1972. *Institution:* 4-year. *Control:* Private. *Enrollment:* Undergraduate, 2704.

Departments that offer bachelor's degrees: 1) Administrative Science, 2) Business Administration, 3) Commerce; 4) Economics, 5) Fine Arts, 6) Gymnastics, 7) Library Science, 8) Literature.

**143. (A) Hansung Women's College** (*See also* No. 142, Hansung University). *Institution:* 4-year.

**144. (A) Hanyang College of Engineering** (*See also* No. 145, Hanyang University). *Institution:* 4-year.

**145. (A) Hanyang University** (Hanyang Dae Hak Kyo), 17, Haeng-dang-dong, Song-dong-gu, Seoul, 133, Korea. *Previous Names:* Dong-Ah Polytechnic Academy (Dong-Ah Kongkwa Hak Won), Konkuk College (Konkuk Kisul Hak Kyo), Hanyang College of Engineering (Hanyang Kongkwa Dae Hak). *Established:* 1939. *Institution:* 4-year. *Control:* Private. *Enrollment:* Undergraduate, 25,168; Graduate, 3000.

Colleges within the University: 1) Economics and Commerce, 2) Education, 3) Engineering, 4) Gymnastics, 5) Home Economics, 6) Law, 7) Liberal Arts, 8) Medicine, 9) Music, 10) Science, 11) Social Science.

Departments that offer bachelor's degrees: 1) Administrative Science, 2) Business Administration, 3) Commerce, 4) Economics, 5) Education, 6) Engineering, 7) Gymnastics, 8) Home Economics, 9) Jurisprudence, 10) Literature, 11) Medicine, 12) Music, 13) Nursing, 14) Physical Science, 15) Political Science.

| Graduate Schools: | Master's | Doctor's |
|---|---|---|
| General | x | x |
| Business Administration | x | – |
| Education | x | – |
| Environmental Studies | x | – |
| Industry | x | – |
| Public Administration | x | – |

**146. (C) Hanyang Women's Junior College** (Hanyang Yoja Chonmun Dae Hak), 17, Haengdang-dong, Songdong-gu, Seoul, 133, Korea. *Established:* 1974. *Institution:* Junior Vocational. *Control:* Private. *Enrollment:* 2800.

Programs: 1) Commerce, 2) Design, 3) Engineering/Technical, 4) Liberal Arts, 5) Textiles, 6) Tourism.

**147. (D) The Holiness Church's Theological Seminary** (Songgyolkyo Shin Hak Kyo), 147, Anyang-dong, Anyang-shi, Kyonggi-do, 171, Korea. *Established:* 1965. *Institution:* Specialized Institution—Miscellaneous. *Control:* Private. *Enrollment:* Day—1180; Evening—800. *Ministry of Education:* Accreditation.

Diploma Program: Theology.

**148. (A) Honam College** (Honam Dae Hak), 148, Ssangchon-dong, So-gu, Kwangju-shi, Chon-Nam, 505, Korea. *Established:* 1981. *Institution:* 4-year. *Control:* Private. *Enrollment:* Undergraduate, 3952.

Departments that offer bachelor's degrees: 1) Administrative Science, 2) Business Administration, 3) Commerce, 4) Economics, 5) Fine Arts, 6) Gymnastics, 7) Home Economics, 8) Jurisprudence, 9) Literature.

**149. (D) Honam Theological Seminary** (Changnohoe Honam Shin Hak Kyo), 108, Yanglim-dong, So-gu, Kwangju-shi, Chon Nam, 500, Korea. *Established:* 1960. *Institution:* Specialized Institution—Miscellaneous. *Control:* Private. *Enrollment:* Day—320; Evening—120. *Ministry of Education:* Accreditation

Diploma Program: Theology.

**150. (C) Hongik Technical Junior College** (Hongik Kongop Chonmun Dae Hak), 72-1, Sangsu-dong, Mapo-gu, Seoul, 121, Korea. *Established:* 1971. *Institution:* Junior Vocational. *Control:* Private. *Enrollment:* 2480.

Programs: 1) Design, 2) Industrial Arts, 3) Tourism.

**151. (A) Hong Ik University** (Hong Ik Dae Hak Kyo), 72-1, Sangsu-dong, Mapo-gu, Seoul, 121, Korea. *Established:* 1948. *Institution:* 4-year. *Control:* Private. *Enrollment:* Undergraduate, 9308; Graduate, 1046.

Colleges within the University: 1) Economics and Commerce, 2) Education, 3) Engineering, 4) Fine Arts, 5) Liberal Arts.

Departments that offer bachelor's degrees: 1) Business Administration, 2) Commerce, 3) Economics, 4) Education, 5) Engineering, 6) Fine Arts, 7) Literature.

| Graduate Schools: | Master's | Doctor's |
| --- | --- | --- |
| General | x | x |
| Education | x | — |
| Environmental Studies | x | — |
| Industrial Arts | x | — |

**152. (A) Hoseo College** (Hoso Dae Hak), San 120-1, Anso-dong, Chonan-shi, Chung Nam, 330, Korea. *Previous Name:* Cheonwon Technical Junior College (Chonwon Kongop Chonmun Dae Hak). *Established:* 1978. *Institution:* 4-year. *Control:* Private. *Enrollment:* Undergraduate: 3016.

Departments that offer bachelor's degrees: 1) Administrative Science, 2) Business Administration, 3) Economics, 4) Engineering, 5) Literature, 6) Music, 7) Theology.

**153. (C) Hyejeon Junior College** (Hyejon Chonmun Dae Hak), San 16, Namjang-ri, Hongsong-up, Chung Nam, 350, Korea. *Established:* 1981. *Institution:* Junior Vocational. *Control:* Private. *Enrollment:* 1520.

Programs: 1) Commerce, 2) Education, 3) Engineering/Technical, 4) Liberal Arts,
5) Textiles, 6) Tourism.

154. (A) **Hye Wha College** (*See also* No. 97, Dong Guk University). *Institution:* 4-year.

155. (A) **Hyop Song Theological School** (*See also* No. 251, Methodist Theological Seminary). *Institution:* 4-year.

156. (A) **Hyosung Women's University** (Hyosong Yoja Dae Hak Kyo), Bongdok-dong, Nam-gu, Taegu, 634, Korea. *Established:* 1952. *Institution:* 4-year. *Control:* Private. *Religious Affiliation:* Catholic. *Enrollment:* Undergraduate, 8720; Graduate, 198.

Colleges within the University: 1) Arts, 2) Education, 3) Home Economics, 4) Law and Economics, 5) Liberal Arts and Science, 6) Pharmacology.

Departments that offer bachelor's degrees: 1) Administrative Science, 2) Business Administration, 3) Commerce, 4) Economics, 5) Education, 6) Fine Arts, 7) Home Economics, 8) Jurisprudence, 9) Library Science, 10) Literature, 11) Music, 12) Pharmacology.

| Graduate Schools: | Master's | Doctor's |
|---|---|---|
| General | x | x |

157. (C) **Incheon Junior College** (Inchon Chonmun Dae Hak), 235, Towha-dong, Nam-gu, Inchon, 160-01, Korea. *Established:* 1969. *Institution:* Junior Vocational. *Control:* Private. *Enrollment:* 5240.

Programs: 1) Commerce, 2) Design, 3) Education, 4) Engineering/Technical, 5) Fine Arts, 6) Health Careers, 7) Liberal Arts.

158. (C) **Incheon Nursing Junior College** (Inchon Kanho Chonmun Dae Hak), 472, Kansok-dong, Nam-gu, Inchon, 160-01, Korea. *Established:* 1972. *Institution:* Junior Vocational. *Control:* Private. *Enrollment:* 1040.

Programs: 1) Education, 2) Engineering/Technical, 3) Health Careers, 4) Nursing.

159. (B) **Incheon National Teacher's College** (Inchon Kyoyuk Dae Hak), 203, Sung-ui-dong, Nam-gu, Inchon, 160-01, Korea. *Established:* 1962. *Institution:* 4-year teacher's college. *Control:* National. *Enrollment:* 1307.

Department that offers a bachelor's degree: Primary Education.

160. (A) **Incheon University** (Inchon Dae Hak), Tohwa-dong, Nam-gu, Inchon, 160-01, Korea. *Established:* 1979. *Institution:* 4-year. *Control:* Private. *Enrollment:* Undergraduate, 6188; Graduate, 150.

Departments that offer bachelor's degrees: 1) Administrative Science, 2) Business Administration, 3) Commerce, 4) Engineering, 5) Fine Arts, 6) Gymnastics, 7) Home Economics, 8) Jurisprudence, 9) Literature, 10) Physical Science, 11) Political Science.

| Graduate Schools: | Master's | Doctor's |
|---|---|---|
| Education | x | — |

161. (C) **Indug Technical Junior College** (Indog Kongop Chonmun Dae Hak), San 76, Wolkye-dong, Tobong-gu, Seoul, 132-01, Korea. *Established:* 1972. *Institution:* Junior Vocational. *Control:* Private. *Enrollment:* 2160.

Programs: 1) Commerce, 2) Design, 3) Engineering/Technical, 4) Liberal Arts.

162. (A) **Inha College of Engineering** (*See also* No. 164, Inha University). *Institution:* 4-year.

**163. (C)** **Inha Industrial Junior College** (Inha Kongop Chonmun Dae Hak), 253, Yonghyon-dong, Nam-gu, Inchon, 160, Korea. *Established:* 1958. *Institution:* Junior Vocational. *Control:* Private. *Enrollment:* 3920.

Program: 1) Aviation, 2) Design, 3) Engineering/Technical, 4) Tourism.

**164. (A)** **Inha University** (Inha Dae Hak Kyo), 253, Yonghyon-dong, Nam-gu, Inchon, 160, Korea. *Previous Names:* Inha College of Engineering (Inha Kongkwa Dae Hak). *Established:* 1952. *Institution:* 4-year. *Control:* Private. *Enrollment:* Undergraduate, 17,784; Graduate, 1220.

Colleges within the University: 1) Economics and Commerce, 2) Education, 3) Engineering, 4) Home Economics, 5) Law and Political Science, 6) Liberal Arts, 7) Science.

Departments that offer bachelor's degrees: 1) Administrative Science, 2) Business Administration, 3) Commerce, 4) Economics, 5) Education, 6) Engineering, 7) Home Economics, 8) Jurisprudence, 9) Literature, 10) Physical Science, 11) Political Science.

| Graduate Schools: | Master's | Doctor's |
|---|---|---|
| General | x | x |
| Business Administration | x | – |
| Education | x | – |

**165. (A)** **Inje College** (Inje Dae Hak), Kaegum-dong, Pusanjin-gu, Pusan, 601, Korea. *Previous Names:* Injae Medical College (Injae Ui-kwa Dae Hak), Injae College (Injae Dae Hak). *Established:* 1979. *Institution:* 4-year. *Control:* Private. *Enrollment:* Undergraduate, 1064.

Departments that offer bachelor's degrees: 1) Dentistry, 2) Medicine.

**166. (A)** **Inje Medical College** (*See also* No. 165, Inje College). *Institution:* 4-year.

**167. (A)** **International College** (*See also* No. 168, International University). *Institution:* 4-year.

**168. (A)** **International University** (Kukje Dae Hak), Chungjong-ro, Sodaemun-gu, Seoul, 120, Korea. *Previous Names:* Hankuk Night College (Hankuk Yagan Dae Hak), International College (Kukje Dae Hak). *Established:* 1947. *Institution:* 4-year. *Control:* Private. *Enrollment:* Undergraduate, 1872.

Departments that offer bachelor's degrees: 1) Business Administration, 2) Commerce, 3) Economics, 4) Jurisprudence, 5) Literature.

**169. (C)** **Jangan Vocational Junior College** (Changan Shilop Chonmun Dae Hak), 460, San-ri, Pongdam-myon, Whasong-gun, Kyonggi-do, 170-21, Korea. *Established:* 1979. *Institution:* Junior Vocational. *Control:* Private. *Enrollment:* 3040.

Programs: 1) Commerce, 2) Design, 3) Education, 4) Home Economics, 5) Liberal Arts, 6) Textiles.

**170. (C)** **Jeju Junior College** (Cheju Chonmun Dae Hak), 2235, Yongpyong-dong, Cheju-shi, Cheju-do, 590, Korea. *Established:* 1972. *Institution:* Junior Vocational. *Control:* Private. *Enrollment:* 2040.

Programs: 1) Commerce, 2) Education, 3) Engineering/Technical, 4) Home Economics, 5) Tourism.

**171. (B)** **Jeju National Teacher's College** (Cheju Kyoyuk Dae Hak), 4810, Hwabuk-dong, Cheju-shi, Cheju-do, 590, Korea. *Previous Name:* Cheju National Teacher's School

(Cheju Sabum Hak Kyo). *Established:* 1953. *Institution:* 4-year teacher's college. *Control:* National. *Enrollment:* 232.

Department that offers a bachelor's degree: Primary Education.

172. (A) **Jeju National University** (Cheju Dae Hak Kyo), Ara-dong, Cheju-shi, Cheju-do, 590, Korea. *Established:* 1955. *Institution:* 4-year. *Control:* National. *Enrollment:* Undergraduate, 7852; Graduate, 158.

Colleges within the University: 1) Agriculture, 2) Education, 3) Liberal Arts, 4) Oceanography, 5) Social Science.

Departments that offer bachelor's degrees: 1) Administrative Science, 2) Agriculture Science, 3) Business Administration, 4) Commerce, 5) Education, 6) Jurisprudence, 7) Literature.

| Graduate Schools: | Master's | Doctor's |
|---|---|---|
| General | x | x |
| Education | x | – |

173. (C) **Jeju Nursing Junior College** (Cheju Kanho Chonmun Dae Hak), 9-2, Tonam-dong, Cheju-shi, Cheju-do, 590, Korea. *Established:* 1972. *Institution:* Junior Vocational. *Control:* Private. *Enrollment:* 400.

Programs: 1) Health Careers, 2) Nursing.

174. (A) **Jeon Ju Christian College** (*See also* No. 177, Jeon Ju University). *Institution:* 4-year.

175. (A) **Jeon Ju College** (*See also* No. 177, Jeon Ju University). *Institution:* 4-year.

176. (C) **Jeonju Technical Junior College** (Chonju Kongop Chonmun Dae Hak), 72-Namnosong-dong, Chonju-shi, Chon Puk, 520, Korea. *Established:* 1976. *Institution:* Junior Vocational. *Control:* Private. *Enrollment:* 1800.

Programs: 1) Commerce, 2) Communication, 3) Engineering/Technical.

177. (A) **Jeon Ju University** (Chon Ju Dae Hak Kyo), 2-ka, Hyoja-dong, Chonju-shi, Chonbuk, 520, Korea. *Previous Names:* Jeon Ju Christian College (Chon Ju Yong-Saeng Dae Hak), Jeon Ju College (Chon Ju Dae Hak). *Established:* 1964. *Institution:* 4-year. *Control:* Private. *Enrollment:* Undergraduate, 7904; Graduate, 12.

Colleges within the University: 1) Arts, 2) Education, 3) Liberal Arts, 4) Social Science.

Departments that offer bachelor's degrees: 1) Administrative Science, 2) Business Administration, 3) Commerce, 4) Education, 5) Fine Arts, 6) Gymnastics, 7) Jurisprudence, 8) Literature.

| Graduate Schools: | Master's | Doctor's |
|---|---|---|
| General | x | – |

178. (A) **Jeonju Woosuk College** (*See also* No. 179, Jeonju Woosuk University). *Institution:* 4-year.

179. (A) **Jeonju Woosuk University** (Chonju Usok Dae Hak), Korang-ri, Chochon-myon, Wanju-gun, Chonbuk, 520-75, Korea. *Previous Names:* Jeonju Woosuk Women's College (Chonju Usok Yoja Dae Hak), Jeonju Woosuk College (Chonju Usok Dae Hak). *Established:* 1979. *Institution:* 4-year. *Control:* Private. *Enrollment:* Undergraduate: 4576.

Departments that offer bachelor's degrees: 1) Business Administration, 2) Commerce, 3) Education, 4) Home Economics, 5) Literature, 6) Pharmacology, 7) Physical Science.

180. **(A) Jeonju Woosuk Women's College** (*See also* No.179, Jeonju Woosuk University). *Institution:* 4-year.

181. **(C) Jesus Nursing Junior College** (Yesu Kanho Chonmun Dae Hak), 168-1, Chungwhasan-dong, 1 ka, Chonju-shi, Chon Puk, 520, Korea. *Established:* 1979. *Institution:* Junior Vocational. *Control:* Private. *Religious Affiliation:* Presbyterian. *Enrollment:* 240.

    Program: Nursing.

182. **(C) Jinju Agriculture and Forestry Junior College** (Chinju Nonglim Chonmun Dae Hak), 150, Chilam-dong, Chinju-shi, Kyong Nam, 620, Korea. *Established:* 1973. *Institution:* Junior Vocational. *Control:* National. *Enrollment:* 1120.

    Program: Engineering/Technical.

183. **(B) Jinju National Teacher's College** (Chinju Kyoyuk Dae Hak), 380, Shinan-dong, Chinju-shi, Kyong Nam, 620, Korea. *Previous Name:* Chinju National Teacher's School (Chinju Sabum Hak Kyo). *Established:* 1946. *Institution:* 4-year teacher's college. *Control:* National. *Enrollment:* 848.

    Department that offers a bachelor's degree: Primary Education.

184. **(C) Jinju Nursing Health Junior College** (Chinju Kanho Pokon Chonmun Dae Hak), 1142, Sangbongso-dong, Chinju-shi, Kyong Nam, 620, Korea. *Established:* 1972. *Institution:* Junior Vocational. *Control:* Private. *Enrollment:* 880.

    Program: 1) Health Careers, 2) Nursing.

185. **(C) Jinju Vocational Junior College** (Chinju Shilop Chonmun Dae Hak), 101, Hadae-dong, Chinju-shi, Kyong Nam, 620, Korea. *Established:* 1977. *Institution:* Junior Vocational. *Control:* Private. *Enrollment:* 1460.

    Programs: 1) Commerce, 2) Education, 3) Engineering/Technical, 4) Home Economics, 5) Liberal Arts.

186. **(C) Jisan Nursing Health Junior College** (Chisan Kanho Pokon Chonmun Dae Hak), San 8, Pukok-dong, Tonglae-gu, Pusan, 607, Korea. *Established:* 1963. *Institution:* Junior Vocational. *Control:* Private. *Enrollment:* 1040.

    Programs: 1) Health Careers, 2) Nursing.

187. **(C) Junggyeong Technical Junior College** (Chunggyong Kongop Chonmun Dae Hak), 155-3, Chayang-dong, Tong-gu, Taejon-shi, Chung Nam, 300, Korea. *Previous Name:* Dejeon Vocational Junior College (Taejon Shilop Chonmun Hak Kyo). *Established:* 1974. *Institution:* Junior Vocational. *Control:* Private. *Enrollment:* 3080.

    Programs: 1) Commerce, 2) Design, 3) Education, 4) Engineering, 5) Textiles.

188. **(A) Kangreung National University** (Kangrung Dae Hak), Chodang-dong, Kangrung-shi, Kangwon-do, 210, Korea. *Established:* 1969. *Institution:* 4-year. *Control:* National. *Enrollment:* Undergraduate: 2964.

    Departments that offer bachelor's degrees: 1) Business Administration, 2) Commerce, 3) Fine Arts, 4) Literature, 5) Music, 6) Physical Science.

189. **(A) Kangweon National University** (Kangwon Dae Hak Kyo), 22, Hyoja-dong, Chun Chon, Kangwon-do, 200, Korea. *Previous Names:* Chunchon Agricultural College (Chunchon Nong op Hak Kyo), Kwangweon University (Kwangwon Dae Hak Kyo). *Established:* 1947. *Institution:* 4-year. *Control:* National. *Enrollment:* Undergraduate, 14,040; Graduate, 728.

Colleges within the University: 1) Agriculture, 2) Business Administration, 3) Education, 4) Engineering, 5) Law, 6) Literature and Humanities, 7) Natural Science.

Departments that offer bachelor's degrees: 1) Administrative Science, 2) Agriculture Science, 3) Business Administration, 4) Commerce, 5) Economics, 6) Education, 7) Engineering, 8) Gymnastics, 9) Jurisprudence, 10) Literature, 11) Pharmacology, 12) Physical Science, 13) Political Science.

| Graduate Schools: | Master's | Doctor's |
|---|---|---|
| General | x | x |
| Business and Public Administration | x | – |
| Education | x | – |

**190. (A) Keimyung University** (Kyemyong Dae Hak Kyo), Taemyong-dong, Nam-gu, Taegu, 634, Korea. *Established:* 1954. *Institution:* 4-year. *Control:* Private. *Religious Affiliation:* Christian, Non-denominational. *Enrollment:* Undergraduate, 16,856; Graduate, 841.

Colleges within the University: 1) Arts, 2) Business Administration, 3) Education, 4) Foreign Languages, 5) Liberal Arts, 6) Medicine, 7) Music, 8) Science and Engineering, 9) Social Science.

Departments that offer bachelor's degrees: 1) Administrative Science, 2) Business Administration, 3) Commerce, 4) Economics, 5) Education, 6) Engineering, 7) Fine Arts, 8) Gymnastics, 9) Home Economics, 10) Jurisprudence, 11) Library Science, 12) Literature, 13) Medicine, 14) Music, 15) Physical Science, 16) Theology.

| Graduate Schools: | Master's | Doctor's |
|---|---|---|
| General | x | x |
| Education | x | – |
| International Trade | x | – |

**191. (A) King Sejong College** (*See also* No. 192, King Sejong University). *Institution:* 4-year.

**192. (A) King Sejong University** (Sejong Dae Hak), San 2, Kunja-dong, Songdong-gu, Seoul, 133, Korea. *Previous Names:* Sudo Teacher's College for Women (Sudo Yoja Sabom Dae Hak), King Sejong College (Sejong Dae Hak). *Established:* 1947. *Institution:* 4-year. *Control:* Private. *Enrollment:* Undergraduate, 5148; Graduate, 294.

Departments that offer bachelor's degrees: 1) Business Administration, 2) Commerce, 3) Economics, 4) Fine Arts, 5) Gymnastics, 6) Home Economics, 7) Literature, 8) Music.

| Graduate Schools: | Master's | Doctor's |
|---|---|---|
| General | x | x |
| Business Administration | x | – |

**193. (A) Kong Ju National Teacher's College** (Kong Ju Sabom Dae Hak), Changgi, Kongju-gun, Chungnam, 301, Korea. *Established:* 1948. *Institution:* 4-year. *Control:* National. *Enrollment:* Undergraduate, 4004; Graduate, 20.

Department that offers a bachelor's degree: Education.

| Graduate Schools: | Master's | Doctor's |
|---|---|---|
| General | x | – |

**194. (B) Kong Ju National Teacher's College** (Kong Ju Kyoyuk Dae Hak), 376, Pongh-wang-dong, Kongju-up, Kongju-gun, Chung Nam, 301, Korea. *Previous Name:* Kong Ju Women Teacher's School (Kong Ju Yoja Sabum Hak Kyo). *Established:* 1938. *Institution:* 4-year teacher's college. *Control:* National. *Enrollment:* 1039.

Department that offers a bachelor's degree: Primary Education.

**195. (B) Kong Ju Women Teacher's School** (*See also* No. 194, Kong Ju National Teacher's College). *Institution:* 4-year teacher's college.

**196. (A) Konkuk College** (*See also* No. 197, Kon Kuk University). *Institution:* 4-year.

**197. (A) Kon Kuk University** (Kon Kuk Dae Hak Kyo), 93-1 Mojin-dong, Songdong-gu, Seoul, 133, Korea. *Previous Name:* Cho-Sun Political Science School (Cho-Sun Chongchi Hak Kyo). *Established:* 1946. *Institution:* 4-year. *Control:* Private. *Enrollment:* Undergraduate, 17,784; Graduate, 1860.

Colleges within the University: 1) Agriculture, 2) Animal Husbandry, 3) Engineering, 4) Home Economics, 5) Law and Economics, 6) Liberal Arts and Science.

Departments that offer bachelor's degrees: 1) Administrative Science, 2) Agriculture Science, 3) Business Administration, 4) Commerce, 5) Economics, 6) Education, 7) Engineering, 8) Fine Arts, 9) Gymnastics, 10) Home Economics, 11) Jurisprudence, 12) Library Science, 13) Literature, 14) Physical Science, 15) Political Science, 16) Veterinary Medicine.

| Graduate Schools: | Master's | Doctor's |
| --- | --- | --- |
| General | x | x |
| Education | x | – |
| Industry | x | – |
| Public Administration | x | – |

**198. (A) Kookmin College** (*See also* No. 199, Kookmin University). *Institution:* 4-year.

**199. (A) Kookmin University** (Kukmin Dae Hak Kyo), Chongrung-dong, Songbuk-gu, Seoul, 132, Korea. *Established:* 1946. *Institution:* 4-year. *Control:* Private. *Enrollment:* Undergraduate, 7956; Graduate, 660.

Colleges within the University: 1) Economics and Commerce, 2) Education, 3) Engineering, 4) Law and Political Science, 5) Liberal Arts, 6) Modeling.

Departments that offer bachelor's degrees: 1) Administrative Science, 2) Business Administration, 3) Commerce, 4) Economics, 5) Education, 6) Engineering, 7) Jurisprudence, 8) Literature, 9) Political Science.

| Graduate Schools: | Master's | Doctor's |
| --- | --- | --- |
| General | x | x |
| Business Administration | x | – |
| Education | x | – |

**200. (A) Korea Baptist Theological College** (Chimnye Shinhak Dae Hak), Mok-dong, Chung-gu, Taejon, Chungnam, 300, Korea. *Established:* 1954. *Institution:* 4-year. *Control:* Private. *Enrollment:* Undergraduate, 1248; Graduate, 190.

Departments that offer bachelor's degrees: 1) Education, 2) Theology.

| Graduate Schools: | Master's | Doctor's |
| --- | --- | --- |
| General | x | – |
| Theology | x | – |

**201. (D) Korea Bible School** (Han Kuk Song So Shin Hak Kyo), 205, Sanggye-dong, Tobong-gu, Seoul, 134-02, Korea. *Established:* 1955. *Institution:* Specialized Institution—Miscellaneous. *Control:* Private. *Enrollment:* Day—440; Evening—200.

Diploma Program: Theology.

**202. (A) Korea Christian College** (Christ Shinhak Dae Hak), San 61-1, Dungchon-dong, Kangso-gu, Seoul, 150, Korea. *Established:* 1973. *Institution:* 4-year. *Control:* Private. *Enrollment:* Undergraduate, 208.

Department that offers a bachelor's degree: Theology.

**203. (D) Korea Christianity Theological Seminary** (Tae Han Kidogkyo Shin Hak Kyo), 5-198, Hyochang-dong, Yongsan-gu, Seoul, 140, Korea. *Established:* 1958. *Institution:* Specialized Institution—Miscellaneous. *Control:* Private. *Enrollment:* Day—240; Evening—240.

Diploma Program: Theology.

**204. (F) Korea Correspondence College** (*See also* No. 205, Korea Correspondence University). *Institution:* Radio and Correspondence.

**205. (F) Korea Correspondence University** (Hankuk Pangsong Tongshin Dae Hak), 169, Tongsung-dong, Chongno-gu, Seoul, 110, Korea. *Previous Name:* Korea Correspondence College (Hankuk Pangsong Tongshin Dae Hak). *Established:* 1972. *Institution:* Radio and Correspondence. *Control:* National. *Enrollment:* 119,624.

Departments that offer bachelor's degrees: 1) Administrative Science, 2) Agriculture Science, 3) Business Administration, 4) Economics, 5) Education–Elementary, 6) Engineering (Computer Science and Statistics), 7) Home Economics, 8) Jurisprudence, 9) Library Science.

Diploma Program: Education–Early Childhood.

**206. (D) Korea Judo College** (Tae Han Yudo Hak Kyo), 355, Pungnap-dong, Kangdong-gu, Seoul, 134-01, Korea. *Established:* 1953. *Institution:* Specialized Institution—Miscellaneous. *Control:* Private. *Enrollment:* 1900. *Ministry of Education:* Accreditation.

Diploma Program: Physical Education.

**207. (A) Korea Maritime University** (Hankuk Haeyang Dae Hak), Yongdo-gu, Pusan 606, Korea. *Established:* 1945. *Institution:* 4-year. *Control:* National. *Enrollment:* Undergraduate, 2796; Graduate, 52.

Departments that offer bachelor's degrees: 1) Business Administration, 2) Commerce, 3) Engineering, 4) Jurisprudence.

| Graduate Schools: | Master's | Doctor's |
|---|---|---|
| General | x | x |

**208. (A) Korea National College of Physical Education** (Han Kuk Cheyuk Dae Hak), Kongrung-dong, Tobong-gu, Seoul, 130-02, Korea. *Established:* 1976. *Institution:* Specialized Institution—Miscellaneous. *Control:* National. *Enrollment:* 800. *Ministry of Education:* Accreditation.

Department that offers a bachelor's degree: Physical Education.

**209. (A) Korea Social Work College** (*See also* No. 74, Daegu University). *Institution:* 4-year.

**210. (A) Korea Theological Seminary** (*See also* No. 217, Kosin College). *Institution:* 4-year.

**211. (C) Korea Union Agricultural Junior College** (Samyuk Nongop Chonmun Dae Hak), San 133, Whajom-ri, Pyolnae-myon, Yangju-gun, Kyonggi-do, 130-71, Korea. *Established:* 1978. *Institution:* Junior Vocational. *Control:* Private. *Enrollment:* 320.

Program: Engineering/Technical.

**212. (C) Korea Union Nursing Junior College** (Samyuk Kanho Chonmun Dae Hak), 29-1, Hwikyong-dong, Tongdaemun-gu, Seoul, 131, Korea. *Established:* 1960. *Institution:* Junior Vocational. *Control:* Private. *Enrollment:* 120.

Program: Nursing.

**213. (A) Korea University** (Koryo Dae Hak Kyo), 5-ka, Anam-dong, Songbuk-gu, Seoul, 132, Korea. *Previous Names:* Posong College (Posong Chonmun Hak Kyo). *Established:* 1905. *Institution:* 4-year. *Control:* Private. *Enrollment:* Undergraduate, 22,556; Graduate, 3657.

Colleges within the University: 1) Agriculture, 2) Business Administration, 3) Economics and Political Science, 4) Education, 5) Engineering, 6) Law, 7) Liberal Arts, 8) Medicine, 9) Science.

Departments that offer bachelor's degrees: 1) Administrative Science, 2) Agriculture Science, 3) Business Administration, 4) Commerce, 5) Economics, 6) Education, 7) Engineering, 8) Jurisprudence, 9) Literature, 10) Medicine, 11) Nursing, 12) Physical Science, 13) Political Science.

| Graduate Schools: | Master's | Doctor's |
| --- | --- | --- |
| General | x | x |
| Business Administration | x | – |
| Education | x | – |
| Food Development | x | – |

**214. (C) Korea University, Junior College of Public Health and Medical Technology** (Koryo Dae Hak Kyo Pyongsol Pokon Chonmun Dae Hak), San-1, Chongrung 3-dong, Songbuk-gu, Seoul, 132, Korea. *Established:* 1963. *Institution:* Junior Vocational. *Control:* Private. *Enrollment:* 700.

Program: Health Careers.

**215. (D) Korean Nazarene Theological Seminary** (Hankuk Nazaret Shin Hak Kyo), Chonan, Chungam, 330, Korea. *Established:* 1981. *Institution:* Specialized Institution—Miscellaneous. *Control:* Private. *Enrollment:* 120 est.

Diploma Program: Theology.

**216. (A) Korean Union College** (Samyuk Dae Hak), Kongrung-dong, Tobong-gu, Seoul, 130-02, Korea. *Established:* 1906. *Institution:* 4-year. *Control:* Private. *Enrollment:* Undergraduate, 1664; Graduate, 20.

Departments that offer bachelor's degrees: 1) Business Administration, 2) Education, 3) Home Economics, 4) Literature, 5) Nursing, 6) Pharmacology, 7) Theology.

| Graduate Schools: | Master's | Doctor's |
| --- | --- | --- |
| General | x | – |

**217. (A) Kosin College** (Kosin Dae Hak), Amnam-dong, So-gu, Pusan, 600, Korea. *Previous Names:* Korea Theological Seminary (Koryo Shin Hak Kyo). *Established:*

1946. *Institution:* 4-year. *Control:* Private. *Religious Affiliation:* Presbyterian. *Enrollment:* Undergraduate, 1184; Graduate, 196.

Departments that offer bachelor's degrees: 1) Education, 2) Medicine, 3) Music, 4) Theology.

| Graduate Schools: | Master's | Doctor's |
| --- | --- | --- |
| General | x | – |
| Theology | x | – |

218. **(A) Kumoh Institute of Technology** (Kum-oh Kongkwa Dae Hak), Sinpyong-dong, Kumi, Kyongbuk, 641, Korea. *Established:* 1979. *Institution:* 4-year. *Control:* Private. *Enrollment:* Undergraduate, 1976.

Department that offers a bachelor's degree: Engineering.

219. **(C) Kwangju Health Junior College** (Kwangju Pokon Chonmun Dae Hak), 66-1, Yanglim-dong, So-gu, Kwangju-shi, Chon Nam, 500, Korea. *Previous Name:* Supia Womens Vocational Junior College (Supia Yoja Shilop Chonmun Dae Hak). *Established:* 1972. *Institution:* Junior Vocational. *Control:* Private. *Religious Affiliation:* Christian Non-denominational. *Enrollment:* 1920.

Programs: 1) Education, 2) Engineering/Technical, 3) Health Careers, 4) Home Economics.

220. **(C) Kwangju Junior College of Business and Commerce** (Kwangju Kyong-Sang Chonmun Dae Hak), 592, Chinwol-dong, So-gu, Kwangju-shi, Chon Nam, 500, Korea. *Established:* 1980. *Institution:* Junior Vocational. *Control:* Private. *Enrollment:* 1200.

Programs: 1) Administration, 2) Commerce, 3) Education, 4) Liberal Arts.

221. **(E) Kwangju Open College** (Kwangju Kaebang Dae Hak), 592, Chinwol-dong, So-gu, Kwangju-shi, Chon Nam, 500, Korea. *Established:* 1983. *Institution:* 4-year Open. *Control:* National. *Enrollment:* N.A.

Departments that offer bachelor's degree: 1) Commerce, 2) Education, 3) Liberal Arts, 4) Management.

222. **(B) Kwang Ju National Teacher's College** (Kwang Ju Kyoyuk Dae Hak), 1-1, Punghyang-dong, Puk-gu, Kwangju-shi, Chon Nam, 500, Korea. *Previous Name:* Kwang Ju Teacher's School (Kwang Ju Sabum Hak Kyo). *Established:* 1938. *Institution:* 4-year teacher's college. *Control:* National. *Enrollment:* 1347.

Department that offers a bachelor's degree: Primary Education.

223. **(B) Kwang Ju National Teacher's School** (*See also* No. 222, Kwang Ju National Teacher's College). *Institution:* 4-year teacher's college.

224. **(A) Kwangweon University** (*See also* No. 189, Kangweon National University). *Institution:* 4-year.

225. **(A) Kwangwoon College of Electronic Technology** (*See also* No. 226, Kwangwoon University). *Institution:* 4-year.

226. **(A) Kwangwoon University** (Kwang-un Dae Hak), Wolkye-dong, Tobong-gu, Seoul, 132, Korea. *Previous Names:* Dongkuk Junior College (Dongkuk Chonja Chogup Dae Hak), Kwangwoon College of Electronic Technology (Kwang-un Kongkwa Dae Hak). *Established:* 1962. *Institution:* 4-year. *Control:* Private. *Enrollment:* Undergraduate, 4420; Graduate, 70.

Departments that offer bachelor's degrees: 1) Administrative Science, 2) Business Administration, 3) Commerce, 4) Engineering, 5) Literature.

Graduate Schools:                                  Master's  Doctor's

General                                                    x         –

227. (C) **Kyemyung Vocational Junior College** (Kyemyong Shilop Chonmun Dae Hak), 2139, Taemyong-dong, Nam-gu, Taegu, 634, Korea. *Established:* 1962. *Institution:* Junior Vocational. *Control:* Private. *Enrollment:* 3720.

Programs: 1) Commerce, 2) Education, 3) Engineering/Technical, 4) Liberal Arts, 5) Textiles, 6) Tourism.

228. (C) **Kyeong-buk Technical Junior College** (Kyongbuk Kongop Chonmun Dae Hak), 55, Hyomok-dong, Tong-gu, Taegu, 635, Korea. *Established:* 1976. *Institution:* Junior Vocational. *Control:* Private. *Enrollment:* 4160.

Programs: 1) Commerce, 2) Engineering/Technical, 3) Textiles.

229. (C) **Kyeongbuk Vocational Junior College** (Kyongbuk Shilop Chonmun Dae Hak), San 5-4, Pyongsan-dong, Apnyang-myon, Kyongsan-gun, Kyong Puk, 630-14, Korea. *Previous Name:* Daeil Vocational Junior College (Tae-il Shilop Chonmun Dae Hak). *Established:* 1980. *Institution:* Junior Vocational. *Control:* Private. *Enrollment:* 1140.

Programs: 1) Commerce, 2) Education, 3) Liberal Arts, 4) Tourism.

230. (B) **Kyeong-gi Teacher's School** (*See also* No. 303, Seoul National Teacher's College). *Institution:* 4-year teacher's college.

231. (C) **Kyeong-gi Technical Junior College** (Kyong-gi Kongop Chonmun Dae Hak), 172, Kongrung-dong, Tobong-gu, Seoul, 130-02, Korea. *Established:* 1974. *Institution:* Junior Vocational. *Control:* National. *Enrollment:* 1520.

Program: Engineering/Technical.

232. (E) **Kyeong-gi Technical Open College** (Kyong-gi Kongop Kaebang Dae Hak), 172, Kongrung-dong, Tobong-gu, 130-02, Seoul, Korea. *Established:* 1983. *Institution:* 4-year Open. *Control:* National. *Enrollment:* 6880.

Department that offers a bachelor's degree: Engineering.

233. (C) **Kyeong Hee Hotel Management Junior College** (Kyong-Hi Hotel Kyongyong Chonmun Dae Hak), 1, Hoegi-dong, Tongdaemun-gu, Seoul, 131, Korea. *Established:* 1975. *Institution:* Junior Vocational. *Control:* Private. *Enrollment:* 800.

Programs: 1) Commerce, 2) Engineering/Technical, 3) Tourism.

234. (C) **Kyeong Hee University, Nursing Junior College** (Kyong Hi Dae Hak Kyo Pyongsol Kanho Chonmun Dae Hak), 1, Hoegi-dong, Tongdaemun-gu, Seoul, 131, Korea. *Established:* 1970. *Institution:* Junior Vocational. *Control:* Private. *Enrollment:* 360.

Program: Nursing.

235. (C) **Kyeongju Vocational Junior College** (Kyongju Shilop Chonmun Dae Hak), San 42-1, Hyohyon-dong, Kyongju-shi, Kyong Puk, 681, Korea. *Established:* 1981. *Institution:* Junior Vocational. *Control:* Private. *Enrollment:* 1200.

Programs: 1) Commerce, 2) Design, 3) Education, 4) Liberal Arts, 5) Tourism.

236. (C) **Kyeongnam Nursing Junior College** (Kyongnam Kanho Chonmun Dae Hak), 97-1, Chuyak-dong, Chinju-shi, Kyong Nam, 620, Korea. *Established:* 1926. *Institution:* Junior Vocational. *Control:* Provincial. *Enrollment:* 240.

Program: Nursing.

**237. (C) Kyeong Nam Technical Junior College** (Kyong Nam Kongop Chonmun Dae Hak), 167, Chulye-dong, Puk-gu, Pusan, 601, Korea. *Established:* 1970. *Institution:* Junior Vocational. *Control:* Private. *Enrollment:* 4000.

Programs: 1) Commerce, 2) Design, 3) Engineering/Technical, 4) Liberal Arts.

**238. (C) Kyeongwon Technical Junior College** (Kyongwon Kongop Chonmun Dae Hak), San-65, Pokjong-dong, Songnam-shi, Kyonggi-do, 130-14, Korea. *Established:* 1978. *Institution:* Junior Vocational. *Control:* Private. *Enrollment:* 3130.

Programs: 1) Commerce, 2) Design, 3) Education, 4) Engineering/Technical, 5) Home Economics, 6) Textiles.

**239. (A) Kyonggi University** (Kyonggi Dae Hak Kyo), Iui-ri, Suji-myon, Yong-in-gun, Kyonggi-do, 170-40, Korea. *Previous Names:* Choyang Teacher Training College (Choyang Sabom Hak Kyo). *Established:* 1947. *Institution:* 4-year. *Control:* Private. *Enrollment:* Undergraduate, 9932; Graduate, 70.

Departments that offer bachelor's degrees: 1) Administrative Science, 2) Business Administration, 3) Commerce, 4) Economics, 5) Engineering, 6) Fine Arts, 7) Gymnastics, 8) Jurisprudence, 9) Library Science, 10) Literature, 11) Physical Science.

| Graduate Schools: | Master's | Doctor's |
|---|---|---|
| General | x | – |

**240. (A) Kyongsong Agriculture School** (*See also* No. 297, Seoul City University). *Institution:* 4-year.

**241. (C) Kyunggi Nursing Junior College** (Kyonggi Kanho Chonmun Dae Hak), 22, Shinhung-dong, 2 ka, Chung-gu, Inchon, 160, Korea. *Established:* 1973. *Institution:* Junior Vocational. *Control:* Private. *Enrollment:* 520.

Programs: 1) Health Careers, 2) Nursing.

**242. (A) Kyung Hee University** (Kyong Hi Dae Hak Kyo), Tongdaemun-gu, Seoul, 131, Korea. *Previous Name:* Shin Hung College (Shin Hung Chogup Dae Hak). *Established:* 1949. *Institution:* 4-year. *Control:* Private. *Enrollment:* Undergraduate, 18,452; Graduate, 2354.

Colleges within the University: 1) Dentistry, 2) Economics and Political Science, 3) Education, 4) Engineering, 5) Gymnastics, 6) Industry, 7) Law and Public Administration, 8) Liberal Arts and Science, 9) Medicine, 10) Music, 11) Oriental Medicine, 12) Pharmacology.

Departments that offer bachelor's degrees: 1) Administrative Science, 2) Agriculture Science, 3) Business Administration, 4) Commerce, 5) Dentistry, 6) Economics, 7) Education, 8) Engineering, 9) Fine Arts, 10) Gymnastics, 11) Home Economics, 12) Jurisprudence, 13) Literature, 14) Medicine, 15) Nursing, 16) Oriental Medicine, 17) Pharmacology, 18) Physical Science, 19) Political Science.

| Graduate Schools: | Master's | Doctor's |
|---|---|---|
| General | x | x |
| Business Administration | x | – |
| Education | x | – |
| Public Administration | x | – |

APPENDIX A 121

**243. (A) Kyungnam College** (*See also* No. 245, Kyungnam University) *Institution:* 4-year.

**244. (A) Kyungnam Teacher's College** (*See also* No. 277, Pusan Sanub University). *Institution:* 4-year.

**245. (A) Kyungnam University** (Kyongnam Dae Hak Kyo), Wolyong-dong, Masan-shi, Kyongnam, 610, Korea. *Previous Names:* Kookmin College (Kukmin Dae Hak), Hae In College (Hae In Dae Hak), Changwon College (Masan Dae Hak), Kyungnam College (Kyongnam Dae Hak). *Established:* 1947. *Institution:* 4-year. *Control:* Private. *Enrollment:* Undergraduate, 14,612; Graduate, 761.

Colleges within the University: 1) Economics and Commerce, 2) Education, 3) Engineering, 4) Law and Political Science, 5) Liberal Arts and Science.

Departments that offer bachelor's degrees: 1) Administrative Science, 2) Business Administration, 3) Commerce, 4) Economics, 5) Education, 6) Engineering, 7) Jurisprudence, 8) Literature, 9) Physical Science, 10) Political Science.

| Graduate Schools: | Master's | Doctor's |
|---|---|---|
| General | x | x |
| Business Administration | x | – |
| Education | x | – |

**246. (A) Kyungpook National University** (Kyongbuk Dae Hak Kyo), Sankyok-dong, Puk-gu, Taegu, 635, Korea. *Established:* 1952. *Institution:* 4-year. *Control:* National. *Enrollment:* Undergraduate, 22,396; Graduate, 2491.

Colleges within the University: 1) Agriculture, 2) Dentistry, 3) Economics and Commerce, 4) Education, 5) Engineering, 6) Law and Political Science, 7) Liberal Arts and Science, 8) Medicine.

Departments that offer bachelor's degrees: 1) Administrative Science, 2) Agriculture Science, 3) Business Administration, 4) Commerce, 5) Dentistry, 6) Economics, 7) Education, 8) Engineering, 9) Fine Arts, 10) Gymnastics, 11) Home Economics, 12) Jurisprudence, 13) Library Science, 14) Literature, 15) Medicine, 16) Music, 17) Nursing, 18) Physical Science, 19) Political Science, 20) Veterinary Medicine.

| Graduate Schools: | Master's | Doctor's |
|---|---|---|
| General | x | x |
| Business Administration | x | – |
| Education | x | – |
| Public Administration | x | – |
| Public Health | x | – |

**247. (A) Kyung Won College** (Kyong Won Dae Hak), San 65, Pokjong-dong, Songnam, Kyonggi-do, 130-14, Korea. *Established:* 1981. *Institution:* 4-year. *Control:* Private. *Enrollment:* Undergraduate, 4160.

Departments that offer bachelor's degrees: 1) Administrative Science, 2) Business Administration, 3) Commerce, 4) Economics, 5) Jurisprudence, 6) Literature.

**248. (D) Luther Theological Seminary** (Luto Shin Hak Kyo) Yong-in, Kyonggi-do, Korea. *Established:* 1982. *Institution:* Specialized Institution—Miscellaneous. *Control:* Private. *Enrollment:* 40 est.

Diploma Program: Theology

**249. (C) Masan Nursing Junior College** (Masan Kanho Chonmun Dae Hak), 3, Chung-ang-dong, 3 ka, Masan-shi, Kyong Nam, 610, Korea. *Established:* 1979. *Institution:* Junior Vocational. *Control:* Private. *Enrollment:* 400.

Programs: 1) Health Careers, 2) Nursing.

**250. (A) Methodist Theological College** (*See also* No. 251, Methodist Theological Seminary). *Institution:* 4-year.

**251. (D) Methodist Theological Seminary** (Kamrikyo Hyupsong Shin Hak Kyo), Nam Yangju, Kyongi-do, Korea. *Established:* 1983. *Institution:* Specialized Institution—Miscellaneous. *Control:* Private. *Enrollment:* 80 est.

Diploma Program: Theology.

**252. (A) Methodist Theological Seminary** (Kamlikyo Shinhak Dae Hak), Naengchon-dong, Sodaemun-gu, Seoul, 120, Korea. *Previous Names:* Hyop Song Theological School (Hyop Song Shinhak Kyo), Methodist Theological College (Kamlikyo Shinhak Kyo). *Established:* 1905. *Institution:* 4-year. *Control:* Private. *Religious Affiliation:* Methodist. *Enrollment:* Undergraduate, 1248; Graduate, 236.

Departments that offer bachelor's degrees: 1) Education, 2) Theology.

| Graduate Schools: | Master's | Doctor's |
|---|---|---|
| General | x | x |
| Theology | x | x |

**253. (C) Milyang Agriculture and Sericulture Junior College** (Milyang Nongjam Chonmun Dae Hak), 1028, Nae-i-dong, Milyang-up, Milyang-gun, Kyong Nam, 605, Korea. *Previous Name:* Milyang Agriculture and Sericulture School (Milyang Nongjam Hak Kyo). *Established:* 1923. *Institution:* Junior Vocational. *Control:* National. *Enrollment:* 880.

Program: Engineering/Technical.

**254. (C) Milyang Agriculture and Sericulture School** (*See also* No. 253, Milyang Agriculture and Sericulture Junior College), 1028, Nae-i-dong, Milyang-up, Milyang-gun, Kyong Nam, 605, Korea. *Established:* 1923. *Institution:* Junior Vocational. *Control:* National. *Enrollment:* N.A.

Program: Engineering/Technical.

**255. (C) Mogpo Mercantile Marine Junior College** (Mokpo Hae-Yang Chonmun Dae Hak), 571-2, Chukkyo-dong, Mokpo-shi, Chon Nam, 580, Korea. *Established:* 1973. *Institution:* Junior Vocational. *Control:* National. *Enrollment:* 1000.

Programs: 1) Engineering/Technical, 2) Marine.

**256. (A) Mogpo National College** (Mokpo Dae Hak), Yonghae-dong, Mokpo, Chonnam, 580, Korea. *Established:* 1963. *Institution:* 4-year. *Control:* National. *Enrollment:* Undergraduate, 5096.

Departments that offer bachelor's degrees: 1) Administrative Science, 2) Agriculture Science, 3) Business Administration, 4) Commerce, 5) Economics, 6) Education, 7) Gymnastics, 8) Home Economics, 9) Jurisprudence, 10) Literature, 11) Music.

**257. (C) Mogpo Vocational Junior College** (Mokpo Shilop Chonmun Dae Hak), 525, Sang-dong, Mokpo-shi, Chon Nam, 580, Korea. *Established:* 1976. *Institution:* Junior Vocational. *Control:* Private. *Enrollment:* 2200.

Programs: 1) Commerce, 2) Education, 3) Engineering/Technical, 4) Health Careers.

**258. (A) Mok Won College** (*See also* No. 259, Mokwon Methodist College). *Institution:* 4-year.

**259. (A) Mokwon Methodist College** (Mokwon Dae Hak), 24, Mok-dong, Chung-gu, Taejon, Chungnam, 300, Korea. *Previous Names:* Tae Jon Methodist Theological College (Kamlikyo Taejon Shinhak Kyo), Mok Won College (Mok Won Dae Hak). *Established:* 1957. *Institution:* 4-year. *Control:* Private. *Religious Affiliation:* Methodist. *Enrollment:* Undergraduate, 5048; Graduate, 120.

Departments that offer bachelor's degrees: 1) Administrative Science, 2) Business Administration, 3) Commerce, 4) Economics, 5) Education, 6) Fine Arts, 7) Jurisprudence, 8) Literature, 9) Music, 10) Theology.

| Graduate Schools: | Master's | Doctor's |
|---|---|---|
| General | x | – |
| Theology | x | – |

**260. (C) Myeongji Vocational Junior College** (Myonggi Shilop Chonmun Dae Hak), 50-3, Namgajwa-dong, Sodaemun-gu, Seoul, 122, Korea. *Established:* 1975. *Institution:* Junior Vocational. *Control:* Private. *Enrollment:* 3200.

Programs: 1) Commerce, 2) Design, 3) Education, 4) Engineering/Technical, 5) Fine Arts, 6) Liberal Arts.

**261. (A) Myong Ji College** (*See also* No. 262, Myong Ji University). *Institution:* 4-year.

**262. (A) Myong Ji University** (Myong Ji Dae Hak Kyo), Nam-ri, Yong-In, Kyonggi-do, Korea. *Previous Names:* Seoul Higher Education for Women (Seoul Kodung Kajong Hak Kyo), Seoul Mun-Lee College (Seoul Mun-Lee Silkwa Dae Hak), Myong Ji College (Myong Ji Dae Hak). *Established:* 1948. *Institution:* 4-year. *Control:* Private. *Religious Affiliation:* Christian, Non-denominational. *Enrollment:* Undergraduate, 8996; Graduate, 284.

Colleges within the University: 1) Economics and Commerce, 2) Engineering, 3) Law and Public Administration, 4) Liberal Arts, 5) Science.

Departments that offer bachelor's degrees: 1) Administrative Science, 2) Business Administration, 3) Commerce, 4) Engineering, 5) Gymnastics, 6) Home Economics, 7) Jurisprudence, 8) Library Science, 9) Literature, 10) Physical Science.

| Graduate Schools: | Master's | Doctor's |
|---|---|---|
| General | x | x |

**263. (A) Myong Jin School** (*See also* No. 97, Dong Guk University). *Institution:* 4-year.

**264. (A) National Fisheries University of Pusan** (Pusan Susan Dae Hak), 599, Teayon-dong, Pusanjin-gu, Pusan, 608, Korea. *Established:* 1941. *Institution:* 4-year. *Control:* National. *Enrollment:* Undergraduate, 5252; Graduate, 250.

Departments that offer bachelor's degrees: 1) Business Administration, 2) Commerce, 3) Engineering, 4) Fisheries, 5) Physical Science.

| Graduate Schools: | Master's | Doctor's |
|---|---|---|
| General | x | x |

**265. (C) National Medical Center, Nursing Junior College** (Kuglip Uiryowon Kanho Chonmun Dae Hak), 18-79, Ulji-ro 6 ka, Chung-gu, Seoul, 100, Korea. *Established:* 1958. *Institution:* Junior Vocational. *Control:* National. *Enrollment:* 120.

Program: Nursing.

**266. (C) O-San Technical Junior College** (O-San Kongop Chonmun Dae Hak), 18, Chonghak-ri, O-San-up, Whasong-gun, Kyonggi-do, 170-83, Korea. *Established:* 1978. *Institution:* Junior Vocational. *Control:* Private. *Enrollment:* 1600.

Programs: 1) Commerce, 2) Education, 3) Engineering/Technical.

**267. (A) Pai Chai College** (Pae Chae Dae Hak), Toma-dong, Chung-gu, Taejon, Chungnam, 300-01, Korea. *Previous Names:* Paichai Tae Jeon Junior College (Pae-Chae Tae-Jon Chogup Dae Hak). *Established:* 1978. *Institution:* 4-year. *Control:* Private. *Enrollment:* Undergraduate, 2808.

Departments that offer bachelor's degrees: 1) Agriculture Science, 2) Business Administration, 3) Education, 4) Literature, 5) Physical Science.

**268. (A) Paichai Tae Jeon Junior College** (*See also* No. 267, Pai Chai College). *Institution:* 4-year.

**269. (D) Piason Biblical Theological Seminary** (Pioson Songso Shin Hak Kyo), San 84, Yong-i-ri, Won-gok-myon, Ansong-gun, Kyonggi-do, 180-21, Korea. *Established:* 1980. *Institution:* Specialized Institution—Miscellaneous. *Control:* Private. *Enrollment:* Day—600. *Ministry of Education:* Accreditation.

Diploma Program: Theology.

**270. (C) Pohang Fisheries Junior College** (*See also* No. 272, Pohang Vocational Junior College). *Institution:* Junior Vocational.

**271. (C) Pohang Nursing Junior College** (Pohang Kanho Chonmun Dae Hak), 315, Yonghung 1-dong, Pohang-shi, Kyong Puk, 680, Korea. *Established:* 1979. *Institution:* Junior Vocational. *Control:* Private. *Enrollment:* 320.

Program: Nursing.

**272. (C) Pohang Vocational Junior College** (Pohang Shilop Chonmun Dae Hak), 254-70, Songdo-dong, Pohang-shi, Kyong Puk, 680, Korea. *Previous Name:* Pohang Fisheries Junior College (Pohang Susan Shonmun Dae Hak). *Established:* 1952. *Institution:* Junior Vocational. *Control:* Private. *Enrollment:* 2320.

Programs: 1) Commerce, 2) Education, 3) Engineering/Technical.

**273. (A) Posong College** (*See also* No. 213, Korea University). *Institution:* 4-year.

**274. (A) Presbyterian College and Theological Seminary** (Changnohoe Shinhak Dae Hak), Songdong-gu, Seoul, 133, Korea. *Established:* 1952. *Institution:* 4-year. *Control:* Private. *Religious Affiliation:* Presbyterian. *Enrollment:* Undergraduate, 1352; Graduate, 266.

Departments that offer bachelor's degrees: 1) Education, 2) Music, 3) Theology.

| Graduate Schools: | Master's | Doctor's |
| --- | --- | --- |
| General | x | x |
| Theology | x | – |

**275. (A) Pusan College of Foreign Studies** (Pusan Oe-kuk-o Dae Hak), Uam-dong, Nam-gu, Pusan, 608, Korea. *Established:* 1981. *Institution:* 4-year. *Control:* Private. *Enrollment:* Undergraduate, 3952.

Departments that offer bachelor's degrees: 1) Business Administration, 2) Commerce, 3) Engineering, 4) Literature.

**276. (A) Pusan Sanub College** (*See also* No. 277, Pusan Sanub University). *Institution:* 4-year.

**277. (A) Pusan Sanub University** (Pusan Sanop Dae Hak Kyo), Taeyon-dong, Nam-gu, Pusan, 608, Korea. *Previous Names:* Kyungnam Teacher's College (Kyongnam Sabom Dae Hak), Pusan Industrial College (Pusan Sanop Dae Hak). *Established:* 1955. *Institution:* 4-year. *Control:* Private. *Enrollment:* Undergraduate, 8892; Graduate, 100.

Colleges within the University: 1) Arts, 2) Economics and Commerce, 3) Education, 4) Law and Political Science, 5) Liberal Arts, 6) Science.

Departments that offer bachelor's degrees: 1) Administrative Science, 2) Business Administration, 3) Commerce, 4) Economics, 5) Education, 6) Fine Arts, 7) Gymnastics, 8) Home Economics, 9) Jurisprudence, 10) Library Science, 11) Literature, 12) Music, 13) Pharmacology, 14) Physical Science, 15) Political Science.

| Graduate Schools: | Master's | Doctor's |
|---|---|---|
| International Trade | x | – |

**278. (A) Pusan Women's University** (Pusan Yoja Dae Hak), Yonsan-dong, Tongnae-gu, Pusan, 607, Korea. *Established:* 1970. *Institution:* 4-year. *Control:* Private. *Enrollment:* Undergraduate, 4464; Graduate, 12.

Departments that offer bachelor's degrees: 1) Business Administration, 2) Education, 3) Fine Arts, 4) Gymnastics, 5) Home Economics, 6) Library Science, 7) Literature, 8) Music.

| Graduate Schools: | Master's | Doctor's |
|---|---|---|
| General | x | – |

**279. (C) Railroad High School** (*See also* No. 280, Railroad Junior College). *Institution:* Junior Vocational.

**280. (C) Railroad Junior College** (Choldo Chonmun Dae Hak), 63-3 ka, Hangang-ro, Yongsan-gu, Seoul, 140, Korea. *Previous Name:* Railroad High School (Choldo Kodung Hak Kyo). *Established:* 1967. *Institution:* Junior Vocational. *Control:* National. *Enrollment:* 300.

Program: Railroad.

**281. (C) Railroad-Nursing Junior College** (Choldo Kanho Chonmun Dae Hak), 63-3 ka, Hangang-ro, Youngsan-gu, Seoul, 140, Korea. *Established:* 1977. *Institution:* Junior Vocational. *Control:* National. *Enrollment:* 110.

Program: Nursing.

**282. (C) Red Cross Nursing Junior College** (Chokshipja Kanho Chonmun Dae Hak), 85-15, Pyong-dong, Chongro-gu, Seoul, 110, Korea. *Established:* 1945. *Institution:* Junior Vocational. *Control:* Private. *Enrollment:* 400.

Program: Nursing.

**283. (A) Sacred Heart College for Women** (Songshim Yoja Dae Hak), San 43-1, Yokkok-dong, Puchon, Kyonggi-do 150-71, Korea. *Established:* 1964. *Institution:* 4-year. *Control:* Private. *Enrollment:* Undergraduate, 3648; Graduate, 90.

Departments that offer bachelor's degrees: 1) Business Administration, 2) Commerce, 3) Home Economics, 4) Literature, 5) Nursing.

| Graduate Schools: | Master's | Doctor's |
|---|---|---|
| General | x | – |

**284. (C) Samcheok Technical Junior College** (Samchok Kongop Chonmun Dae Hak), 253, Kyo 3-ri, Samchok-up, Samchok-gun, Kangwon-do, 240, Korea. *Previous Name:* Samcheok Vocational School (Samchok Chigop Hak Kyo). *Established:* 1939. *Institution:* Junior Vocational. *Control:* National. *Enrollment:* 1760.

Program: 1) Development of Resources, 2) Engineering/Technical.

**285. (C) Samcheok Vocational School** (*See also* No. 284, Samcheok Technical Junior College). *Institution:* Junior Vocational.

**286. (A) Sangji College** (Sangji Dae Hak), 1082, Pongsan-dong, Wonju, Kangwon-do, 220, Korea. *Established:* 1963. *Institution:* 4-year. *Control:* Private. *Enrollment:* Undergraduate, 4680.

Departments that offer bachelor's degrees: 1) Administrative Science, 2) Agriculture Science, 3) Business Administration, 4) Commerce, 5) Education, 6) Fine Arts, 7) Home Economics, 8) Jurisprudence.

**287. (C) Sangji University, Vocational Junior College** (Sangji Dae Hak Pyongsol Shilop Chonmun Dae Hak), San 41, Usan-dong, Wonju-shi, Kangwon-do, 220, Korea. *Established:* 1973. *Institution:* Junior Vocational. *Control:* Private. *Enrollment:* 2080.

Programs: 1) Commerce, 2) Education, 3) Liberal Arts, 4) Tourism.

**288. (C) Sangji Vocational Junior College** (Sangji Shilop Chonmun Dae Hak), 393, Yulse-dong, Andong-shi, Kyong Puk, 660, Korea. *Established:* 1969. *Institution:* Junior Vocational. *Control:* Private. *Enrollment:* 1680.

Programs: 1) Commerce, 2) Design, 3) Education, 4) Engineering/Technical, 5) Textiles.

**289. (A) Sang Myung Women's University** (Sang Myong Yoja Dae Hak), Hongji-dong, Chongno-gu, Seoul, 110, Korea. *Established:* 1965. *Institution:* 4-year. *Control:* Private. *Enrollment:* Undergraduate, 4576; Graduate, 150.

Departments that offer bachelor's degrees: 1) Administrative Science, 2) Business Administration, 3) Education, 4) Library Science, 5) Literature.

| Graduate Schools: | Master's | Doctor's |
|---|---|---|
| General | x | — |

**290. (C) Sangju Agricultural Junior College** (Sangju Nongop Chonmun Dae Hak), 140, Namsong-dong, Sangju-up, Sangju-gun, Kyong Puk, 642, Korea. *Established:* 1973. *Institution:* Junior Vocational. *Control:* National. *Enrollment:* 1120.

Programs: 1) Commerce, 2) Engineering/Technical.

**291. (C) Seogang Vocational Junior College** (Sokang Shilop Chonmun Dae Hak), 789-1, Unam-dong, Puk-gu, Kwangju-shi, Chon Nam, 500-05, Korea. *Established:* 1979. *Institution:* Junior Vocational. *Control:* Private. *Enrollment:* 2600.

Programs: 1) Commerce, 2) Design, 3) Education, 4) Engineering/Technical, 5) Liberal Arts, 6) Tourism.

**292. (C) Seohae Technical Junior College** (Sohae Kongop Chonmun Dae Hak), 663, Sorung-dong, Kunsan-shi, Chon Puk, 511, Korea. *Previous Name:* Gunsan Technical Junior College (Kunsan Kongop Chonmun Dae Hak). *Established:* 1977. *Institution:* Junior Vocational. *Control:* Private. *Enrollment:* 1960.

Programs: 1) Commerce, 2) Design, 3) Engineering/Technical, 4) Tourism.

**293. (C) Seo-il Technical Junior College** (So-il Kongop Chonmun Dae Hak), 49-3, Myonmok 6-dong, Tongdaemun-gu, Seoul, 130-01, Korea. *Previous Name:* Baeseong Vocational Junior College (Paesong Shilop Chonmun Hak Kyo). *Established:* 1976. *Institution:* Junior Vocational. *Control:* Private. *Enrollment:* 1600.

Programs: 1) Design, 2) Education, 3) Engineering/Technical.

**294. (C) Seongji Technical Junior College** (Songji Kongop Chonmun Dae Hak), 38-1, Uam-dong, Nam-gu, Pusan, 608, Korea. *Established:* 1976. *Institution:* Junior Vocational. *Control:* Private. *Enrollment:* 2800.

Programs: 1) Commerce, 2) Design, 3) Engineering/Technical.

**295. (C) Seongshim Junior College of Foreign Studies** (Songshim Oekuko Chomun Dae Hak), 249, Pansong-dong, Haeundae-gu, Pusan, 607-05, Korea. *Established:* 1983. *Institution:* Junior Vocational. *Control:* Private. *Enrollment:* 560.

Program: Liberal Arts.

**296. (C) Seongshin Nursing Junior College** (Songshin Kanho Chonmun Dae Hak), 97, Sanjong-dong, Mokpo-shi, Chon Nam, 580, Korea. *Previous Name:* Golromban Nursing Junior College (Golromban Kanho Chonmun Hak Kyo). *Established:* 1971. *Institution:* Junior Vocational. *Control:* Private. *Religious Affiliation:* Catholic. *Enrollment:* 240.

Program: Nursing.

**297. (A) Seoul City University** (Seoul Shirip Dae Hak), Chonnong-dong, Tongdaemun-gu, Seoul, 131, Korea. *Previous Names:* Kyongsong Agriculture School (Kyongsong Nong-op Hak Kyo), Seoul Industry College (Seoul San-op Dae Hak). *Established:* 1918. *Institution:* 4-year. *Control:* Municipal. *Enrollment:* Undergraduate, 4420; Graduate, 150.

Departments that offer bachelor's degrees: 1) Administrative Science, 2) Business Administration, 3) Commerce, 4) Economics, 5) Engineering, 6) Fine Arts, 7) Jurisprudence, 8) Literature, 9) Music.

| Graduate Schools: | Master's | Doctor's |
|---|---|---|
| General | x | x |
| Urban Administrative Science | x | – |

**298. (C) Seoul Health Junior College** (Seoul Pokon Chonmun Dae Hak), 50, Chungmu-ro, 3 ka, Chung-gu, Seoul, 100, Korea. *Established:* 1967. *Institution:* Junior Vocational. *Control:* Private. *Enrollment:* 1440.

Programs: 1) Engineering/Technical, 2) Health Careers.

**299. (A) Seoul Higher Education for Women** (*See also* No. 262, Myong Ji University). *Institution:* 4-year.

**300. (A) Seoul Industry College** (*See also* No. 297, Seoul City University). *Institution:* 4-year.

**301. (C) Seoul Junior College of Arts** (Seoul Yesul Chonmun Dae Hak), 8-19, Yejang-dong, Chung-gu, Seoul, 100, Korea. *Established:* 1962. *Institution:* Junior Vocational. *Control:* Private. *Enrollment:* 1440.

Programs: 1) Design, 2) Fine Art.

**302. (A) Seoul Mun-Lee College** (*See also* No. 262, Myong Ji University). *Institution:* 4-year.

**303. (B) Seoul National Teacher's College** (Seoul Kyoyuk Dae Hak), 636, Socho-dong, Kangnam-gu, Seoul, 134-04, Korea. *Previous Name:* Kyeong-gi Teacher's School (Kyong-gi Sabum Hak Kyo). *Established:* 1946. *Institution:* 4-year teacher's college. *Control:* National. *Enrollment:* 1517.

Department that offers a bachelor's degree: Primary Education.

**304. (A) Seoul National University** (Seoul Dae Hak Kyo), San 56-1, Shinrim-dong, Kwan ak-gu, Seoul, 151, Korea. *Established:* 1945. *Institution:* 4-year. *Control:* National. *Enrollment:* Undergraduate, 25,600; Graduate, 6484.

Colleges within the University: 1) Agriculture, 2) Business Administration, 3) Dentistry, 4) Education, 5) Engineering, 6) Fine Arts, 7) Home Economics, 8) Humanistics, 9) Law, 10) Medicine, 11) Music, 12) Natural Science, 13) Pharmacy, 14) Social Science, 15) Veterinary Medicine.

Departments that offer bachelor's degrees: 1) Administrative Science, 2) Agriculture Science, 3) Business Administration, 4) Commerce, 5) Dentistry, 6) Economics, 7) Education, 8) Engineering, 9) Fine Arts, 10) Home Economics, 11) Jurisprudence, 12) Literature, 13) Medicine, 14) Music, 15) Nursing, 16) Pharmacology, 17) Physical Science, 18) Political Science, 19) Veterinary Medicine.

| Graduate Schools: | Master's | Doctor's |
| --- | :---: | :---: |
| General | x | x |
| Environmental Studies | x | – |
| Public Administration | x | – |
| Public Health | x | – |

**305. (C) Seoul Nursing Junior College** (Seoul Kanho Chonmun Dae Hak), 287-89, Hongje-dong, Sodaemun-gu, Seoul, 120, Korea. *Established:* 1964. *Institution:* Junior Vocational. *Control:* Private. *Enrollment:* 400.

Program: Nursing.

**306. (A) Seoul Theological Seminary** (Seoul Shinhak Dae Hak), 101, Sosa-dong, Pochon, Kyonggi-do, 150-71, Korea. *Established:* 1959. *Institution:* 4-year. *Control:* Private. *Enrollment:* Undergraduate, 1456; Graduate, 156.

Departments that offer bachelor's degrees: 1) Education, 2) Music, 3) Theology.

| Graduate Schools: | Master's | Doctor's |
| --- | :---: | :---: |
| General | x | x |
| Theology | x | – |

**307. (A) Seoul Women's University** (Seoul Yoja Dae Hak), Kongrung-dong, Tobong-gu, Seoul, 130-02, Korea. *Established:* 1961. *Institution:* 4-year. *Control:* Private. *Enrollment:* Undergraduate, 3380; Graduate, 70.

Departments that offer bachelor's degrees: 1) Business Administration, 2) Education, 3) Fine Arts, 4) Home Economics, 5) Library Science, 6) Literature.

| Graduate Schools: | Master's | Doctor's |
| --- | :---: | :---: |
| General | x | – |

**308. (A) Severance Union Medical Clinic** (*See also* No. 357, Yonsei University). *Institution:* 4-year.

**309. (C) Shingu Industrial Junior College** (*See also* No. 310, Shingu Vocational Junior College). *Institution:* Junior Vocational.

**310. (C) Shingu Vocational Junior College** (Shingu Shilop Chonmun Dae Hak), 191, Tandae-dong, Songnam-shi, Kyonggi-do, 130-14, Korea. *Previous Name:* Shingu Industrial Junior College (Shingu Sanop Chonmun Hak Kyo). *Established:* 1974. *Institution:* Junior Vocational. *Control:* Private *Enrollment:* 3689.

Programs: 1) Commerce, 2) Education, 3) Engineering/Technical 4) Health Careers, 5) Textiles.

**311. (C) Shinheung Institute** (*See also* No. 312, Shinheung Vocational Junior College). *Institution:* Junior Vocational.

**312. (C) Shinheung Vocational Junior College** (Shinhung Shilop Chonmun Dae Hak), 115, Howon-dong, Uijongbu-shi, Kyonggi-do, 130-03, Korea. *Previous Name:* Shinheung Institute (Shinhung Hak Won). *Established:* N.A. *Institution:* Junior Vocational. *Control:* Private. *Enrollment:* 2160.

Programs: 1) Commerce, 2) Education, 3) Health Careers.

**313. (A) Shin Hung College** (*See also* No. 242, Kyung Hee University). *Institution:* 4-year.

**314. (C) Shinil Vocational Junior College** (Shinil Shilop Chonmun Dae Hak), San 395, Manchon-dong, Susong-gu, Taegu, 634, Korea. *Established:* 1979. *Institution:* Junior Vocational. *Control:* Private. *Enrollment:* 2640.

Programs: 1) Commerce, 2) Education, 3) Engineering/Technical, 4) Health Careers, 5) Home Economics, 6) Nursing, 7) Tourism.

**315. (A) Sogang University** (Sogang Dae Hak Kyo), Sinsu-dong, Mapo-gu, Seoul, 121, Korea. *Established:* 1960. *Institution:* 4-year. *Control:* Private. *Religious Affiliation:* Catholic. *Enrollment:* Undergraduate, 7072; Graduate, 1200.

Colleges within the University: 1) Economics and Commerce, 2) Liberal Arts, 3) Science and Engineering.

Departments that offer bachelor's degrees: 1) Business Administration, 2) Economics, 3) Engineering, 4) Literature, 5) Physical Science, 6) Political Science.

| Graduate Schools: | Master's | Doctor's |
|---|---|---|
| General | x | x |
| Business Administration | x | − |

**316. (C) Sokcho Junior College** (*See also* No. 106, Dongwoo Junior College). *Institution:* Junior Vocational.

**317. (C) Songwon Vocational Junior College** (Songwon Shilop Chonmun Dae Hak), 199-1, Kwangchon-dong, So-gu, Kwangju-shi, Chon Nam, 500, Korea. *Established:* 1974. *Institution:* Junior Vocational. *Control:* Private. *Enrollment:* 2800.

Programs: 1) Commerce, 2) Design, 3) Education, 4) Engineering/Technical.

**318. (A) Sook Myung Women's University** (Suk Myong Yoja Dae Hak Kyo), San 2-1, Chongpa-dong, Yongsan-gu, Seoul, 140, Korea. *Established:* 1938. *Institution:* 4-year. *Control:* Private. *Enrollment:* Undergraduate, 6756; Graduate, 673.

Colleges within the University: 1) Economics and Political Science, 2) Fine Arts, 3) Home Economics, 4) Liberal Arts, 5) Music, 6) Pharmacology, 7) Science.

Departments that offer bachelor's degrees: 1) Administrative Science, 2) Business Administration, 3) Commerce, 4) Economics, 5) Fine Arts, 6) Gymnastics, 7) Home Economics, 8) Jurisprudence, 9) Library Science, 10) Literature, 11) Music, 12) Pharmacology, 13) Physical Science, 14) Political Science.

| Graduate Schools: | Master's | Doctor's |
|---|---|---|
| General | x | x |
| Education | x | – |

**319. (C) Soongcheon Nursing Junior College** (Sunchon Kanho Chonmun Dae Hak), 130, Maegok-dong, Suchon-shi, Chon Nam, 540, Korea. *Established:* 1979. *Institution:* Junior Vocational. *Control:* Private. *Enrollment:* 360.

Program: Nursing.

**320. (C) Sooncheon Technical Junior College** (Sunchon Kongop Chonmun Dae Hak), 9-1, Tokwol-dong, Sunchon-shi, Chon Nam, 540, Korea. *Previous Name:* Woojin Institute (Ujin Hak Won). *Established:* 1977. *Institution:* Junior Vocational. *Control:* Private. *Enrollment:* 1560.

Programs: 1) Commerce, 2) Design, 3) Education, 4) Engineering/Technical, 5) Liberal Arts.

**321. (A) Soonchunhyang College** (Sunchonhyang Dae Hak), Asan-gun, Chungnam 330-62, Korea. *Previous Names:* Soonchunhyang Medical College (Sunchonhyang Ui-kwa Dae Hak). *Established:* 1978. *Institution:* 4-year. *Control:* Private. *Enrollment:* Undergraduate, 2936.

Departments that offer bachelor's degrees: 1) Business Administration, 2) Dentistry, 3) Literature, 4) Medicine, 5) Physical Science.

**322. (A) Soonchunhyang Medical College** (*See also* No. 321, Soonchunhyang College). *Institution:* 4-year.

**323. (C) Soong-eui Womens Junior College** (Sungui Yoja Chonmun Dae Hak), 8-3, Ye-jang-dong, Chung-gu, Seoul, 100, Korea. *Established:* 1971. *Institution:* Junior Vocational. *Control:* Private. *Enrollment:* 2340.

Programs: 1) Commerce, 2) Education, 3) Fine Arts, 4) Home Economics, 5) Liberal Arts, 6) Tourism.

**324. (A) Soong Jun University** (Sung Jon Dae Hak Kyo), 135, Sangdo-dong, Kwan-ak-gu, Seoul, 151, Korea. *Established:* 1954. *Institution:* 4-year. *Control:* Private. *Religious Affiliation:* Presbyterian. *Enrollment:* Undergraduate, 7696; Graduate, 723.

Colleges within the University: 1) Engineering, 2) Law and Economics, 3) Liberal Arts and Science.

Departments that offer bachelor's degrees: 1) Administrative Science, 2) Business Administration, 3) Commerce, 4) Economics, 5) Engineering, 6) Jurisprudence, 7) Literature, 8) Physical Science.

| Graduate Schools: | Master's | Doctor's |
|---|---|---|
| General | x | x |
| Industry | x | – |

**325. (A) Soong Jun University at Taejon** (*See also* No. 135, Han Nam College). *Institution:* 4-year.

326. (D) **Sudo Baptist Theological Seminary** (Sudo Chimrye Shin Hak Kyo), Ansong, Kyongi-do, Korea. *Established:* 1984. *Institution:* Specialized Institution—Miscellaneous. *Control:* Private. *Enrollment:* 40 est.

Diploma Program: Theology.

327. (A) **Sudo Teacher's College for Women** (*See also* No. 192, King Sejong University). *Institution:* 4-year.

328. (A) **Suncheon Junior College** (*See also* No. 329, Suncheon National College). *Institution:* 4-year.

329. (A) **Suncheon National College** (Sunchon Dae Hak), Maegok-dong, Sunchon, Chonnam, 540, Korea. *Previous Name:* Suncheon Junior College (Sunchon Nong-Lim Chonmun Hak Kyo). *Established:* 1973. *Institution:* 4-year. *Control:* National. *Enrollment:* Undergraduate, 3068.

Departments that offer bachelor's degrees: 1) Administrative Science, 2) Business Administration, 3) Education, 4) Engineering.

330. (A) **Sung Kyun Kwan University** (Song Kyun Kwan Dae Hak Kyo), Myongryundong, Chongno-gu, Seoul, 110, Korea. *Background:* Founded in 992 as sole national institute of higher learning in Koryo Dynasty based on Confucian doctrine and principles. Re-established in 1400 on present site under Yi Dynasty. Became a college in 1937 and private university in 1953. *Established:* 992. *Institution:* 4-year. *Control:* Private. *Enrollment:* Undergraduate, 17,076; Graduate, 1940.

Colleges within the University: 1) Agriculture, 2) Economics and Commerce, 3) Education, 4) Engineering, 5) Home Economics, 6) Law, 7) Liberal Arts, 8) Pharmacology, 9) Science, 10) Social Science, 11) *Yukyo* (Confucianism).

Departments that offer bachelor's degrees: 1) Administrative Science, 2) Agriculture Science, 3) Business Administration, 4) Commerce, 5) Economics, 6) Education, 7) Engineering, 8) Home Economics, 9) Jurisprudence, 10) Library Science, 11) Literature, 12) Pharmacology, 13) Physical Science, 14) Political Science.

| Graduate Schools: | Master's | Doctor's |
| --- | --- | --- |
| General | x | x |
| Business Administration and Public Administration | x | – |
| Education | x | – |
| International Trade | x | – |

331. (A) **Sungshin Women's University** (Songshin Yoja Dae Hak Kyo), 3-ka, Tongsondong, Songbuk-gu, Seoul, 132, Korea. *Established:* 1965. *Institution:* 4-year. *Control:* Private. *Enrollment:* Undergraduate, 6004; Graduate, 370.

Colleges within the University: 1) Arts, 2) Education, 3) Liberal Arts, 4) Science, 5) Social Science.

Departments that offer bachelor's degrees: 1) Business Administration, 2) Economics, 3) Education, 4) Fine Arts, 5) Home Economics, 6) Literature, 7) Music, 8) Political Science.

| Graduate Schools: | Master's | Doctor's |
| --- | --- | --- |
| General | x | x |
| Industry | x | – |

332. **(C) Supia Nursing Junior College** (*See also* No. 52, Christian Hospital, Nursing Junior College). *Institution:* Junior Vocational.

333. **(C) Supia Women's Vocational Junior College** (*See also* No. 219, Kwangju Health Junior College). *Institution:* Junior Vocational.

334. **(A) Suweon Catholic College** (Suwon Catholic Dae Hak). *Address:* N.A. *Established:* 1984. *Institution:* 4-year. *Control:* Private. *Enrollment:* Undergraduate, N.A.

     Departments that offer bachelor's degrees: N.A.

335. **(C) Suweon Nursing Junior College** (Suwon Kanho Chonmun Dae Hak), 249, Shinpung-dong, Suwon-shi, Kyonggi-do, 170, Korea. *Established:* 1973. *Institution:* Junior Vocational. *Control:* Private. *Enrollment:* 440.

     Programs: 1) Education, 2) Nursing.

336. **(C) Suweon Technical Junior College** (Suwon Kongop Chonmun Dae Hak), San 2-18, Wa-u-ri, Pongdam-myon, Whasong-gun, Kyonggi-do, 170-21, Korea. *Established:* 1977. *Institution:* Junior Vocational. *Control:* Private. *Enrollment:* 2480.

     Programs: 1) Commerce, 2) Design, 3) Engineering/Technical.

337. **(A) Suwon University** (Suwon Dae Hak), Pongdam-myon, Hwasong-gun, Kyonggi-do, Korea. *Established:* 1981. *Institution:* 4-year. *Control:* Private. *Enrollment:* Undergraduate, 4160.

     Departments that offer bachelor's degrees: 1) Economics, 2) Engineering, 3) Gymnastics, 4) Literature.

338. **(A) Tae Jon Methodist Theological College** (*See also* No. 259, Mokwon Methodist College). *Institution:* 4-year.

339. **(C) Tongyeing Fisheries Junior College** (Tongyong Susan Chonmun Dae Hak), 445, Inpyong-dong, Chungmu-shi, Kyong Nam, 603, Korea. *Established:* 1917. *Institution:* Junior Vocational. *Control:* National. *Enrollment:* 1120.

     Program: Engineering/Technical.

340. **(A) Ulsan University** (Ulsan Dae Hak), Ulsan-shi, Kyongnam 690, Korea. *Established:* 1970. *Institution:* 4-year. *Control:* Private. *Enrollment:* Undergraduate, 8580; Graduate, 60.

     Departments that offer bachelor's degrees: 1) Business Administration, 2) Economics, 3) Engineering, 4) Fine Arts, 5) Home Economics, 6) Literature, 7) Physical Science.

     | Graduate Schools: | Master's | Doctor's |
     |---|---|---|
     | General | x | – |

341. **(C) Ulsan Institute of Technology, Technical Junior College** (Ulsan Kongkwa Dae Hak Pongsol Kongop Chonmun Dae Hak), San 29, Mugo-dong, Ulsan-shi, Kyong Nam, 690, Korea. *Established:* 1973. *Institution:* Junior Vocational. *Control:* Private. *Enrollment:* 1960.

     Programs: 1) Commerce, 2) Engineering/Technical.

342. **(C) Wonju Nursing Junior College** (Wonju Kanho Chonmun Dae Hak), San 2-1, Hungop-ri, Hungop-myon, Wonsong-gun, Kangwon-do, 220-13, Korea. *Established:* 1967. *Institution:* Junior Vocational. *Control:* National. *Enrollment:* 499.

     Programs: 1) Education, 2) Nursing.

**343. (C) Wonkwang Health Junior College** (Wonkwang Pokon Chonmun Dae Hak), 344-2, Shinyong-dong, I-ri-shi, Chon Puk, 510, Korea. *Established:* 1976. *Institution:* Junior Vocational. *Control:* Private. *Enrollment:* 2080.

Programs: 1) Education, 2) Health Careers, 3) Nursing.

**344. (A) Won Kwang University** (Won Kwang Dae Hak Kyo), Sinyong-dong, Iri-shi, Chonbuk, 510, Korea. *Established:* 1953. *Institution:* 4-year. *Control:* Private. *Religious Affiliation:* Buddhist. *Enrollment:* Undergraduate: 16,748; Graduate, 744.

Colleges within the University: 1) Agriculture, 2) Business Administration, 3) Dentistry, 4) Education, 5) Engineering, 6) Home Economics, 7) Law, 8) Liberal Arts and Science, 9) Medicine, 10) Oriental Medicine, 11) Pharmacology, 12) Social Science.

Departments that offer bachelor's degrees: 1) Administrative Science, 2) Agriculture Science, 3) Business Administration, 4) Commerce, 5) Dentistry, 6) Economics, 7) Education, 8) Engineering, 9) Fine Arts, 10) Gymnastics, 11) Home Economics, 12) Jurisprudence, 13) Literature, 14) Medicine, 15) Oriental Medicine, 16) Pharmacology, 17) Physical Science.

| Graduate Schools: | Master's | Doctor's |
|---|---|---|
| General | x | x |
| Education | x | – |

**345. (C) Woojin Institute** (*See also* No. 320, Sooncheon Technical Junior College). *Institution:* Junior Vocational.

**346. (C) Yeonam Junior College of Animal and Horticulture** (Yonam Chuksan-Wonye Chonmun Dae Hak), San 3-1, Suhyang-ri, Songwhan-up, Chonwon-gun, Chung Nam, 330-81, Korea. *Established:* 1982. *Institution:* Junior Vocational. *Control:* Private. *Enrollment:* 640.

Program: Engineering/Technical.

**347. (C) Yeonam Technical Junior College** (Yonam Kongop Chonmun Dae Hak), San 100, Kajwa-dong, Chinju-shi, Kyong Nam, 620-15, Korea. *Established:* 1984. *Institution:* Junior Vocational. *Control:* Private. *Enrollment:* 240.

Program: Engineering/Technical.

**348. (C) Yeongjin Vocational Junior College** (Yongjin Shilop Chonmun Dae Hak), 218, Pokhyon-dong, Puk-gu, Taegu, 635, Korea. *Established:* 1974. *Institution:* Junior Vocational. *Control:* Private. *Enrollment:* 3440.

Programs: 1) Commerce, 2) Design, 3) Education, 4) Engineering/Technical, 5) Home Economics.

**349. (C) Yeongju Junior College** (*See also* No. 353, Yeonju Junior College of Business and Commerce). *Institution:* Junior Vocational.

**350. (C) Yeongnam Technical Junior College** (Yongnam Kongop Chonmun Dae Hak), 1737, Taemyong-dong, Nam-gu, Taegu, 634, Korea. *Established:* 1968. *Institution:* Junior Vocational. *Control:* Private. *Enrollment:* 4000.

Programs: 1) Commerce, 2) Design, 3) Engineering/Technical, 4) Home Economics, 5) Textiles.

**351. (D) Yeongnam Theological Seminary** (Yongnam Shin Hak Kyo), 230, Tongsan-dong, Chung-gu, Taegu, 630, Korea. *Established:* 1957. *Institution:* Specialized Institution—Miscellaneous. *Control:* Private. *Enrollment:* Day—410.

Diploma Program: Theology.

352. **(C) Yeongseong School** (*See also* No. 7, Anyang Technical Junior College). *Institution:* Junior Vocational.

353. **(C) Yeonju Junior College of Business and Commerce** (Yongju Kyong-Sang Chonmun Dae Hak), 630, Hyuchon 2-dong, Yongju-shi, Kyong Puk, 650, Korea. *Previous Name:* Yeongju Junior College (Yongju Chonmun Hak Kyo). *Established:* 1971. *Institution:* Junior Vocational. *Control:* Private. *Enrollment:* 2240.

Programs: 1) Commerce, 2) Education, 3) Engineering/Technical, 4) Liberal Arts, 5) Tourism.

354. **(C) Yeosu Fisheries Junior College** (Yosu Susan Chonmun Dae Hak), 195, Kuk-dong, Yosu-shi, Chon Nam, 542, Korea. *Established:* 1915. *Institution:* Junior Vocational. *Control:* National. *Enrollment:* 1120.

Programs: 1) Commerce, 2) Communication, 3) Engineering/Technical.

355. **(C) Yesan Agricultural Junior College** (Yesan Nongop Chonmun Dae Hak), 527, Yesan-ri, Yesan-up, Yesan-gun, Chung Nam, 340, Korea. *Established:* 1974. *Institution:* Junior Vocational. *Control:* National. *Enrollment:* 1280.

Programs: 1) Commerce, 2) Engineering/Technical.

356. **(A) Yeungnam University** (Yongnam Dae Hak Kyo), Kyongsan-gun, Kyongbuk, 632, Korea. *Established:* 1967. *Institution:* 4-year. *Control:* Private. *Enrollment:* Undergraduate, 23,060; Graduate, 1844.

Colleges within the University: 1) Agriculture (including Animal Science), 2) Arts, 3) Economics and Commerce, 4) Education, 5) Engineering, 6) Home Economics, 7) Law and Political Science, 8) Liberal Arts, 9) Medicine, 10) Music, 11) Pharmacology, 12) Science.

Departments that offer bachelor's degrees: 1) Administrative Science, 2) Agriculture Science, 3) Business Administration, 4) Commerce, 5) Economics, 6) Education, 7) Engineering, 8) Fine Arts, 9) Home Economics, 10) Jurisprudence, 11) Literature, 12) Medicine, 13) Music, 14) Pharmacology, 15) Physical Science, 16) Political Science.

| Graduate Schools: | Master's | Doctor's |
|---|---|---|
| General | x | x |
| Business Administration | x | — |
| Education | x | — |
| Environmental Studies | x | — |

357. **(A) Yonsei University** (Yonsei Dae Hak Kyo), 134, Sinchon-dong, Sodaemun-gu, Seoul, 120, Korea. *Background:* Founded in 1885 as Severance Union Medical Clinic. Merged with Chosun Christian College, established in 1915, to form present university in 1957. *Established:* 1885. *Institution:* 4-year. *Control:* Private. *Religious Affiliation:* Methodist. *Enrollment:* Undergraduate, 21,932; Graduate, 4974.

Colleges within the University: 1) Dentistry, 2) Economics and Commerce, 3) Education, 4) Engineering, 5) Home Economics, 6) Law, 7) Liberal Arts, 8) Medicine, 9) Music, 10) Nursing, 11) Science, 12) Social Science, 13) Theology.

Departments that offer bachelor's degrees: 1) Administrative Science, 2) Business Administration, 3) Dentistry, 4) Economics, 5) Education, 6) Engineering, 7) Home Economics, 8) Jurisprudence, 9) Library Science, 10) Literature, 11) Medicine, 12) Music, 13) Nursing, 14) Physical Science, 15) Political Science, 16) Theology.

| Graduate Schools: | Master's | Doctor's |
|---|---|---|
| General | x | x |
| Business Administration | x | – |
| Education | x | – |
| Health Science and Management | x | – |
| Industry | x | – |
| Public Administration | x | – |
| Theology | x | – |

358. (C) **Yuhan Technical Junior College** (Yuhan Kongop Chonmun Dae Hak), 61-6, Yokgok-dong, Puchon-shi, Kyonggi-do, 150-71, Korea. *Established:* 1978. *Institution:* Junior Vocational. *Control:* Private. *Enrollment:* 2880.

Programs: 1) Commerce, 2) Design, 3) Engineering/Technical.

# Appendix B

## Transliteration and Pronunciation Guide

There is no single standard method to transliterate Korean words into the English alphabet. This is also true regarding an individual Korean name. For instance, the most popular Korean surname is written alternately: Kim, Kimm, Gim, Kiem.

There are two commonly used romanization systems, McCune-Reischauer and that of the Korean Ministry of Education. These two approaches are not familiar to many Korean people. As a result, Koreans use any combination of these two systems or others. It is important, therefore, as in the transliteration of any language, to pay more attention to a word's pronunciation than to the exact spelling of it in the Western alphabet.

The most common surnames used by Koreans with the variations in transliteration are listed below:

|  |  |  |
|---|---|---|
| 1. Kim | – Kim, Gim, Kimm, Kiem |
| 2. Park | – Park, Pak, Bak, Bark |
| 3. Lee | – Lee, Rhee, Yi, Li |
| 4. Chung | – Chung, Jung, Chong, Jeong |
| 5. Chang | – Chang, Jang |
| 6. Chun | – Chon, Cheon, Chun |
| 7. Shin | – Sin, Shin |
| 8. Choi | – Choi, Choe, Che |
| 9. Song | – Sung, Song |
| 10. Kwun | – Kwun, Gwon, Kwon |

In receiving mail from the Republic of Korea, letters normally sent to the United States use the McCune-Reischauer or the Ministry of Education system of transliteration. Some words commonly used in addresses are:

|  |  |
|---|---|
| 1. Ga or Ka | – Street or avenue in cities |
| 2. Ri or Li | – Street or section in township |
| 3. Dong | – Section in cities |
| 4. Gu or Ku | – District in cities |
| 5. Myun | – Township |
| 6. Gun or Kun | – County |
| 7. Do | – Province |

Examples and explanations of the transliteration of addresses follow:

1. Kim, So Youn (Surname first)
   123 Yeonhi-dong, Seodaemun-ku, Seoul (ROK)

2. So Youn Kim (Surname last)
   10 Ibuk-ri, Samsun-myun, Kwanju-gun,
   Chol La Nam Do (Chon Nam) (ROK)

In the case of the provinces found on the map of the Republic of Korea, abbreviations are sometimes used as with states in the United States. For example, the proper noun, Michigan, can be abbreviated as Mich. or MI. Because the province name will be a part of the institutional address, and may be abbreviated, the full romanization of the name of the province as well as the abbreviation is also found on the map.

# Appendix C

## Useful Addresses

Note: The following list includes the addresses and locations of the Korean embassy and consulates in the United States. It also includes the addresses of the Ministry of Education of the Republic of Korea and three organizations that may be helpful in providing information regarding educational questions and requests for materials.

### Korean Embassy and Consulates

Embassy of the Republic of Korea
2320 Massachusetts Avenue, N.W
Washington, DC 20008

### Korean Consulates:

101 Benson Boulevard
304
Anchorage, AK 99503

Suite 500
229 Peachtree Street, N.E.
Atlanta, GA 30303

Suite 610
500 N. Michigan Avenue
Chicago, IL 60611

2756 Pali Highway
Honolulu, HI 96817

Suite 745
1990 Post Oak Boulevard
Houston, TX 77056

Suite 1101
Lee Power Building
5455 Wilshire Boulevard
Los Angeles, CA 90036

460 Park Avenue
5th Floor
New York, NY 10022

3500 Clay Street
San Francisco, CA 94118

1125 United Airlines Building
2033 6th Avenue
Seattle, WA 98121

### Addresses in Korea

Ministry of Education
77 Sejong-ro,
Seoul, 110 Republic of Korea

Korea Council for University
   Education
27-2 Yonido-dong, Young-gungpo-Ku
Seoul, 150, Korea

Korean Federation of Education
1-Ka Shinmun-ro, Chongro-Ku
Seoul, 110, Korea

Korea Research Foundation
199-1 Donsoong-dong, Chongro-Ru
Seoul, 110, Korea

# Glossary

Note: The transliteration of common words used in Korean academic transcripts or credentials follows. Terms are alphabetized in two lists, giving the English with the corresponding Korean word and vice versa. McCune-Reischauer is the guideline used for transliteration.

## I. English

| English | Korean |
|---|---|
| Air and Correspondence School | *Bangsong Tongshin Hakkyo* |
| Bachelor's degree | *Haksa* |
| Biology | *Saengmulhak* |
| Buddhism | *Pulgyo* |
| Chemistry | *Hwahak* |
| Civic school | *Kongmin Hakkyo* |
| College | *Taehak* or *Daehak* |
| Commerce | *Sanggyonghak* |
| Confucianism | *Yugyo* |
| Dental college | *Chikwa Taehak* |
| Department | *-gwa* or *-kwa* |
| Diploma | *Chorupchung* |
| Drama | *Yongguk* |
| Economics | *Kyongjehak* |
| Education | *Kyoyukhak* |
| Elementary (primary) school | *Kukmin Hakkyo* |
| Engineering | *Kongkwa* |
| English | *Yong-o* |
| English literature | *Yongmunhak* |
| Fine arts | *Yesul* |
| Graduate school | *Taehagwon* |
| Graduation degree | *Hagui* |
| Higher | *Kodung* |
| High school | *Kodung Hakkyo* |
| Higher civic school | *Kodung Kongmin Hakkyo* |
| Higher technical school | *Kodung Kisul Hakkyo* |
| History | *Sahak* |
| Humanities | *Munkwa* |
| Journalism | *Sinmunhak* |
| Junior college | *Chogup Taehak* |
| Junior high school | *Chung Hakkyo* |
| Junior technical college | *Kongop Chonmun Taehak* |
| Junior vocational college or junior college | *Chonmun Taehak* |
| Kindergarten | *Yuchiwon* |

138

| | |
|---|---|
| Laboratory | *Silhomsil* |
| Master's degree | *Suksa* |
| Mathematics | *Suhak* |
| Medical college | *Uikwa Taehak* |
| Ministry of Education | *Mungyobu* |
| Miscellaneous schools at all levels | *Kakchong Hakkyo* |
| Music | *Umak* |
| Nursing junior college | *Kanho Chonmun Taehak* |
| Nursing school | *Kanho Hakkyo* |
| Ph.D. | *Paksa* |
| Pharmacy | *Yakgwa* |
| Philosophy | *Cholhak* |
| Physical education | *Cheyuk* |
| Physics | *Mullihak* |
| Political science | *Chongchihak* |
| Psychology | *Shimnihak* |
| Research | *Yongku* |
| School | *Hakkyo* |
| Science | *Kwahak* |
| Special schools | *Tuksu Hakkyo* |
| Statistics | *Tonggyehak* |
| Teacher's college | *Kyoyuk Taehak* |
| Teacher training | *Sabom* |
| Technical school | *Kisul Hakkyo* |
| Theology | *Shinhak* |
| University | *Taehakkyo* |
| Vocational/technical high school | *Silop Kodung Hakkyo* |

**II.  Korean**

| | |
|---|---|
| *Bangsong Tongshin Hakkyo* | Air and Correspondence School |
| *Cheyuk* | Physical education |
| *Chikwa Taehak* | Dental college |
| *Chogup Taehak* | Junior college |
| *Cholhak* | Philosophy |
| *Chongchihak* | Political science |
| *Chonmun Taehak* | Junior vocational college or junior college |
| *Chorupchung* | Diploma |
| *Chung Hakkyo* | Junior high school |
| *-gwa* or *-kwa* | Department |
| *Hagui* | Graduation degree |
| *Hakkyo* | School |
| *Haksa* | Bachelor's degree |
| *Hwahak* | Chemistry |
| *Kakchong Hakkyo* | Miscellaneous schools at all levels |

**English**

| | |
|---|---|
| *Kanho Chonmun Taehak* | Nursing junior college |
| *Kanho Hakkyo* | Nursing school |
| *Kisul Hakkyo* | Technical school |
| *Kodung* | Higher |
| *Kodung Hakkyo* | High school |
| *Kodung Kisul Hakkyo* | Higher technical school |
| *Kodung Kongmin Hakkyo* | Higher civic school |
| *Kongkwa* | Engineering |
| *Kongmin Hakkyo* | Civic school |
| *Kongop Chonmun Taehak* | Junior technical college |
| *Kukmin Hakkyo* | Elementary school |
| *-kwa* or *-gwa* | Department |
| *Kwahak* | Science |
| *Kyongjehak* | Economics |
| *Kyoyukhak* | Education |
| *Kyoyuk Taehak* | Teacher's college |
| *Mullihak* | Physics |
| *Munkwa* | Humanities |
| *Mungyobu* | Ministry of Education |
| *Paksa* | Ph.D. |
| *Pulgyo* | Buddhism |
| *Sabom* | Teacher training |
| *Saengmulhak* | Biology |
| *Sahak* | History |
| *Sanggyonghak* | Commerce |
| *Shinhak* | Theology |
| *Shimnihak* | Psychology |
| *Silhomsil* | Laboratory |
| *Silop Kodung Hakkyo* | Vocational/technical high school |
| *Sinmunhak* | Journalism |
| *Suhak* | Mathematics |
| *Suksa* | Master's degree |
| *Taehagwon* | Graduate school |
| *Taehak* or *Daehak* | College |
| *Taehakkyo* | University |
| *Tonggyehak* | Statistics |
| *Tuksu Hakkyo* | Special schools |
| *Uikwa Taehak* | Medical college |
| *Umak* | Music |
| *Yakgwa* | Pharmacy |
| *Yesul* | Fine arts |
| *Yongguk* | Drama |
| *Yongku* | Research |
| *Yongmunhak* | English literature |
| *Yong-o* | English |
| *Yuchiwon* | Kindergarten |
| *Yugyo* | Confucianism |

# Useful References

Educational Facilities Bureau. *Investment Strategies for the Improvement of Vocational High School Education.* Final Report of the Feasibility Study Team. Seoul: Ministry of Education, 1985.

*A Handbook of Korea.* 5th ed. Seoul: Korea Overseas Information Service, Ministry of Culture and Information, 1983.

*Inter-Institutional Cooperation in Higher Education.* Seoul: The Korean Council for University Education, 1984.

*The Korean Economy, Review and Prospects.* 5th ed. Seoul: Korea Exchange Bank, Research Department.

National Institute of Educational Research and Training. *Education in Korea,* Series from 1976-1984. Seoul: Ministry of Education, 1976-1984.

*The Pursuit of Excellence in Higher Education.* Daegu, Korea: Keimyung University, 1980.

# Index

# NATIONAL COUNCIL ON THE EVALUATION OF FOREIGN EDUCATIONAL CREDENTIALS

The Council is an interassociational group that serves as a forum for developing consensus and recognition of certificates, diplomas, and degrees awarded throughout the world. It also assists in establishing priorities for research and publication of country, regional, or topical studies. One of its main purposes is to review and modify admissions and placement recommendations drafted by World Education Series authors or others who might ask for such review. (The practices followed in fulfilling this purpose are explained on page 85.)

Chairperson—Karlene N. Dickey, Associate Dean of Graduate Studies and Research, Stanford University, Standford, CA 94305.

Vice Chairperson/Secretary—Stan Berry, Director of Admissions, Washington State University, Pullman, WA 99163.

## MEMBER ORGANIZATIONS AND THEIR REPRESENTATIVES:

American Association of Collegiate Registrars and Admisions Officers—Chairperson of World Education Series Committee, Alan M. Margolis, Registrar, Queens College, Flushing NY 11367; G. James Haas, Associate Director of Admissions, Indiana University, Bloomington, IN 47405; Kitty Villa, Acting Assistant Director, International Office, University of Texas at Austin, Austin, TX 78712.

American Association of Community and Junior Colleges—Philip J. Gannon, President, Lansing Community College, Lansing, MI 48901.

American Council on Education—Joan Schwartz, Senior Program Associate, Office on Educational Credit and Credentials, ACE, Washington, DC 20036.

College Entrance Examination Board—Sanford C. Jameson, Director, Office of International Education, CEEB, Washington, DC 20036.

Council of Graduate Schools—Andrew J. Hein, Assistant Dean, Graduate School, University of Minnesota, Minneapolis, MN 55455.

Institute of International Education—Martha Renaud, Director, Placement and Special Services Division, IIE, New York, NY 10017.

National Association for Foreign Student Affairs—David Horner, Office of International Students and Scholars, Michigan State University, E. Lansing, MI 48824; Joann Stedman, Director, Foreign Student Services, Columbia University, New York, NY 10027; Valerie Woolston, Director, International Education Services, University of Maryland, College Park, MD 20742.

## OBSERVER ORGANIZATIONS AND THEIR REPRESENTATIVES:

USIA—Joseph Bruns, Chief, Student Support Services Division, Office of Academic Programs, USIA, Washington, DC 20547.

AID—Hattie Jarmon, Education Specialist, Office of International Training, U.S. Department of State/AID, Washington, DC 20523.

State of New York Education Department—Mary Jane Ewart, Associate in Comparative Education, State Education Department, The University of the State of New York, Albany, NY 12230.